W9-DFT-317

Methods in Human Growth Research

In order to gain an understanding of the dynamics of human individual and average growth patterns it is essential that the right methods are selected. There are a variety of methods available to analyse individual growth patterns, to estimate variation in different growth measures in populations and to relate genetic and environmental factors to individual and average growth. This volume provides an overview of modern techniques for the assessment and collection of growth data and methods of analysis for individual and population growth data. The book contains the basic mathematical and statistical tools required to understand the concepts of the methods under discussion and worked examples of analyses, but it is neither a mathematical treatise, nor a recipe book for growth data analysis. Aimed at junior and senior researchers involved in the analysis of human growth data, this book will be an essential reference for anthropologists, auxologists and paediatricians.

ROLAND C. HAUSPIE is Professor of Human Biology in the Laboratory of Anthropogenetics at the Free University of Brussels. His main research interests are in human growth and development, especially the influence of genetic and environmental factors on growth and development.

NOËL CAMERON is Professor of Human Biology in the Department of Human Sciences at Loughborough University. His current research interests include factors affecting the development of peak bone mass during adolescence, factors influencing the acquisition of obesity during childhood and adolescence, and the determinants of sedentary behaviour in adolescents.

LUCIANO MOLINARI is a senior researcher at the University Children's Hospital in Zurich. He is in charge of the statistical analysis of the Zürich Longitudinal Studies and in the Electroencephalography Department, and is a statistical consultant within the Children's Hospital as a whole.

Cambridge Studies in Biological and Evolutionary Anthropology

Series editors

Selected titles also in the series

25 *Human Growth in the Past* Robert D. Hoppa & Charles M. FitzGerald (eds.) 0 521 63153 X
26 *Human Paleobiology* Robert B. Eckhardt 0 521 45160 4
27 *Mountain Gorillas* Martha M. Robbins, Pascale Sicotte & Kelly J. Stewart (eds.) 0 521 76004 7
28 *Evolution and Genetics of Latin American Populations* Francisco M. Salzano & Maria C. Bortolini 0 521 65275 8
29 *Primates Face to Face* Agustin Fuentes & Linda D. Wolfe (eds.) 0 521 79109 X
30 *Human Biology of Pastoral Populations* William R. Leonard & Michael H. Crawford (eds.) 0 521 78016 0
31 *Paleodemography* Robert D. Hoppa & James W. Vaupel (eds.) 0 521 80063 3
32 *Primate Dentition* Davis R. Swindler 0 521 65289 8
33 *The Primate Fossil Record* Walter C. Hartwig (ed.) 0 521 66315 6
34 *Gorilla Biology* Andrea B. Taylor & Michele L. Goldsmith (eds.) 0 521 79281 9
35 *Human Biologists in the Archives* D. Ann Herring & Alan C. Swedlund (eds.) 0 521 80104 4
36 *Human Senescence Life Span* Douglas E. Crews 0 521 57173 1
37 *Patterns of Growth and Development in the Genus* Homo. Jennifer L. Thompson, Gail E. Krovitz & Andrew J. Nelson (eds.) 0 521 82272 6
38 *Neanderthals and Modern Humans* Clive Finlayson 0 521 82087 1

Methods in Human Growth Research

Edited by

Roland C. Hauspie
Free University of Brussels, Belgium

Noël Cameron
Loughborough University, UK

Luciano Molinari
Kinderspital, Zürich, Switzerland

CAMBRIDGE
UNIVERSITY PRESS

PUBLISHED BY THE PRESS SYNDICATE OF THE UNIVERSITY OF CAMBRIDGE
The Pitt Building, Trumpington Street, Cambridge, United Kingdom

CAMBRIDGE UNIVERSITY PRESS
The Edinburgh Building, Cambridge, CB2 2RU, UK
40 West 20th Street, New York, NY 10011–4211, USA
477 Williamstown Road, Port Melbourne, VIC 3207, Australia
Ruiz de Alarcón 13, 28014 Madrid, Spain
Dock House, The Waterfront, Cape Town 8001, South Africa

http://www.cambridge.org

First published 2004

Printed in the United Kingdom at the University Press, Cambridge

Typeface Times 10/12.5 pt. *System* LaTeX 2_ε [TB]

A catalogue record for this book is available from the British Library

Library of Congress Cataloguing in Publication data
Methods in human growth research / edited by Roland C. Hauspie, Noël Cameron, Luciano Molinari.
p. cm. – (Cambridge studies in biological and evolutionary anthropology ; 39)
Includes bibliographical references and index.
ISBN 0 521 82050 2
1. Human growth – Research. 2. Human growth – Longitudinal studies.
3. Human growth – Statistical methods. 4. Population research. 5. Physical anthropology – Methodology. I. Hauspie, Roland. II. Cameron, Noël.
III. Molinari, Luciano, 1944– IV. Series.
GN62.9.M48 2004
559.9′01 – dc22 2003055593

ISBN 0 521 82050 2 hardback

Contents

Contributors

Elizabeth Barden
Division of Gastroenterology and Nutrition
The Children's Hospital of Philadelphia
34th Street and Civic Center Boulevard
Philadelphia, PA 19104-4399, USA

Adam Baxter-Jones
College of Kinesiology
University of Saskatchewan
87 Campus Drive
Saskatoon, SK S7N 5B2, Canada

Darrell Bock
Department of Psychology
University of Chicago
5848 South University Avenue
Chicago, IL 60637, USA

Noël Cameron
Department of Human Sciences
Lougborough University
Loughborough LE11 3TU, UK

Julio Cesar Carrillo
Auxology Unit
Anthropology Department
National University of Colombia
Bogota A.A. 144490, Colombia

Stefan A. Czerwinski
Department of Community Health
Lifespan Health Research Center
Wright State University School of Medicine
3171 Research Blvd.
Kettering, OH 45420-4014, USA

Stephen du Toit
Scientific Software International
7383 North Lincoln Avenue
Chicago, IL 60647-1704, USA

Edward A. Frongillo
Division of Nutritional Sciences
Cornell University
B17 Savage Hall
Ithaca, NY 14853-6301, USA

Theo Gasser
Department of Biostatistics
University of Zürich
Sumatrastrasse 30
Zürich, CH-8006, Switzerland

Daniel Gervini
Department of Biostatistics
University of Zürich
Sumatrastrasse 30
Zürich, CH-8006, Switzerland

Jacques A. Hagenaars
Department of Methodology and Statistics
Faculty of Social and Behavioural Sciences
Tilburg University
P.O.Box 90153
Tilburg, 5000 LE, The Netherlands

Roland Hauspie
Laboratory of Anthropogenetics
Free University Brussels
Pleinlaan 2
Brussels, B-1050, Belgium

John Himes
Division of Epidemiology
School of Public Health
University of Minnesota,
1300 South 2nd Street
Minneapolis, MN 55454-1015, USA

Marie-José Ireton
Auxology Unit
Anthropology Department
National University of Colombia
Bogota A.A. 144490, Colombia

Robert Mirwald
College of Kinesiology
University of Saskatchewan
87 Campus Drive
Saskatoon, SK S7N 5B2, Canada

Luciano Molinari
Abteilung EEG und AWE
Kinderspital Zürich
Steinwiesstrasse 75
Zürich, CH-8006, Switzerland

Bradford Towne
Lifespan Health Research Center
Wright State University School of Medicine
3171 Research Blvd.
Kettering, OH 45420-4014, USA

Gino Verleye
Faculty of Political and Social Sciences
Rijksuniversiteit Gent
Universiteitsstraat 8
Gent, B-9000, Belgium

Jeroen K. Vermunt
Department of Methodology and Statistics
Faculty of Social and Behavioural Sciences
Tilburg University
P.O.Box 90153
Tilburg, 5000 LE, The Netherlands

Babette Zemel
Division of Gastroenterology and Nutrition
The Children's Hospital of Philadelphia
34th Street and Civic Center Boulevard
Philadelphia, PA 19104-4399, USA

Foreword

The study of human growth is one of the most fascinating areas of human biology, but also one of the most difficult from a methodological point of view. However, the application of appropriate methods for collecting and analysing cross-sectional, longitudinal and mixed-longitudinal growth data is a prerequisite for a correct understanding of the dynamics of the individual and average growth patterns in humans. The purpose of this book is two-fold: first, it provides an extensive and critical review of modern techniques to assess and collect data on growth and maturation, and of methods to analyse individual and population growth data; second, it gives a practical help in choosing the most appropriate method for specific questions.

The chapters in this volume are arranged into four parts. Part I deals with the characteristics of growth studies and the methodological issues involved. Himes (Chapter 1) introduces the very purpose of studying human growth focusing on some important types of postnatal growth studies and their chief contributions. The author emphasizes on the nature of growth and maturation and the underlying principles and applications related to child health. Molinari and Gasser, in Chapter 2, examine the human growth curve in terms of distance, velocity and acceleration, and caution against the incorrect use of velocity charts, which could lead to serious misinterpretation of the growth data. They introduce the basic concepts necessary to analyse individual series of growth measurements. In Chapter 3, Frongillo gives directions for the collection of representative samples for growth studies and discusses the use of growth data to assess, monitor and survey disease in epidemiological settings. This chapter specifically discusses sampling schemes, characteristics to evaluate sampling, types of samples, and implications of the type of sampling for analysis, sample size considerations and examples of sampling strategies. Cameron (Chapter 4) deals with measuring growth using primarily anthropometric techniques and external dimensions of the body such as height, weight and measures of subcutaneous fat. In Chapter 5 he discusses the way in which maturity may be assessed, concentrating on the assessment of the process of maturation from birth through childhood and adolescence, i.e. the period of time in which maturation interacts with growth. Zemel and Barden (Chapter 6) discuss the techniques for assessing body composition, including an extensive discussion of those techniques most feasible and practical for use with children. They also review considerations in the methodological assumptions, and the analysis and presentation of body composition results for children.

Part II covers the methods for modelling individual growth. Gasser, Gervini and Molinari (Chapter 7) give an overview of non-parametric approaches, which have the potential to describe in a flexible way special features of the growth curve. They concentrate on kernel estimators, shape-invariant modelling and structural analysis. In Chapter 8, Hauspie and Molinari discuss the use of parametric models to describe the human growth curve after birth, and emphasize some of the potentials and limitations of these approaches. The chapter covers a number of non-linear growth models suitable for fitting infant growth, adolescent growth, and the complete growth cycle. The authors clearly illustrate the application of the various models on the basis of examples and further focus on aspects of goodness of fit and the consequences of violation of the basic assumptions required for curve fitting. Bock and du Toit (Chapter 9) deal with the estimation methods in non-linear models, such as conditioned maximum-likelihood estimation and empirical Bayes estimation. They also deal with non-homogeneous and autocorrelated residuals. The results are also illustrated with two well-known growth models applied to data from the Fels Longitudinal Growth Study.

Methods for the assessment of population growth are considered in Part III. This includes some special topics, that will be of particular relevance for clinical auxologists. Frongillo (Chapter 10) reviews methods for constructing human growth references addressing questions like: 'What are the purposes and uses of growth references?', 'How are growth references presented?', 'What design features are needed for studies to construct growth references?', 'What analytical methods are used to construct growth references?', 'What are some recent examples of how growth references were constructed?' and 'What are future challenges for the construction of growth references?'. Verleye *et al.* (Chapter 11) introduce structural equation modelling in the analysis of human growth data. This rather new approach in auxology combines the benefits of multiple regression and principal components analysis and allows the study of complex systems of interrelationships between biometric variables and genetic and/or environmental or demographic factors. Baxter-Jones and Mirwald (Chapter 12) explore the advantages of multilevel modelling in the analysis of longitudinal growth data. This technique allows the identification of independent inter-group effects while simultaneously controlling the effects of growth and maturation within each individual. Czerwinski and Towne (Chapter 13) provide an overview of the methods currently available for studying the genetic epidemiology of normal human growth and development, including most of the technological innovations over the past two decades that have enabled researchers to investigate issues that were previously intractable. In Chapter 14, Cameron discusses the methods currently available for prediction in both cross-sectional and longitudinal settings. In the former case, the prediction consists of estimating a body dimension or body feature from one or several other features measured at the same time, while in the latter case prediction consists of estimating bodily characteristics at some future time on the basis of information obtained at the present time. The typical example of predicting adult height, as well as its limitations when applied to pathological conditions, is also discussed. Finally, in Chapter 15, Vermunt and Hagenaars give suggestions for the analysis of ordinal longitudinal data. These methods can be applied to variables of developmental stages, which may be classified into ordinal categories, and behavioural variables, which are repeatedly measured by discrete ordinal scales.

All contributors have succeeded in making understandable the concepts that underlie the described methodologies, and have produced a volume that responds to a long-lasting need among researchers. This book will be of particular value to workers in clinical science, nutrition, anthropology and public health.

Professor Giulio Gilli
Turin, Italy

Acknowledgements

Georges-Louis Leclerc Count of Buffon (1707–88) was a great French naturalist, and famous from his monumental life's work, *Histoire naturelle*. In 1777, Buffon published the growth records of the son of Count Philibert Guéneau de Montbeillard (1720–85), a friend and admirer of Buffon's work, in one of the supplements to his series of Natural History volumes.[1] This was the first longitudinal growth study on data collected on an individual from birth to 18 years of age between 1759 and 1777.[2] The cover shows the famous height data of Montbeillard's son, together with a mathematical model fitted to the distance values, the yearly increments and the instantaneous velocity curve. In the background is a picture of Buffon after an original engraving.

[1] J. M. Tanner, *A History of the Study of Human Growth*. Cambridge: Cambridge University Press, 1981.

[2] N. Cameron, *Human Growth and Development*. New York: Academic Press, 2002.

Part I

Growth data and growth studies: characteristics and methodological issues

1 *Why study child growth and maturation?*

John H. Himes
University of Minnesota

Introduction

On a recent visit with family to an amusement park, a sign welcoming customers to one of the gyrating rides announced a vetting criterion for proceeding and partaking of the thrills: 'Riders Must Be At Least This Tall.' The sign pictured a monkey pointing to an elongated horizontal banana about 120 cm from the ground. I wondered at the time about the actual scientific rationale for such a height requirement, *inter alia*, estimating body weight for mechanical counterbalance, estimating age for appropriate behaviour, estimating sitting height for reaching the head rest, estimating minimum waist girth for the safety belt. In any case, this rather pedestrian application of child growth data, probably for some human engineering or safety purpose, is an example of the widely divergent applications of data pertaining to growth and maturation of children.

Even though the actual rationales and applications of growth data vary tremendously, to some degree, they all are grounded in and emanate from auxological studies. Careful measurement, development of relevant theory and statistical analysis are required to answer the myriad research questions related to particular applications. It is, then, the collective scientific, biomedical and practical importance of the many applications that provides the justification for studying child growth and maturation.

In this chapter, I shall briefly present some important types of postnatal growth studies and their chief contributions. At selected junctures I shall elaborate on issues pertaining to the nature of growth and maturation, with particular emphasis on underlying principles and applications related to child health. With no attempt to be exhaustive, the commentaries reflect personal experience, recurrent musings and pet peeves in the field. Of course, which of the

Methods in Human Growth Research, eds. R. C. Hauspie, N. Cameron and L. Molinari. Published by Cambridge University Press.

studies and issues are the most important depends on to whom the question is put. Certainly, there are many other types of growth studies and applications of results that are important in their own spheres, viz. foetal growth, neurological maturation, human engineering and clinical syndromes. Many specific kinds of growth studies and analysis are discussed elsewhere in this volume.

Definitions and foci

The terms 'growth' and 'maturation' have, at times, been used broadly to include all developmental processes, for example, cognitive growth and social maturation (Salkind, 2002); nevertheless, I shall use stricter definitions and limit my focus to physical aspects. Growth will be considered as increases in size or mass, primarily resulting at the cellular level from hypertrophy and hyperplasia, and from interstitial accretion (Thompson, 1971). Accordingly, cells, tissues, organelles, organs, organ systems and organisms all grow. I shall focus primarily on growth at the organismic level, at the level of the whole child and, to a lesser degree, the growth of selected organ systems.

Often, attained size or body dimensions are used as measures of child growth and, consistent with the above definition, they are considered cumulative measures of past changes in size or mass. Not surprisingly then, much attention has been given to the variation in stature and weight of children at a chronological age, and to the velocity of growth, standardized relative to passage of time, e.g. centimetres per year, kilograms per year.

Maturation will be considered as the progressive achievement of adult status, and it comprises a subset of developmental changes that include morphology, function and complexity (Todd, 1937). So, the progressive acquisition of secondary sexual characteristics, such as breast development in girls and facial hair in boys, are common examples of maturation during adolescence. In the case of sexual maturation, progress is usually assessed by attainment of stages described mostly in qualitative terms (Reynolds and Wines, 1948, 1951; Tanner, 1962). Size and morphological variables can be scaled continuously relative to achievement of adult status and they therefore become maturational in meaning; for example, percentage of adult stature attained (Roche *et al.*, 1983) and scoring schemes for skeletal maturity (Roche *et al.*, 1975; Tanner *et al.*, 1983). A major distinction between size per se and maturation is that all children eventually achieve the same adult maturity, whereas adult size and body dimensions vary considerably. I shall focus mostly on maturation of morphology and function rather than skills and behaviour. For example, no mention will be made of maturational changes in motor development and physical performance.

The relative rate of maturation sometimes has been referred to as the tempo of growth (Tanner, 1962), and accordingly, children may be identified as 'early maturing' or 'late maturing' compared with population norms. Also, the chronological age or timing at which a maturational event occurs relative to some population norm is a measure of the rate of maturational progress, e.g. age of menarche.

Applications of growth and maturational data may differ according to the appropriate level of inference. A clinician may measure a child's weight and compare it to reference data to estimate a centile or *Z*-score to assist with diagnosis of the individual child. An epidemiologist may measure the weight of many children, compare them to reference data, and calculate mean *Z*-scores or the proportion above or below a *Z*-score cut-off and make inferences about a whole group or community. Although both applications may be appropriate and valid, individual and group inferences have different minimum requirements for the quality of data, and the nature of the variables used (World Health Organization, 1995).

For example, the observed number of the first 28 permanent teeth erupted has been suggested as an indicator of biological maturation during the ages when the permanent dentition usually emerges (5–13 years). Erupted teeth can be counted reliably, the number increases monotonically, the count is sensitive to environmental influences, and it has reasonable validity relative to other measures of maturation (Filipsson *et al.*, 1978; Hagg and Taranger, 1985). Nevertheless, the variation in chronological age within a given tooth count in healthy children is so large (SD = 0.5–1.3 years) relative to the mean change in age associated with the eruption of an additional tooth ($\approx$0.24 year) that tooth counts cannot distinguish between meaningful grades of maturational progress for individual children. In this instance, chronological age is important because it represents the average maturational progress. Nevertheless, the number of erupted permanent teeth remains an acceptable measure of maturation for group comparisons of children, although it is seldom used outside orthodontia.

Understanding human biology and physiology

A fundamental impetus to study growth and maturation is to understand the basic biological processes and relationships. Understandably, considerable attention has been given to investigating hormonal control of growth and maturation. Hormones influence growth and maturation, beginning *in utero* and continuing throughout childhood and adolescence. Studies of normal children and pathological conditions have provided fundamental understanding of physiological

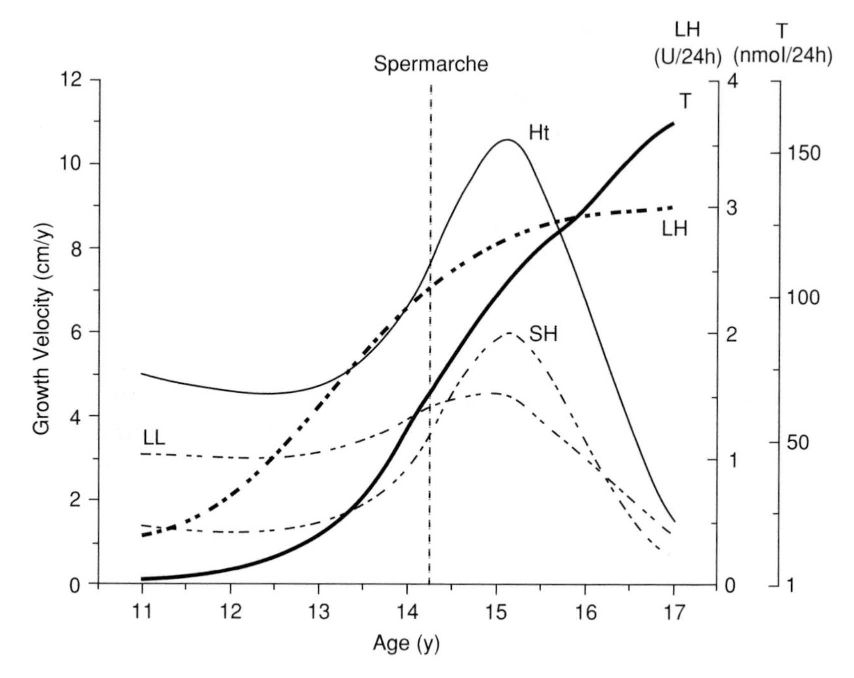

Figure 1.1 Growth velocities for height (Ht), sitting height (SH) and leg length (LL) and 24-hour urinary concentrations of luteinizing hormone (LH) and testosterone (T) for boy no. 31 (redrawn from the data of Nielsen *et al.*, 1985).

mechanisms, established normal developmental patterns, and aided in developing therapies for endocrine-related disorders (Styne, 2001).

Figure 1.1 presents peri-adolescent growth velocities for a single boy relative to spermarche and urinary concentrations of luteinizing hormone (LH) and testosterone (T) (Nielsen *et al.*, 1986a). Velocities were calculated by applying the model of Preece and Baines (1978) to serial measurements, and spermarche was based on spermaturia (Nielsen *et al.*, 1986b). Relative to normative data, the boy would be considered as rather late maturing because the timing of his adolescent spurts are delayed approximately 12–14 months.

The figure is a busy but a good example of the types of information obtained from such studies, e.g. patterns and levels of hormone excretion, magnitude and timing of growth spurts in stature and body segments, relationships among onset of spermarche, peak velocities and hormone excretion. In addition, this particular study provided median values for LH and T excretion and pubic hair stages at the ages of maximum acceleration, peak velocity and maximum deceleration in the body segments, as well as average age intervals between chief features of the curves.

Genetic studies of growth and maturation are important for identifying the hereditary control and expression of these processes, and for quantifying the extent of genetic determination. With growth and maturation, the phenotypes studied are usually measured on a continuous scale so the discrete effect of a gene or set of genes cannot be easily distinguished. Just as important, however, is to identify those aspects that are not under genetic control and therefore considered to be influenced primarily by disease, nutrition or other aspects of the environment.

Genetic studies encompass an enormous range from molecular probes and genome-wide scans to identify specific genomic regions associated with growth (Rao and Majumder, 2001) to population studies of the genetics of body proportions (Bogin *et al.*, 2001); from twin studies of muscle fibre types (Bouchard *et al.*, 1986), to clinical tools for accounting for the stature of parents when evaluating that of a child (Himes *et al.*, 1985). Medical geneticists are very interested in normal and abnormal growth in a wide variety of anthropometric dimensions because of the many presentations of dysmorphic features in genetic and congenital disorders (Meaney and Farrer, 1986; Hall *et al.*, 1989). Some methods for the study of the genetics of growth are detailed in Chapter 13 in the present volume.

The description of the normal progress of childhood growth has been central for almost all applications of growth and maturational data, and there is a long history of such studies (Boyd *et al.*, 1980; Tanner, 1981). Normal patterns and features of the growth curve have been described and, accordingly, auxologic jargon has been coined, e.g. preadolescent fat wave, adiposity rebound, mid-growth spurt, age at take-off, saltatory growth. Usually, the attempt to clarify concepts and standardize scientific discourse through this new language has been successful.

An important approach to describing normal growth has been the mathematical modelling of the growth curve to describe, summarize and quantify features and patterns, usually based on long-term longitudinal data for individuals. The early approaches considered serial data for small segments of the curve, such as infancy (Jenss and Bayley, 1937) or the adolescent spurt (Deming, 1957). Subsequently, the entire growth curve from infancy through young adulthood was fitted and parameterized, using single or linked non-linear functions for different portions of the curve (Preece and Baines, 1978; Bock and Thissen, 1980; Karlberg, 1989). These approaches have been major steps forward in understanding basic relationships among growth and maturational features and have allowed the summarization of large amounts of growth data into meaningful parameters and patterns. Several of the chapters in the present volume explain and elaborate such approaches, and provide examples of their applications (e.g. Gasser and colleagues, Hauspie and Molinari, Bock and du Toit).

A persistent aspect of growth studies has been the notion of variation, normal and abnormal. Variation in size, rates of growth and timing of maturational stages and events has been studied at many levels: across populations, across groups within populations, within populations over time, across individuals within populations, and within individuals.

The focus on variation arose historically for several reasons and it has many valid rationales. Some early workers considered anthropometric dimensions to characterize population types and, implicitly, congenital determinism (Martin, 1928). Consequently, variation across populations was seen primarily in racial or genetic terms. Nevertheless, it was appreciated rather early that phenotypic variation in human growth has both genetic and environmental components, and that growth is sensitive to major environmental influences, e.g. war and poverty (Sanders, 1934). Accordingly, differences among populations and even among groups within populations might be attributed to such environmental influences. Similarly, variation within a population over time may reflect temporal or secular changes in the same environment, or in new environments for immigrant populations. Human growth has been a traditional area of emphasis within physical anthropology, and growth variation among individuals and populations has been studied in terms of evolutionary selection and environmental adaptation (Bogin, 1999).

A significant focus on variation in individual growth arises from the fact that growth and maturation are sensitive to nutrition and disease, and conversely, that growth may be used as an indicator of health and well-being. Growth variation within individuals, then, is of particular interest because of the clinical and health applications, but also because repeated measurements on the same individual over short periods of time allow evaluation of the reliability and errors associated with measurement protocols themselves (Himes, 1989).

Growth charts, references and standards

An important research thrust has been to provide high-quality growth reference data or standards that provide a basis of comparisons of individuals and groups for a wide range of purposes. The technical aspects of studies to develop normative growth reference data or standards have evolved so that the best results require rigorous planning, sampling, measurement, data management and statistical analysis. For example, the development of the 2000 CDC Growth Charts (Kuczmarski *et al.*, 2000) for the USA began with planning meetings in 1988. Glib comments about sluggish government bureaucracy aside, such efforts are huge undertakings, with many technical issues to be resolved (Roche, 1992); much expertise and many hands are required. Even the logistical issues attending an effort like the new World Health Organization Multicentre Growth Reference

Study (de Onis *et al.*, 2001), which will involve 10 000 children in six countries and many collaborators, should give investigators pause before casually embarking on such efforts. Issues related to the technical development, presentation, implementation, interpretation and theoretical appropriateness of growth reference data and standards are many and complicated, and have been much discussed (Goldstein and Tanner, 1980; Johnston and Ouyang, 1991; Cole *et al.*, 1998; Cameron, 1999; Himes, 1999a).

Figure 1.2 presents the weight-for-age centiles for girls, 2–20 years of age, from the CDC Growth Charts (Kuczmarski *et al.*, 2000) as an example. This typical presentation describes the basic pattern of attained weight, the cumulative sum of prior growth and the distributions of weight attained at each age. The selected centiles were calculated within age groups and were smoothed mathematically across the ages, assuming that this is the best single representation of the growth process. The centiles allow comparisons of individuals or groups with the distribution from the reference population, often with implications of normality or health. Obviously, the inferences made from such reference curves depend on the referent population used for developing them. Age- and gender-specific *Z*-scores may be calculated from the actual data values to allow evaluation of extreme measurements beyond the selected centiles, and to facilitate comparisons across genders and ages, and comparisons with other growth variables. Such attained-size or distance growth charts are the most commonly used type of reference, and they are a fundamental tool in paediatrics and public health.

Ideally, such charts are developed for several anthropometric dimensions and age groups from the same reference population. For example, the CDC Growth Charts include separate gender-specific charts for weight-for-age (birth–36 months; 2–20 years); length-for-age (birth–36 months); weight-for-length (birth–36 months, 45–103 cm length); head circumference-for-age (birth–36 months); stature-for-age (2–20 years); weight-for-stature (77–121 cm stature); body mass index (2–20 years).

While it is customary to consider growth in a body measurement in terms of age or the passage of time, it may also be viewed relative to, or conditioned upon, a different body measurement. Consequently, the weight-for-stature charts, considering body weight distributions relative to specific statures, allow for evaluations of weight variation largely excluding effects of stature (Figure 1.3). Because age is not used when applying the charts, this approach assumes that the relationship of weight as a function of stature is the same throughout the age range considered, so that children who are unusually short or tall for their ages are evaluated without bias. Actually, the age-specific weight-for-stature slopes systematically increase as age increases during this period but the practical effects of this violated assumption are probably small (Roche, 1992).

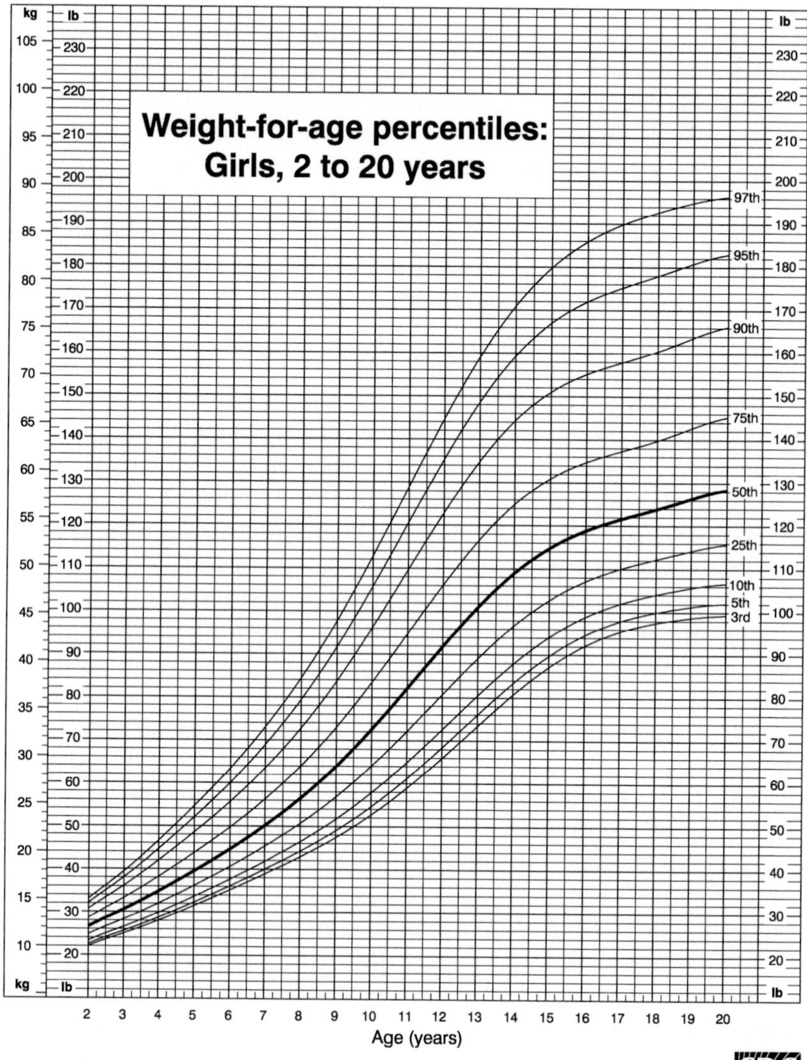

Published May 30, 2000.
SOURCE: Developed by the National Center for Health Statistics in collaboration with the National Center for Chronic Disease Prevention and Health Promotion (2000).

Figure 1.2 Weight-for-age percentiles for US girls, 2–20 years (from Kuczmarski *et al.*, 2000).

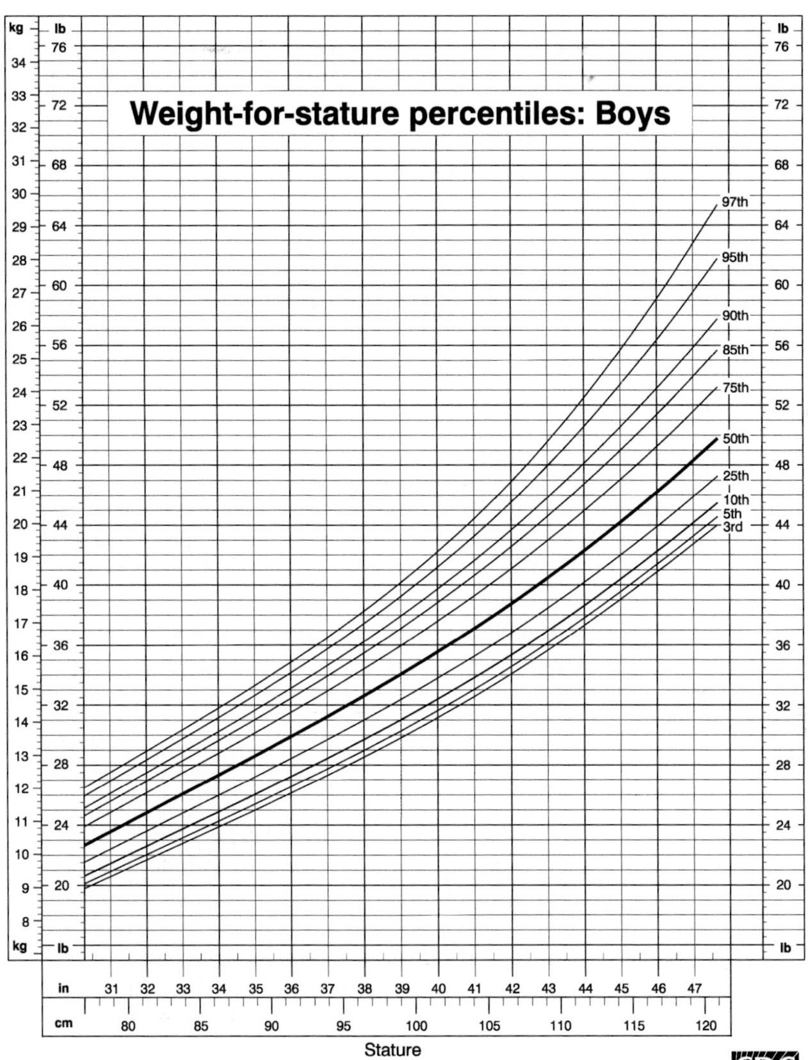

Figure 1.3 Weight-for-stature percentiles for US boys (from Kuczmarski *et al.*, 2000).

Weight-for-length and weight-for-stature are useful because weight is relatively more labile and sensitive to short-term effects of nutrition and infection than length or stature. A similar rationale was used for the QUAC (Quaker arm circumference) stick (although not strictly reference data) that relates arm circumference to stature (Arnhold, 1969; Sommer and Loewenstein, 1975). This was originally developed for use during famine and crisis situations where rapid assessment is needed and children's ages are often unknown. More complete reference data for arm circumference-for-stature have been recently provided (Mei *et al.*, 1997). Head circumference reference data conditioned upon recumbent length have been developed for preterm infants to account for the relationship between these two correlated measurements (Roche *et al.*, 1997).

Reference data for ratios of two measurements (or functions of measurements) are related to conditional reference data, in as much as they summarize the relationship between two dimensions, although they are usually presented relative to age. The most common reference data of this type currently are for body mass index (BMI), which describes the ratio of weight (kg) : stature $(m)^2$. While the term BMI was first coined by Ancel Keys in 1972 (Keys *et al.*, 1972), the ratio appeared in the literature more than a century ago (Quetelet, 1869).

BMI has become the preferred anthropometric weight–stature indicator for children, adolescents and adults describing overweight and obesity (Himes and Dietz, 1994; World Health Organization, 1995, 1998). The rationale of using BMI is based on its associations with total body fat and its associations with concurrent and subsequent risk factors, morbidity and mortality. BMI is not free from technical issues regarding the exact power transformation of stature that is best (Cole, 1991), and specificity of BMI to body fat (Himes and Bouchard, 1989). Nevertheless, the convenience of having a single ratio or index that is reasonably good for everyone at all ages has caught on in the scientific and practice communities. Other growth-related ratios used are head circumference : chest circumference for nutrition assessment (Malina, 1991), waist : hip (and other trunk versus extremity ratios for skinfolds) for fat distribution (Bouchard and Johnston, 1988) and sitting height : stature for endocrine evaluations (Styne, 2001).

The reference data addressed above concern the attained size and proportions of children at an age. Presentations are usually based on large representative samples of children, measured on a single occasion, and centiles are calculated within small chronological age groups. Such reference data are appropriate for assessments of single observations of individuals and for groups with a small range of ages.

The attained-size reference data are often used inappropriately, however, to follow the serial measurements of individual children. While such practice may be useful for evaluating some patterns and trajectories of growth, it can be misleading during adolescence when maturation-related growth and the variation in maturational timing are large (Himes, 1999b). Accordingly, tempo of growth was incorporated into distance growth charts to evaluate the growth of individuals in light of typical patterns and variation in size related to children maturing at rates considered relatively early, average and late (Bayer and Bayley, 1959; Tanner *et al.*, 1966). The basic data for these tempo-specific standards are usually from long-term longitudinal studies on a relatively small number of individuals, and the relative tempo designations are usually based on the age of the maximum velocity in statural growth during adolescence (peak height velocity).

The actual growth, viz. change, in body dimensions is best evaluated relative to reference data that describe the velocity of growth, usually presented as increments relative to a time interval that is useful for measurement and for timely intervention, e.g. 1 month, 3 months, 6 months, 1 year. From a purely statistical perspective, the minimum time between measurements that is appropriate for meaningfully evaluating the growth of an individual child varies across age and growth variables, and depends upon the reliability of measurements, the expected velocity of growth and the observed variation in growth (Himes, 1999c). From a clinical perspective, the appropriate interval between serial growth measurements should be determined by the nature of the targeted disease or treatment, but it should not be less than the statistical limits if the chief purpose is to detect growth.

For incremental growth charts it is important to distinguish between appropriate applications for groups and individuals, especially during adolescence. Figure 1.4 presents the median stature velocities for only those American boys with peak height velocities at the average age (Tanner and Davies, 1985) plotted on the cross-sectional incremental reference data from the Fels Longitudinal Study in Ohio, USA (Roche and Himes, 1980; Baumgartner *et al.*, 1986).

The Roche and Himes reference centiles were calculated on the distribution of increments for all boys and smoothed across age using low-term Fourier transformations. The pattern of growth of an individual boy with average timing of his spurt would be expected to approximate the bold longitudinal curve which has a much sharper increase in growth velocity than the cross-sectional reference data because the latter include the full mix of early-, average- and late-maturing boys at any given age. Consequently, while the average growth increment calculated for groups can be usefully compared to these cross-sectional incremental reference data, no single individual would be expected to follow them. Clinical longitudinal growth charts, taking into account the effects of maturational timing during adolescence, have been developed for several populations (Tanner and Whitehouse, 1976; Hauspie and Wachholder, 1986;

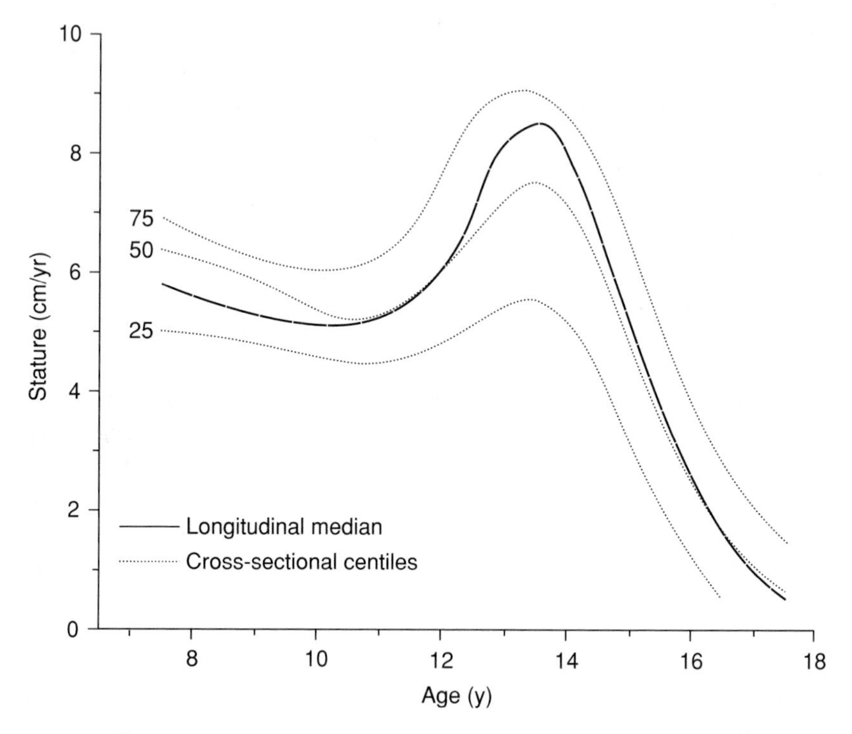

Figure 1.4 Longitudinal median stature velocity for average-maturing boys (Tanner and Davies, 1985) and selected cross-sectional centiles for stature increments (Roche and Himes, 1980).

Hoey *et al.*, 1987). In order to be used effectively, they require frequent measurements during adolescence.

At ages before adolescence, the relative effects of variation in maturational timing are much less and incremental reference data based only on chronological age are usually appropriate for individuals and groups. Such data are available for several growth variables, and are particularly useful during infancy and the first years of life when growth proceeds rapidly and there is great interest in evaluating children for adverse effects of nutrition or disease (Guo *et al.*, 1988, 1991).

Preterm or premature infants (born before 37 weeks gestation) are a special case where the descriptor 'premature' actually refers to the maturation or completion of the average gestational period (40 weeks). These infants are also biologically premature, e.g. in pulmonary function, neurological development and skeletal growth. They follow a different growth curve than infants born at term and they require their own reference data relative to chronological age if they are to be evaluated accurately relative to their preterm peers (Casey *et al.*, 1991; Guo *et al.*, 1996).

Actual centile reference data for evaluating the timing of sexual maturation stages are very few (van Wieringen, 1971; Tanner and Whitehouse, 1976; Taranger *et al.*, 1976). The timing is scaled as chronological age and the centiles for entrance into a pubertal stage are then centiles of ages. In practice, mean or median ages at pubertal stages or events, e.g. menarche or spermarche, are usually used for evaluating relative timing. The best of these data provide the mean or median ages, estimated from dichotomous responses in surveys of whether or not the stage is present or an event has occurred. Analysis with probits or logits on the cumulative distribution of maturational status by age groups fits theoretical curves to the distributions and provides estimates of the ages at which 50% of the individuals achieved the stage or event studied (Neter *et al.*, 1996). Such age estimates should be interpreted as estimates of the age at the entrance into the stage or the onset of the event and not the mean age of all those who are at a particular stage. The latter is influenced by the duration of the particular stage and the shape of the distribution of ages at entry into the stage.

Sun *et al.* (2002) recently published nationally representative median ages for entry into sexual maturity stages for US children in three race/ethnicity groups (Table 1.1). Because these are population data, differences in median timing among groups are due to the combined effects of genetics and environmental factors. Such data are useful to judge normality in timing of stages, durations of stages and differences in relative timing among population groups.

For example, within the US population, non-Hispanic black girls have the earliest median onset of puberty (pubic hair stage 2, breast stage 2) at about 9.4 years, compared with non-Hispanic white girls and Mexican-American girls. In boys, a similar racial/ethnic pattern in initial timing exists, and the median ages for onset of pubescent genital changes (stage 2) actually precede those for the median ages for onset of pubic hair and breast development in girls. This apparent male precedence in onset of puberty may be an artefact of the observational methodology to assess initial enlargement of the testes.

Testicular size is an interesting example of a measurement that has been both misused and used appropriately as a measure of maturation in boys. Testicular size is usually reported as a volume (ml). Linear dimensions of a testicle can be measured and used to calculate the volume (assuming an ellipsoid model) (Steeno, 1989) but currently, testicular size is usually estimated by palpation and comparison with three-dimensional models of known volumes (Zachmann *et al.*, 1974). Testicular size increases dramatically during puberty, primarily reflecting the response of the seminiferous tubules to the sharp increases in follicle-stimulating hormone (FSH), and increase in testicular size is a corollary

Table 1.1 *Median ages of entry into each stage and fiducial limits in years for boys and girls by race*

	Age at entry for boys							Age at entry for girls					
	Non-Hispanic white		Non-Hispanic black		Mexican-American			Non-Hispanic white		Non-Hispanic black		Mexican-American	
Stage[a]	Median	FL[b]	Median	FL[b]	Median	FL[b]		Median	FL[b]	Median	FL[b]	Median	FL[b]
Pubic hair							*Pubic hair*						
PH2	11.98[c]	11.69–12.29	11.16[c]	10.89–11.43	12.30[c]	12.06–12.56	PH2	10.57[c]	10.29–10.85	9.43[c]	9.05–9.74	10.39	—
PH3	12.65	12.37–12.95	12.51[c]	12.26–12.77	13.06[c]	12.79–13.36	PH3	11.80[c]	11.54–12.07	10.57[c]	10.30–10.83	11.70[c]	11.14–12.27
PH4	13.56	13.27–13.86	13.73	13.49–13.99	14.08	13.83–14.32	PH4	13.00[c]	12.71–13.30	11.90[c]	11.38–12.42	13.19[c]	12.88–13.52
PH5	15.67	15.30–16.05	15.32	14.99–15.67	15.75	15.46–16.03	PH5	16.33[c]	15.86–16.88	14.70[c]	14.32–15.11	16.30[c]	15.90–16.76
Genitalia development							*Breast development*						
G2	10.03	9.61–10.40	9.20[c]	8.62–9.64	10.29[c]	9.94–10.60	B2	10.38[c]	10.11–10.65	9.48[c]	9.14–9.76	9.80	0–11.78
G3	12.32	12.00–12.67	11.78[c]	11.50–12.08	12.53[c]	12.29–12.79	B3	11.75[c]	11.49–12.02	10.79[c]	10.50–11.08	11.43	8.64–14.50
G4	13.52	13.22–13.83	13.40	13.15–13.66	13.77	13.51–14.03	B4	13.29[c]	12.97–13.61	12.24[c]	11.87–12.61	13.07[c]	12.79–13.36
G5	16.01[c]	15.57–16.50	15.00[c]	14.70–15.32	15.76[c]	15.39–16.14	B5	15.47[c]	15.04–15.94	13.92[c]	13.57–14.29	14.70[c]	14.37–15.04

[a] Definitions of the stages are given in Sun *et al.* (2002).
[b] FL indicates fiducial limit. Calculated 98.3% FLs to adjust for multiple comparisons between races for an overall α of 0.05.
[c] Significant pair-wise racial difference, $P < 0.05$.
Source: Sun *et al.* (2002).

to functional maturation and the production of sperm (Styne, 2001). According to our original definitions, however, the size or volume per se of a testicle at a given age constitutes cumulative growth, not maturation. Moreover, there is considerable variation in the size of testicles among adult men (Beres *et al.*, 1989). Consequently, the variation in testicular volume at an age is, strictly speaking, an assessment of testes growth, albeit with some important functional maturation implications.

Alternatively, some thresholds of testicular volume have been identified as indicative of the beginning of puberty (4 ml), and as within the range of normal adulthood (12 ml) (Marshall and Tanner, 1970). More meaningful thresholds in terms of maturation might be determined relative to actual functional status. Nevertheless, reference centiles of age at attaining the selected thresholds may rightly be viewed as maturational because they mark the progress toward a defined adult status, and they provide information regarding the relative speed with which the individual has arrived at a particular threshold. The confusion in using testicular size at an age without reference to the thresholds as a measure of maturation is identical to the confusion that would arise by using stature at an age as a measure of maturation. In both cases the growth measure is correlated with maturation but it has not been scaled in a way to appropriately measure progress toward adult status rather than size alone.

Growth as a bioassay of health and well-being

A traditional *in vitro* laboratory technique for measuring the biological activity of compounds thought to affect growth is to add the compound to culture media including the target tissue. The response of the target tissue to the compound then provides a bioassay of the ability of the compound to influence the growth of that particular tissue. Human somatic growth and maturation provide a similar bioassay for a wide range of factors for individual children and populations of children.

For healthy normal children, the course and pattern of growth and maturation is determined by their genetic complement, if unconstrained by environmental factors. With infection or inadequate nutrition, growth slows in an adaptive response to allow the limited nutriture to support biological activities more critical for life, e.g. respiration, circulation and brain function. When adequate nutrition and health are restored, the growth responds with periods of rapid catch-up or compensatory growth if there are sufficient nutrient stores to support it and sufficient time on the maturational clock for it to be effected (Keller, 1988). This conceptual model indicates that growth is sensitive to nutritional adequacy and health, but within genetic limits.

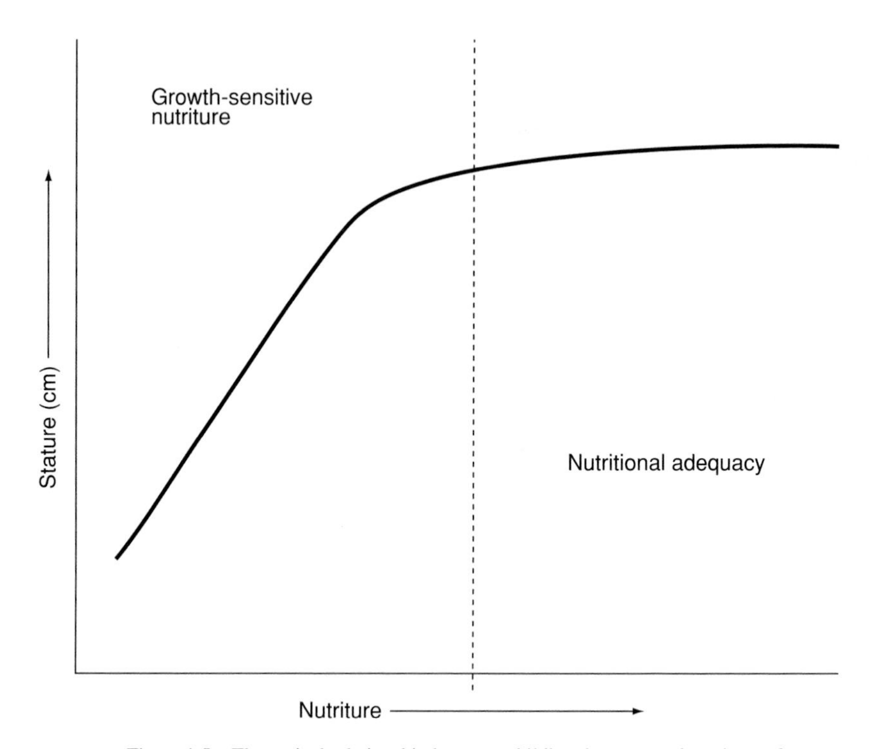

Figure 1.5 Theoretical relationship between childhood stature and nutriture of growth-essential nutrients.

A theoretical example of this general relationship can be seen in Figure 1.5, which presents attained stature relative to nutriture for growth-essential nutrients, e.g. protein and energy. Of course, the extreme of nutritional inadequacy is incompatible with life. As nutriture increases so does stature until it approximates an asymptote, where additional nutriture is no longer associated with additional growth. Relative to growth, the level of nutriture associated with this theoretical asymptote is the point of nutritional adequacy, or the point where stature is no longer constrained by nutriture. Thus, stature is sensitive to differences in nutriture only if it is less than adequate, and stature in situations of adequate nutriture is primarily determined by genetic factors if there are no other constraining factors. Accordingly, variation in stature among those who are adequately nourished and healthy is a poor reflection of whatever variation may exist in their health and nutriture (World Health Organization, 1995). One exception to this latter generalization is the observation that obese children are taller than their non-obese but adequately nourished peers (Wolff, 1955; Himes and Roche, 1986).

	Malnutrition	Adequate Nutriture	Totals
Stunted	a	b	a +b
Not Stunted	c	d	c + d
Totals	a +c	b +d	a + b + c +d

Prevalence = (a+c) / (a+b+c+d)
Sensitivity = a / (a+c)
Specificity = d / (b+d)
Positive predictive value = a / (a+b)

Figure 1.6 Relationships among prevalence, sensitivity, specificity and positive predictive value for stunting relative to true malnutrition.

Growth is sensitive to many factors, including various specific nutrients, infection, psychological deprivation, persistent loud noises, genetic abnormalities and various pathologies (Kelnar *et al.*, 1998). Consequently, the presence of a growth deficiency is not by itself specific to any one diagnosis, although it may be an essential part in arriving at the diagnosis.

The usefulness of a measure of growth like stature as an indicator of health or nutritional status depends on the likelihood that a given short or stunted child or a group of stunted children actually reflect ill health or inadequate nutriture (World Health Organization, 1995). The likelihood chiefly depends on what epidemiologists refer to as the positive predictive value of the growth indicator, which in turn is a function of the sensitivity and specificity of the indicator, and the prevalence of ill health or inadequate nutriture in the population (Kraemer, 1992).

As an example, we shall use stunting as an indicator of malnutrition. The determination of stunting is usually relative to internationally accepted reference standards of recumbent length or stature for age (World Health Organization, 1995). Using a conventional four-fold table crossing stunting with true malnutrition, one can visualize the relationships (Figure 1.6). Sensitivity (the proportion of the truly malnourished who are correctly identified as such by the

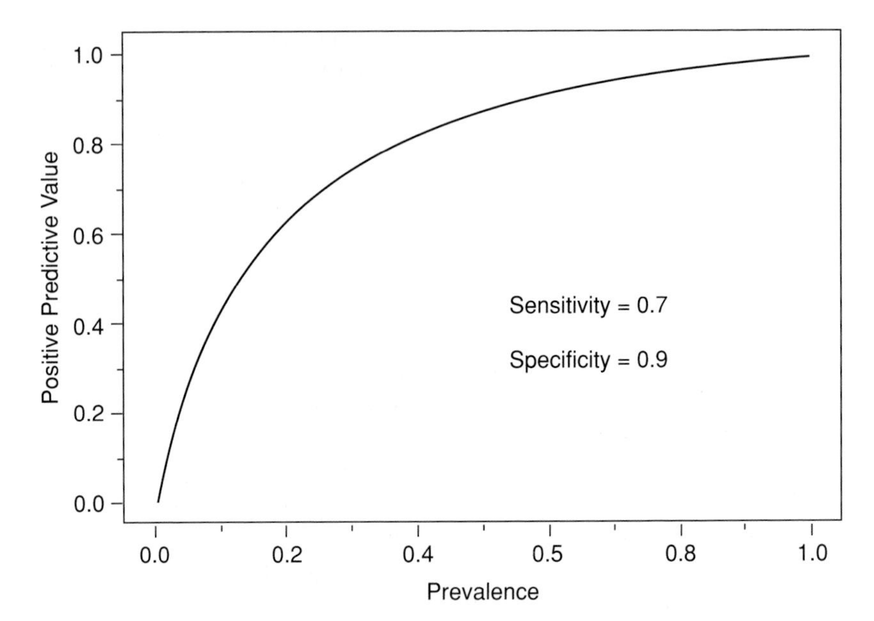

Figure 1.7 Relationship of positive predictive value to prevalence for an indicator with fixed sensitivity and specificity.

indicator) and specificity (the proportion of those truly not malnourished who are correctly identified as such by the indicator) are usually characteristics of the indicator and often independent of the prevalence of malnutrition (Habicht *et al.*, 1982). The positive predictive value (the proportion of those identified as malnourished by the indicator who are truly malnourished), is important because in clinical or screening situations one usually only knows which children are positive for the indicator (in our case, stunted) and not their true nutritional status.

The relationship between prevalence and the positive predictive value for a hypothetical indicator with fixed sensitivity and specificity is presented in Figure 1.7. Clearly, as the prevalence raises so does the positive predictive value. Consequently, at low prevalences such as those in highly developed countries, a small proportion of the children screened positively for malnutrition will actually be malnourished, and the high rate of false positives will lead one to judge stunted children as something other than malnourished; perhaps, short for genetic reasons. In areas where malnutrition is highly prevalent, stunting is a useful measure of nutritional status because of the high likelihood that a given stunted child or group of stunted children is in fact short due to malnutrition.

Clinicians in different countries usually come to the same conclusions in their clinical judgements of stunted children, without understanding the underlying relationships with sensitivity, specificity, and prevalence. Nevertheless, by understanding the principles, one can evaluate the usefulness of specific growth measures and differing definitions for those measures as indicators of health and nutritional status in a wide variety of settings.

Finale: body size, growth, and society

Societal awareness of body size begins early when new mothers are routinely queried about the weight of their newborns, and it continues as the child grows 'big and tall'. Some of these common references undoubtedly have their origins in observations of growth reflecting healthy development and functioning.

Body size, growth and maturational status are entwined with literary references, social and psychological perceptions, and notions of beauty, sexuality and aesthetics; for example, from Francis Bacon's *Apophthegms* (1624) in Spedding *et al.* (1968): 'Nature did never put her precious jewels onto a garret four stories high, and therefore exceeding tall men had ever very empty heads.' We smile at such literary allusions because they may have little to do with scientific evidence and much to do with the personal biases of the authors and prevailing notions of the times. Nevertheless, they reflect and extend social perceptions that may have significant psychological and behavioural impacts, many of them negative. For example, self-esteem and body image appear interdependent in eating disorders such as anorexia nervosa, bulimia and obesity (Thompson, 1995). Short stature may be accompanied by psychological difficulties secondary to peer acceptance and societal views (Sandberg and Voss, 2002). Unhealthy dieting practices in youth may result from idealized notions of beauty promulgated by the media (Field *et al.*, 2001). The onset of puberty is associated with increases in depression and risky behaviours (Wilson *et al.*, 1994; Hayward *et al.*, 1999).

Growth and maturation are ubiquitously shared experiences. Accordingly, they occupy a prominent place in our personal perspectives, our public perceptions, and in our literature and art. The applications of scientific studies of human growth and maturation are similarly far reaching: commonplace and esoteric, profound and prosaic, practical and abstract. Viewed together, they provide compelling justification and continuing impetus for this enduring study of ourselves.

REFERENCES

Arnhold, R. (1969). The arm circumference as a public health index of protein-calorie malnutrition of early childhood. XVII: The QUAC stick: a field measure used by the Quaker Service Team in Nigeria. *Journal of Tropical Paediatrics*, **15**, 243–7.

Baumgartner, R. N., Roche, A. F. and Himes, J. H. (1986). Incremental growth tables: supplementary to previously published charts. *American Journal of Clinical Nutrition*, **43**, 711–22.

Bayer, L. M. and Bayley, N. (1959). *Growth Diagnosis: Selected Methods for Interpreting and Predicting Physical Development from One Year to Maturity*. Chicago, IL: University of Chicago Press.

Beres, J., Papp, G., Pazonyi, I. and Czeizel, E. (1989). Testicular volume variations from 0 to 28 years of age. *International Urology and Nephrology*, **21,** 159–67.

Bock, R. D. and Thissen, D. (1980). Statistical problems of fitting individual growth curves. In *Human Physical Growth and Maturation: Methodologies and Factors*, vol. 30, Series A, *Life Sciences*, eds. F. E. Johnston, A. F. Roche and C. Susanne, pp. 265–90. New York: Plenum Press.

Bogin, B. (1999). *Patterns of Human Growth*, 2nd edn. Cambridge, UK: Cambridge University Press.

Bogin, B., Kapell, M., Varela Silva, M. I., *et al.* (2001). How genetic are human body proportions? In *Perspectives in Human Growth, Development and Maturation*, eds. P. Dasgupta and R. Hauspie, pp. 205–31. Boston, MA: Kluwer Academic Publishers.

Bouchard, C. and Johnston, F. E. (1988). *Fat Distribution during Growth and Later Health Outcomes*. New York: Alan R. Liss.

Bouchard, C., Lesage, R., Lortie, G., *et al.* (1986). Aerobic performance in brothers, dizygotic and monozygotic twins. *Medicine and Science in Sports and Exercise*, **18**, 639–46.

Boyd, E., Scammon, R. E., Savara, B. S. and Schilke, J. F. (1980). *Origins of the Study of Human Growth*. Portland, OR: University of Oregon Health Sciences Center Foundation.

Cameron, N. (1999). The use and abuse of growth charts. In *Human Growth in Context*, eds. F. E. Johnston, B. Zemel and P. B. Eveleth, pp. 65–94. London: Smith-Gordon.

Casey, P. H., Kraemer, H. C., Bernbaum, J., Yogman, M. W. and Sells, J. C. (1991). Growth status and growth rates of a varied sample of low birth weight, preterm infants: a longitudinal cohort from birth to three years of age. *Journal of Pediatrics*, **119**, 599–605.

Cole, T. (1991). Weight–stature indices to measure underweight, overweight, and obesity. In *Anthropometric Assessment of Nutritional Status*, ed. J. H. Himes, pp. 83–111. New York: Wiley–Liss.

Cole, T. J., Freeman, J. V. and Preece, M. A. (1998). British 1990 growth reference centiles for weight, height, body mass index and head circumference fitted by maximum penalized likelihood. *Statistics in Medicine*, **17**, 407–29.

de Onis, M., Victora, C. G., Garza, C., Frongillo, E. A. J. R. and Cole, T. (2001). A new international growth reference for young children. In *Perspectives in Human Growth, Development and Maturation*, eds. P. Dasgupta and R. Hauspie, pp. 45–53. Boston, MA: Kluwer Academic Publishers.

Deming, J. (1957). Application of the Gompertz curve to the observed pattern of growth in length of 48 individual boys and girls during the adolescent cycle of growth. *Human Biology*, **29**, 83–122.

Field, A. E., Camargo, C. A., Jr, Taylor, C. B., *et al.* (2001). Peer, parent, and media influences on the development of weight concerns and frequent dieting among preadolescent and adolescent girls and boys. *Pediatrics*, **107**, 54–60.

Filipsson, R., Hall, K. and Lindsten, J. (1978). Dental maturity as a measure of somatic development in children. *Advances in Metabolic Disorders*, **9**, 425–51.

Goldstein, H. and Tanner, J. M. (1980). Ecological considerations in the creation and the use of child growth standards. *Lancet*, **i**, 582–5.

Guo, S., Roche, A. F. and Moore, W. M. (1988). Reference data for head circumference and 1-month increments from 1 to 12 months of age. *Journal of Pediatrics*, **113**, 490–4.

Guo, S. M., Roche, A. F., Fomon, S. J., *et al.* (1991). Reference data on gains in weight and length during the first two years of life. *Journal of Pediatrics*, **119**, 355–62.

Guo, S. S., Wholihan, K., Roche, A. F., Chumlea, W. C. and Casey, P. H. (1996). Weight-for-length reference data for preterm, low-birth-weight infants. *Archives of Pediatrics and Adolescent Medicine*, **150**, 964–70.

Habicht, J. P., Meyers, L. D. and Brownie, C. (1982). Indicators for identifying and counting the improperly nourished. *American Journal of Clinical Nutrition*, **35**, 1241–54.

Hagg, U. and Taranger, J. (1985). Dental development, dental age and tooth counts. *Angle Orthodontist*, **55**, 93–107.

Hall, J. G., Froster-Iskenius, U. G. and Allanson, J. E. (1989). *Handbook of Normal Physical Measurements*. Oxford, UK: Oxford University Press.

Hauspie, R. and Wachholder, A. (1986). Clinical standards for growth velocity in height of Belgian boys and girls, aged 2 to 18 years. *International Journal of Anthropology*, **1**, 339–48.

Hayward, C., Gotlib, I. H., Schraedley, P. K. and Litt, I. F. (1999). Ethnic differences in the association between pubertal status and symptoms of depression in adolescent girls. *Journal of Adolescent Health*, **25**, 143–9.

Himes, J. H. (1989). Reliability of anthropometric methods and replicate measurements. *American Journal of Physical Anthropology*, **79**, 77–80.

(1999a). Growth reference data for adolescents: maturation-related misclassification and its accommodation. In *Human Growth in Context*, eds. F. E. Johnston, B. Zemel and P. B. Eveleth, pp. 95–100. London: Smith-Gordon.

(1999b). Maturation-related deviations and misclassification of stature and weight in adolescence. *American Journal of Human Biology*, **11**, 499–504.

(1999c). Minimum time intervals for serial measurements of growth in recumbent length or stature of individual children. *Acta Paediatrica*, **88**, 120–5.

Himes, J. H. and Bouchard, C. (1989). Validity of anthropometry in classifying youths as obese. *International Journal of Obesity and Related Metabolic Disorders*, **13**, 183–93.

Himes, J. H. and Dietz, W. H. (1994). Guidelines for overweight in adolescent preventive services: recommendations from an expert committee (The Expert Committee on Clinical Guidelines for Overweight in Adolescent Preventive Services). *American Journal of Clinical Nutrition*, **59**, 307–16.

Himes, J. H. and Roche, A. F. (1986). Subcutaneous fatness and stature: relationship from infancy to adulthood. *Human Biology*, **58**, 737–50.

Himes, J. H., Roche, A. F., Thissen, D. and Moore, W. M. (1985). Parent-specific adjustments for evaluation of recumbent length and stature of children. *Pediatrics*, **75**, 304–13.

Hoey, H. M., Tanner, J. M. and Cox, L. A. (1987). Clinical growth standards for Irish children. *Acta Paediatrica Scandinavica (Supplement)*, **338**, 1–31.

Jenss, R. M. and Bayley, N. (1937). A mathematical method for studying the growth of a child. *Human Biology*, **12**, 556–63.

Johnston, F. E. and Ouyang, Z. (1991). Choosing appropriate reference data for the anthropometric assessment of nutritional status. In *Anthropometric Assessment of Nutritional Status*, ed. J. H. Himes, pp. 337–46. New York: Wiley–Liss.

Karlberg, J. (1989). A biologically-oriented mathematical model (ICP) for human growth. *Acta Paediatrica Scandinavica (Supplement)*, **350**, 70–94.

Keller, W. (1988). The epidemiology of stunting. In *Linear Growth Retardation in Less Developed Countries*, vol. 14, ed. J. C. Waterlow, pp. 17–34. New York: Raven Press.

Kelnar, C. J. H., Savage, M. O., Stirling, H. F. and Saenger, P. (eds.) (1998). *Growth Disorders: Pathophysiology and Treatment*, Philadelphia, PA: Chapman and Hall Medical.

Keys, A., Fidanza, F., Karvonen, M. J., Kimura, N. and Taylor, H. L. (1972). Indices of relative weight and obesity. *Journal of Chronic Disease*, **25**, 329–43.

Kraemer, H. C. (1992). *Evaluating Medical Tests: Objective and Quantitative Guidelines*. Newbury Park, CA: Sage Publications.

Kuczmarski, R. J., Ogden, C. L., Grummer-Strawn, L. M., *et al.* (2000). CDC Growth Charts: United States. *Advance Data*, **314**, 1–27.

Malina, R. M. (1991). Ratios and derived indicators in the assessment of nutritional status. In *Anthropometric Assessment of Nutritional Status*, ed. J. H. Himes, pp. 151–71. New York: Wiley–Liss.

Marshall, W. A. and Tanner, J. M. (1970). Variations in the pattern of pubertal changes in boys. *Archives of Disease in Childhood*, **45**, 13–23.

Martin, R. (1928). *Lehrbuch der Anthropologie in systematischer Darstellung mit besonderer Berücksichtigung der anthropologischen Methoden für Studierende, Artze und Forschungsreisende*. Jena, Germany: G. Fischer.

Meaney, F. J. and Farrer, L. A. (1986). Clinical anthropometry and medical genetics: a compilation of body measurements in genetic and congenital disorders. *American Journal of Medical Genetics*, **25**, 343–59.

Mei, Z., Grummer-Strawn, L. M., de Onis, M. and Yip, R. (1997). The development of a MUAC-for-height reference, including a comparison to other nutritional status screening indicators. *Bulletin of the World Health Organization*, **75**, 333–41.

Neter, J., Kutner, M. H., Nachtsheim, C. J. and Wasserman, W. (1996). *Applied Linear Statistical Models*. Chicago, IL: Irwin.

Nielsen, C. T., Skakkebaek, N. E., Darling, J. A., *et al.* (1986a). Longitudinal study of testosterone and luteinizing hormone (LH) in relation to spermarche, pubic hair, height and sitting height in normal boys. *Acta Endocrinologica Supplement (Copenhagen)*, **279**, 98–106.

Nielsen, C. T., Skakkebaek, N. E., Richardson, D. W., *et al.* (1986b). Onset of the release of spermatozoa (spermarche) in boys in relation to age, testicular growth, pubic hair, and height. *Journal of Clinical Endocrinology and Metabolism*, **62**, 532–5.

Preece, M. A. and Baines, M. J. (1978). A new family of mathematical models describing the human growth curve. *Annals of Human Biology*, **5**, 1–24.

Quetelet, A. (1869). *Physique sociale: ou Essai sur le développement des facultés de l'homme*. Brussels: C. Muquardt.

Rao, D. C. and Majumder, P. P. (2001). Genetics of complex traits with particular attention to fat patterning. In *Perspectives in Human Growth, Development and Maturation*, eds. P. Dasgupta and R. Hauspie, pp. 79–89. Boston, MA: Kluwer Academic Publishers.

Reynolds, E. L. and Wines, J. (1948). Individual differences in physical changes associated with adolescence in girls. *American Journal of Diseases in Children*, **75**, 329–50.

(1951). Physical changes associated with adolescence in boys. *American Journal of Diseases in Children*, **82**, 529–47.

Roche, A. F. (1992). *Executive Summary of the NCHS Growth Chart Workshop 1992*. Hyattsville, MD: National Center for Health Statistics, Centers for Disease Control and Prevention, Public Health Service, US Department of Health and Human Services.

Roche, A. F. and Himes, J. H. (1980). Incremental growth charts. *American Journal of Clinical Nutrition*, **33**, 2041–52.

Roche, A. F., Wainer, H. and Thissen, D. (1975). *Skeletal Maturity: The Knee Joint as a Biological Indicator*. New York: Plenum Press.

Roche, A. F., Tyleshevski, F. and Rogers, E. (1983). Non-invasive measurements of physical maturity in children. *Research Quarterly for Exercise and Sport*, **54**, 364–71.

Roche, A. F., Guo, S. S., Wholihan, K. and Casey, P. H. (1997). Reference data for head circumference-for-length in preterm low-birth-weight infants. *Archives of Paediatrics and Adolescent Medicine*, **151**, 50–7.

Salkind, N. J. (2002). *Child Development*. New York: Macmillan.

Sandberg, D. E. and Voss, L. D. (2002). The psychosocial consequences of short stature: a review of the evidence. *Best Practice and Research, Clinical Endocrinology and Metabolism*, **16**, 449–63.

Sanders, B. S. (1934). *Environment and Growth*. Baltimore, MD: Warwick and York.

Sommer, A. and Loewenstein, M. S. (1975). Nutritional status and mortality: a prospective validation of the QUAC stick. *American Journal of Clinical Nutrition*, **28**, 287–92.

Spedding, J., Ellis, R. L. and Heath, D. D. (eds.) (1968). *The Works of Francis Bacon*, vol. VII, *Literary and Professional Works*, vol. 2, p. 182. New York: Garrett Press.

Steeno, O. P. (1989). Klinische und physikalische Bewertung des infertilen Mannes: Hodenmessung oder Orchiodometrie. *Andrologia*, **21**, 103–12.

Styne, D. (2001). Growth. In *Basic and Clinical Endocrinology*, eds. F. S. Greenspan and D. G. Gardner, pp. 163–200. New York: McGraw Hill.

Sun, S. S., Schubert, C. M., Chumlea, W. C., *et al.* (2002). National estimates of the timing of sexual maturation and racial differences among US children. *Pediatrics*, **110**, 911–19.

Tanner, J. M. (1962). *Growth at Adolescence*. Oxford, UK: Blackwell Scientific Publications.

(1981). *A History of the Study of Human Growth*. Cambridge, UK: Cambridge University Press.

Tanner, J. M. and Davies, P. S. (1985). Clinical longitudinal standards for height and height velocity for North American children. *Journal of Pediatrics*, **107**, 317–29.

Tanner, J. M. and Whitehouse, R. H. (1976). Clinical longitudinal standards for height, weight, height velocity, weight velocity, and stages of puberty. *Archives of Disease in Childhood*, **51**, 170–9.

Tanner, J. M., Whitehouse, R. H. and Takaishi, M. (1966). Standards from birth to maturity for height, weight, height velocity, and weight velocity: British children, 1965. I. *Archives of Disease in Childhood*, **41**, 454–71.

Taranger, J., Engström, I., Lichtenstein, H. and Svennberg-Redegren, I. (1976). Somatic pubertal development. *Acta Paediatrica Scandinavica (Supplement)*, **258**, 121–35.

Tanner, J. M., Whitehouse, R. H., Cameron, N., *et al.* (1983). *Assessment of Skeletal Maturity and Prediction of Adult Height (TW2 Method)*. London: Academic Press.

Thompson, D' A. W. (1971). *On Growth and Form*, abridged edn, ed. J. T. Bonner. Cambridge, UK: Cambridge University Press.

Thompson, J. K. (1995). Assessment of body image. In *Handbook of Assessment Methods for Eating Behaviours and Weight Related Problems: Measures, Theory, and Research*, ed. D. B. Allison, pp. 119–48. Thousand Oaks, CA: Sage Publications.

Todd, T. W. (1937). *Atlas of Skeletal Maturation*. St Louis, MO: Mosby.

van Wieringen, J. C. (1971). *Growth Diagrams 1965 Netherlands: Second National Survey on 0–24-Year-Olds*. Groningen, The Netherlands: Wolters-Noordhoff.

Wilson, D. M., Killen, J. D., Hayward, C., *et al.* (1994). Timing and rate of sexual maturation and the onset of cigarette and alcohol use among teenage girls. *Archives of Paediatrics and Adolescent Medicine*, **148**, 789–95.

Wolff, O. H. (1955). Obesity in childhood: a study of birth weight, the height, and the onset of puberty. *Quarterly Journal of Medicine*, **24**, 109–23.

World Health Organization (1995). *Physical Status: The Use and Interpretation of Anthropometry – Report of a WHO Expert Committee*, WHO Technical Report Series no. 854. Geneva, Switzerland: World Health Organization.

(1998). *Obesity: Preventing and Managing the Global Epidemic*, Report of a WHO Consultation on Obesity, Geneva, 3–5 June 1997. Geneva, Switzerland: World Health Organization.

Zachmann, M., Prader, A., Kind, H. P., Hafliger, H. and Budliger, H. (1974). Testicular volume during adolescence: cross-sectional and longitudinal studies. *Helvetica Paediatrica Acta*, **29**, 61–72.

2 *The human growth curve: distance, velocity and acceleration*

Luciano Molinari
Kinderspital, Zürich

and

Theo Gasser
University of Zürich

Introduction

In this chapter we introduce the basic concepts necessary to analyse *individual series* of growth measurements.

We also indicate where, in this volume, related topics are discussed.

Growth in auxology is defined as the increase in size of a body dimension and stands somewhat in opposition to *development*, which, usually, implies a qualitative change. The distinction is, however, not always clear cut, because the differential growth of different body dimensions can lead to qualitative changes, e.g. in proportion and shape, which well deserve the name of development.

Both in the study of growth, as well as of development, the notion of change is central. This has implications for the design of a growth study: only a minimal amount of information about change can be gathered from cross-sectional studies, namely merely the average change. Short mixed cross-sectional/longitudinal studies provide information about short-term growth trends, often relevant for clinical purposes; however a picture of the growth process as a whole and a correct view of the relationship between different growth phases can only be obtained from time-consuming *longitudinal studies*, where subjects are followed and measured over many years, most desirably from birth, or even earlier, to, or well into, adulthood. Such studies are very delicate in their management, if selection bias is to be avoided, and measurement occasions must be

Methods in Human Growth Research, eds. R. C. Hauspie, N. Cameron and L. Molinari. Published by Cambridge University Press.

designed carefully, essentially with smaller intervals where the expected change is more rapid.

The clinical assessment of the growth of an individual involves answering a number of questions, concerning his/her amount of growth and the corresponding growth velocity: is the growth 'normal' for the age, is it proportionate? A reliable answer, in the clinical context, to such questions can only be obtained if an adequate *reference* is available. How such references, also called *norms* or *standards*, can be established, will be explained in Chapter 10.

Age is, of course, the best single predictor for growth; it is, however, a poor predictor, both in normally developing individuals, and, even more, in pathological situations. Chapter 5 will provide an introduction to the problem of *measuring maturity*, as a precondition for a correct clinical assessment of growth. A complementary methodology, *structural analysis*, useful for scientific work, is presented in Chapter 7.

Growth measurement as represented by a series of observations of certain body dimensions involves a number of 'random' components, such as measurement errors, diurnal or seasonal effects, etc., which may or may not be of interest. *Parametric* and *non-parametric* models to estimate or smooth out, depending on the situation, these effects are the topic of Chapters 8 and 7, respectively.

The data of the First Zürich Longitudinal Study (Prader *et al.*, 1989), one of the best growth studies, will be used for illustration of the substantial points. Extensive use will be made of results obtained by Gasser and co-workers.

Measuring anthropometric variables for a growth study, such as the one considered here, is not trivial. Measuring techniques confront the auxologist with a variety of technical and conceptual problems discussed in Chapter 4.

An excellent, surprisingly modern and readable introduction to the many facets of longitudinal data and their statistical analysis is Goldstein (1979).

Falkner and Tanner (1986), Ulijaszek *et al.* (1998) and Eveleth and Tanner (1991) are comprehensive references, which, although perhaps not in all parts up to date, will help the young scientist to get in touch with related fields not discussed in this book. He/she will then be better able to appreciate the breadth of the field, and to assess the impact of modern statistical methods in different contexts. Plenty of information, mainly from a nutritionist perspective, is in Waterlow and Schürch (1994).

Basic growth description: a gentle introduction

The classical picture (Figure 2.1), from Scammon (1930), quoted for example in Bogin (1999), groups body dimensions into four major types.

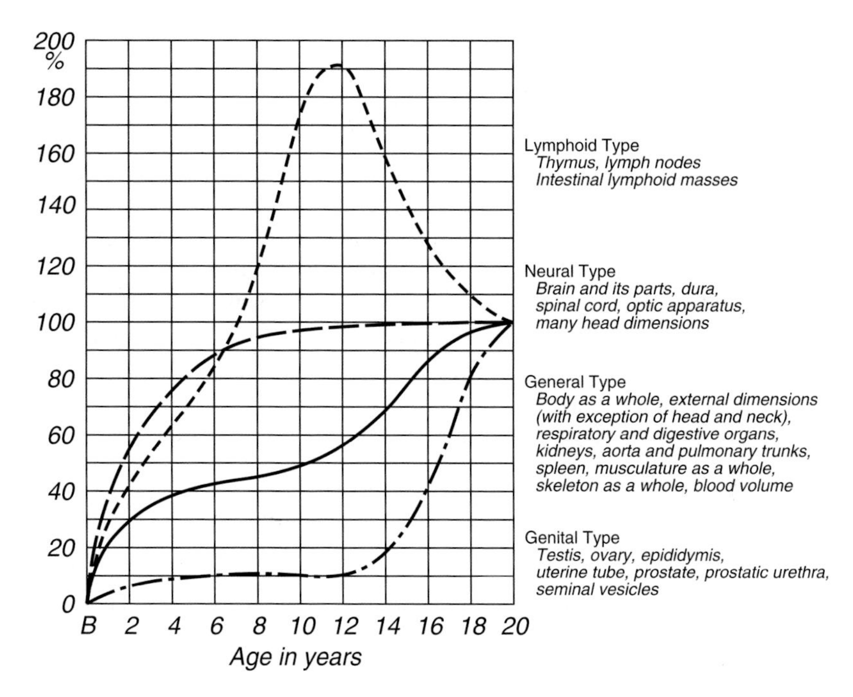

Figure 2.1 Types of postnatal growth curves, as percentage of total growth increment from birth to 20 years (Scammon, 1930).

Three of the curves (corresponding to the genital, general and neural type) are increasing, the general and genital types being S-shaped, due to the pubertal spurt, while the neural type appears to lack this spurt. The lymphoid type curve is increasing until puberty and decreasing thereafter.

These curves describe the qualitative patterns of growth well, and, by emphasizing the similarities as well as the differences between growth in different regions of the body, suggest a number of interesting questions for a quantitative analysis. For example it may be asked, whether it is true that neural type measurements do not have a pubertal spurt. For head circumference, for example, this appears not to be true. Or one can wonder whether all measurements not of the lymphoid type approach asymptotically their adult value, or, alternatively, may grow indefinitely.

To illustrate the basic ideas and definitions, we will make use of measurements of stature done by the Chevalier de Montbeillard on his son: at the suggestion of Buffon he took his height, approximately every six months, from birth in 1759 to age 18 years in 1777. These measurements, (Table 2.1) which have been previously used, among others, by D'Arcy Thompson (1942) and

Table 2.1 *Stature of Montbeillard's son: age in years, months, days; stature in cm*

April 1759	51.4	6,6,19	122.9	13,6	158.6
0,6	65.0	7,0	124.3	14,0	162.9
1,0	73.1	7,3	127.0	14,6,10	169.2
1,6	81.2	7,6	128.9	15,0,2	175.0
2,0	90.0	8,0	130.8	15,6,8	177.5
2,6	92.8	8,6	134.3	16,3,8	181.4
3,0	98.8	9,0	137.0	16,6,6	183.3
3,6	100.4	9,7,12	140.1	17,0,2	184.6
4,0	105.2	10,0	141.9	17,1,9	185.4
4,7	109.5	11,6	141.6	17,5,5	186.5
5,0	111.7	12,0	149.9	17,7,4	186.8
5,7	115.5	12,8	154.1		
6,0	117.8	13,0	155.3		

Source: Sandland and McGilchrist (1979), after Scammon (1927).

Tanner (1955, 1981), form the first longitudinal study known, and, according to Tanner, still one of the best.

To make things simple for the time being, we fit the triple logistic model (Bock and Thissen, 1976) to the measurements as in Table 2.1. The use of this highly parameterized and very flexible model essentially removes measurement errors and small fluctuations, such as seasonal effects, noted by Buffon (Tanner, 1981), yielding pleasant curves, better suited for didactical purposes. In Figure 2.2 the so-called distance, $h(t)$, i.e. height or stature at age t years, velocity, $hv(t) = dh(t)/dt$, i.e. the first derivative with respect to age of the function h, and acceleration, $ha(t) = dhv(t)/dt = d^2h(t)/dt^2$, i.e. the second derivative of the distance or the first derivative of the velocity, are shown. Notice that the clear outlier at age 11.5 years (Table 2.1) was replaced by an interpolated value before fitting the triple logistic. This outlier was curiously discarded, or corrected, without mention, in D'Arcy Thompson (1942) and Tanner (1981).

The preferred tool for the study of individual growth is the velocity curve: it shows in a clear and intuitive way the growth dynamic, the distance curve giving little information. The acceleration, while fascinating, because of its tight relation to force – meaning cause in physics – is not easily interpretable in the context of growth and presents bigger problems of estimation (Chapters 7 and 8). For these reasons, from the very beginning, auxologists have tried to estimate the growth velocity and have worked with that rather than with the distance curve.

The general course of growth in height, as indicated by the velocity curve is as follows:

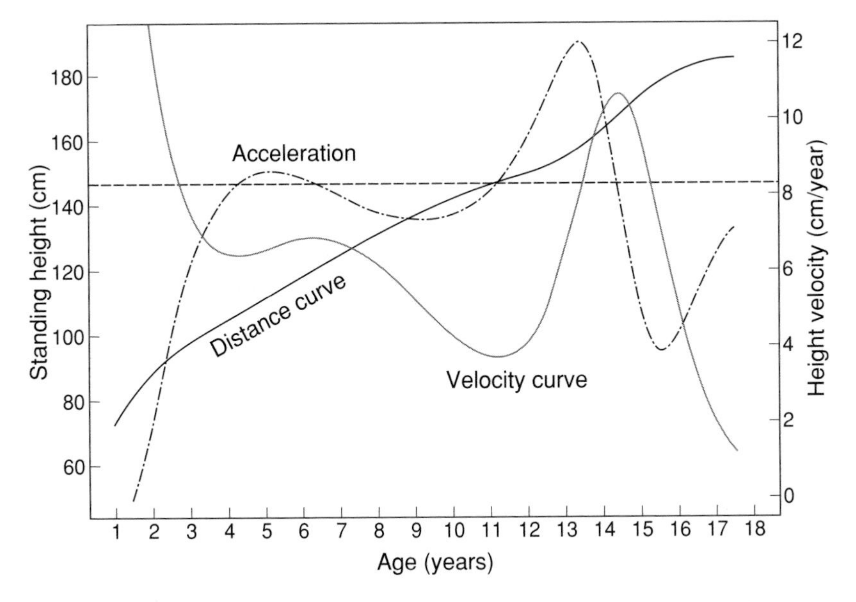

Figure 2.2 Distance, velocity and acceleration curves for standing height. From a triple logistic fit to Montbeillard's son's stature data. The horizontal dashed line indicates the zero level for the acceleration.

- High growth velocity just after birth, rapidly decreasing. The period from (or just before) birth to two years deserves an in-depth analysis, for which, however, narrowly spaced data are scarce. See Bertino *et al.* (1996) for foetal growth and Eiholzer *et al.* (1998) for growth in the first year of life.
- A period of almost constant, or slightly decreasing, growth velocity, from, say, age three to just before puberty, interrupted by what is called the *mid-growth spurt* (Molinari *et al.*, 1980; Tanner and Cameron, 1980). In many subjects this small spurt, of unclear origin, is not a real spurt, defined by a true local maximum of the velocity (zero acceleration), but rather a deviation of the velocity curve from convexity (equivalent to a positive third derivative of $h(t)$). This makes the estimation of the timing and intensity of this spurt relatively error prone, particularly for height, but more reliable when acceleration is used, rather than velocity itself. We note that the mid-growth spurt is visible in most subjects of the First Zürich Longitudinal Study and is more pronounced in other dimensions than height, such as for example shoulder width (Gasser *et al.*, 1985, 1991a,b).
- After a 'latency' phase, of variable duration and at times very short in girls, the pubertal spurt is initiated (see Largo and Prader (1983a,b) for its relation to secondary sexual characters) and growth velocity increases again, achieving its maximum at age at peak height velocity (APHV), about

one year before menarche for girls, or at an age where, on average, about 90% of the final height is reached, both for boys and for girls. The normal range for APHV includes a period of about four years, again for boys and girls.

- After APHV growth velocity decreases, more rapidly than it increased, and, under the influence of the sexual hormones, the epiphyses close and growth almost stops, although it can continue at a very low level for several more years. Notice that cartilage tissues can continue to grow up to a very old age (Zankl *et al.*, 2002).

There is a long auxological tradition going back at least to Count (1943) and summarized by Karlberg *et al.* (1994) and by Milani (2000), and reflected in the mathematical modelling of the growth process by such models as the triple logistic or the JPPS curve (Jolicoeur *et al.*, 1988), by which the process of somatic growth from conception to adulthood consists of three phases, which are under the control of different hormonal systems.

According to Karlberg *et al.* (1994) and Milani (2000), we should roughly distinguish:

- An infancy cycle, beginning before birth and ending (Gasser *et al.*, 1991a,b) before 1.5 to 2 years, steered by thyroid hormones. Some researchers would see this phase rather as an adaptation from intrauterine to extrauterine growth (Smith *et al.*, 1976).
- The childhood cycle, dominating prepubertal growth, but still present during the initial part of puberty, which is under the additional powerful influence of the growth hormones.
- The pubertal phase, in which sexual hormones initiate the pubertal spurt and simultaneously lead, through the closure of the epiphyses, to the termination of growth.

These cycles are not observable, with the exception of the pubertal one, and modelling the full growth by different mathematical models leads to very contradictory results with respect to the infancy and childhood phases (Figure 2.3).

A more modest, but relevant model, distinguishes just a prepubertal and a pubertal cycle and models the effect of puberty on growth by a switch-off function (see the discussion of shape-invariant modelling in Chapter 7).

The statistical analysis of growth data

The standard analysis of longitudinal growth measurements for a sample of subjects consists of three steps:

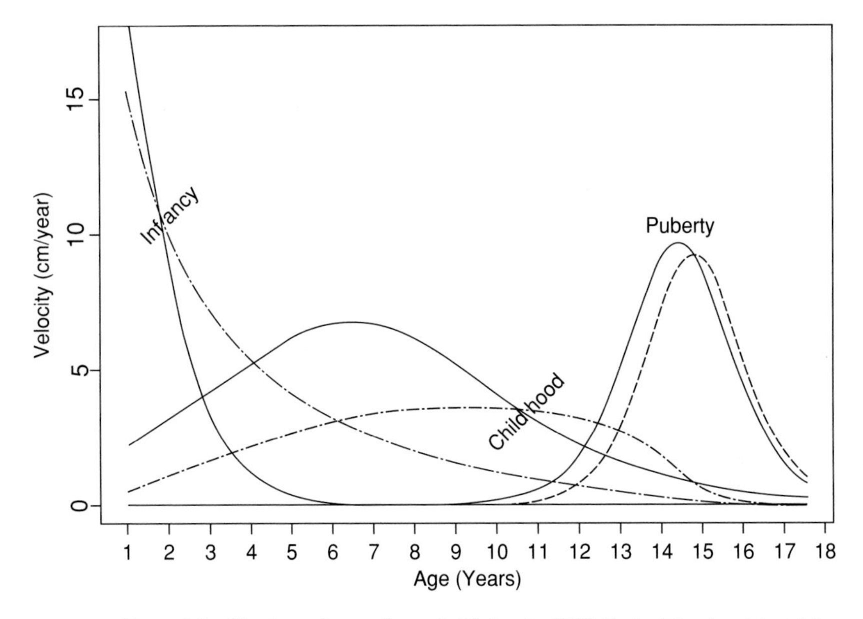

Figure 2.3 The three phases of growth. Fitting the JPSS (dashed lines) and the triple logistic (continuous lines) models to Montbeillard's son's data leads to completely discordant infancy and childhood phases.

1. The distance, velocity and, possibly, acceleration curves are estimated by some smoothing method, parametric (Chapters 8 and 9) or non-parametric (Chapter 7).
2. A number of 'parameters' (characteristic ages, *milestones*), and the corresponding distances, velocities and accelerations are obtained from the individual curves.
3. These parameters are used in further analyses to describe and compare populations, subgroups and different somatic variables.

In this introductory chapter we shall use the following milestones (Gasser, 1985):

- T_3 = age at maximal growth velocity (zero acceleration) during the mid-growth spurt.
- T_6 = age at minimal height velocity (AMHV) (zero acceleration) at the onset of the pubertal spurt.
- T_7 = age at maximal acceleration during the pubertal spurt.
- T_8 = age at peak height velocity (APHV) (maximal velocity) during the pubertal spurt.
- T_9 = age at maximal deceleration during the pubertal spurt.

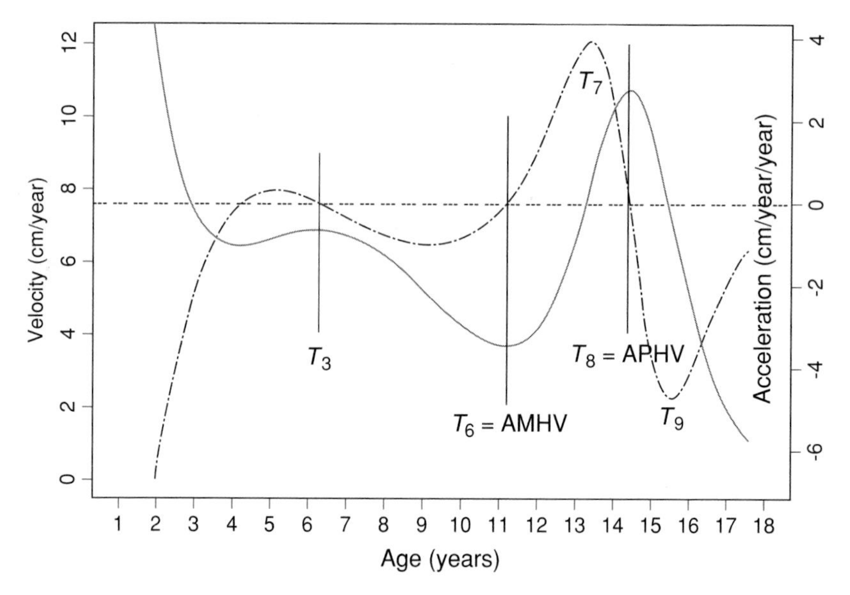

Figure 2.4 Basic milestone ages exemplified at the triple logistic fit of Montbeillard's son's stature data. AMHV, age at minimal height velocity (before puberty); APHV, age at peak height velocity (during puberty). See text for the definition of other milestones.

These parameters are shown in Figure 2.4.

Besides the corresponding distances, velocities and accelerations, further derived quantities are often useful, among which we mention the percentages of adult height reached at ages T_3 to T_9 and:

- $T_{9-6} = T_9 - T_6$: the duration of the pubertal spurt and the corresponding gain in stature $HT_{9-6} = HT_9 - HT_6$.
- $VT_{8-6} = VT_8 - VT_6$: the difference between the maximal growth velocity during the pubertal spurt and the pre-spurt velocity. A possible measure of the intensity of the pubertal spurt.
- $AT_{9-7} = AT_9 - AT_7$: used as a measure of the intensity of the pubertal spurt.

Relative velocities, i.e. velocities standardized with respect to adult size, introduced by Gasser *et al.* (1991a) and used extensively in Sheehy *et al.* (1999) facilitate the comparison of growth across body dimensions, gender and subgroups of subjects. The parameters APHV, AMHV, etc. are at first merely of a descriptive nature and their physiological meaning, e.g. their relation to certain endocrinological events, remains to be established.

When drawing inferences about their biological relevance, for example in relation to gender differences or their correlation to developmental parameters, the researcher must be aware that many steps in the analysis involve considerable statistical uncertainty, due to substantial measurement errors and large within-individual variability.

Standard errors can be difficult to assess reliably, although modern statistical methods, such as the *bootstrap* (Efron and Tibshirani, 1993), may help. It is advisable to repeat the analysis for independent subgroups; in most cases it is natural to use the two genders for this. Patterns, which are not at least qualitatively similar for the two genders, should be looked at with suspicion, unless there are sufficient biological or strong statistical arguments for the observed differences.

Statistical model

The following considerations are in terms of stature, but they apply unchanged to all other body measurements under study.

The usual statistical model assumes the measurements of stature, $H(t)$, to consist of the sum of a 'true' age-dependent stature, denoted by $h(t)$ above (p. 31), unknown, but fixed for any given subject, and of a 'random' part, $\varepsilon(t)$. The random part includes, first of all, the error of measurement, but also short-term growth effects, such as daily variations in stature, seasonal variation in growth velocity, catch-up and catch-down growth (Prader *et al.*, 1963; Tanner, 1986), which are not the object of the study. In other words, what is considered to be the true individual stature $h(t)$ may depend on the purpose of the study; this has implications for the choice of the analysis method (essentially non-parametric or parametric), which can or cannot allow deviations from a predefined growth model.

While the errors of measurement may safely be considered to be statistically independent from occasion to occasion, this will possibly not be the case for the other effects. This correlation has, to our knowledge, always been neglected within longitudinal studies with age intervals of three or more months.

The model is therefore

$$H(t) = h(t) + \varepsilon(t),$$

where t stands for age, or, within a longitudinal study, with fixed design, and 'target' ages t_i,

$$H(t_i) = h(t_i) + \varepsilon(t_i), \quad i = 1, \ldots, m,$$

with m the number of measurements and $\varepsilon(t_i)$ independent, mean zero, residuals, with an age-dependent standard deviation $\sigma_\varepsilon(t)$.

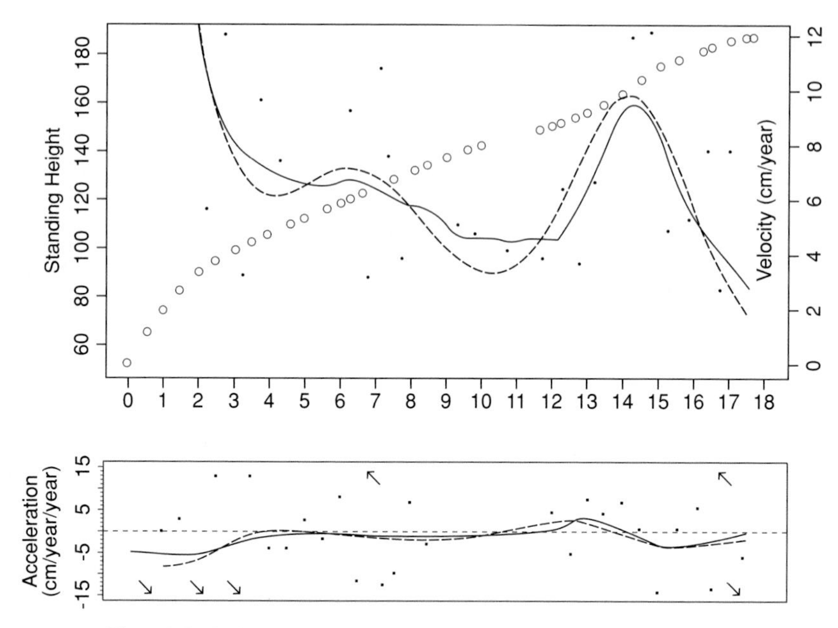

Figure 2.5 Velocity (upper plot) and acceleration (lower plot) for Montbeillard's son. Continuous lines: kernel estimates. Dashed lines: triple logistic fit. Notice the random appearance of the raw accelerations (yearly increments of velocities). Arrows indicate raw values of the acceleration falling outside the plot.

Yearly growth velocities, $hv(t)$, and accelerations, $ha(t)$, can first be estimated (denote an estimate by a ^) by the raw velocities and accelerations, i.e. by the corresponding difference quotients:

$$hv((t_{i+1}+t_i)/2) = (H(t_{i+1}) - H(t_i))/(t_{i+1}-t_i) \tag{2.1}$$

$$ha(t_i) = [hv((t_{i+1}+t_i)/2) - hv((t_i+t_{i-1})/2)]/$$
$$[(t_{i+1}+t_i)/2 - (t_i+t_{i-1})/2]$$

In Figure 2.5 we plot Montbeillard's son's data (Table 2.1, excluding the outlier) and include the corresponding triple logistic and a non-parametric fit (kernel estimator; Chapter 7). Raw velocities and accelerations are also plotted. While raw velocities still give a very rough, general idea of the growth velocity, raw accelerations appear as essentially random. This indicates the necessity of the use of more sophisticated methods for the estimation of the true $hv(t)$ and $ha(t)$.

A scientifically relevant problem is to obtain a valid estimate of the average growth, not distorted by large between-individual variability in growth timing and duration. Many years ago Tanner *et al.* (1966) centred individual height velocities with respect to the pubertal spurt to produce so called 'individual-type' centiles and 'repeated whole-year increments'. In the same spirit 'structural averages' (Gasser *et al.*, 1990; and Chapter 7) are obtained by estimating individual, non-linear, time-scale transformations, which align the individual curves with respect to multiple milestones.

Use of the velocity in a clinical setting

It is worthwhile to consider briefly the implications of Eqn 2.1. For simplicity let us assume that the target ages be equally spaced, $\Delta t = t_{i+1} - t_i$. Then the standard deviation of the observed raw velocity, for a given subject, is $\sigma_\varepsilon \sqrt{2}/\Delta t$, with σ_ε the standard deviation of the measurement error, assumed here to be independent of age. For the raw acceleration the standard deviation is $\sigma_\varepsilon \sqrt{6}/(\Delta t)^2$.

We assume in the following that the concept of Z-score,

$$Z = (\text{observation} - \text{population mean})/\text{population standard deviation},$$

is familiar to the reader.

A paediatrician is now examining a 7-year-old child for growth anomalies and will refer a patient for further examination if the raw growth velocity, based on height measurements taken at age 6 and 7, falls outside the range from the 3rd to the 97th centile of the appropriate reference. Assume a standard deviation of the measurement error of $\sigma_\varepsilon = 0.3$ cm (and, consequently, a standard deviation for the raw velocities of 0.42 cm/year) and a reference standard deviation for the raw velocities of 1 cm/year at age 6.5 years.

Consider now three different subjects with different true growth velocities ('true' referring to the growth velocity we would observe, were we able to measure height without error), average (0 Z-score), on the 6th and on the 2nd centile, respectively, and calculate, using simple properties of the Gaussian distribution (we use throughout the more neutral term 'Gaussian' to indicate the 'normal' distribution): (a) a 95% range for the observed raw velocities expressed in Z-scores and in percentiles; (b) the probability that the subject will be referred or not referred. The results are given in the first three lines of Table 2.2.

The situation is much worse if the raw velocity is calculated from half-yearly measurements, its standard deviation then being 0.85 cm/year, comparable with the population standard deviation. The results are reported in the last three lines of Table 2.2. It should now be obvious that raw velocities based on half-yearly velocities are nearly useless in clinical situations, except in extreme cases. The same is true, of course, for accelerations, even for those based on yearly measurements; but reference values for accelerations have not, to our knowledge, been published.

Table 2.2 *Probabilities of being referred or not referred for different subjects, based on their true growth velocity; probabilities of the wrong assessment are in bold*

Subject: Z-score of true growth velocity	Z-score and percentile range for the observed velocity	Probability that subject will be referred	Probability that subject will not be referred
Velocity based on yearly measurements			
$Z = 0$	$Z = -0.83$ to 0.83 Centile: 20th to 80th	**0**	1
6th centile:[a] $Z = -1.55$	$Z = -2.39$ to -0.72 Centile: 0.9th to 23rd	**0.22**	0.78
2nd centile: $Z = -2.05$	$Z = -2.89$ to -1.22 Centile: 0.2th to 11th	0.66	**0.34**
Velocity based on half-yearly measurements			
$Z = 0$	$Z = -1.66$ to 1.66 Centile: 5th to 95th	**0.027**	0.97
6th centile: $Z = -1.55$	$Z = -3.21$ to 0.11 Centile: 0.06th to 54th	**0.35**	0.65
2nd centile: $Z = -2.05$	$Z = -3.72$ to -0.39 Centile: 0.01th to 34th	0.58	**0.42**

[a] Example: the third line of the table presents the case of a subject, for whom the true yearly velocity (height increment) is on the 6th centile (Z-score -1.55) of the appropriate reference, and who, therefore, should not be referred. His observed growth velocity will be, expressed as Z-score, with 95% chance, between -2.39 and -0.72, corresponding to the 0.9th and the 23rd centiles, respectively. The chance of this subject of being referred (the wrong decision!) will be 22%.

In other words raw velocities and accelerations have, as screening tools, both a poor sensitivity and a poor specificity.

It is somewhat paradoxical that the tool, which, for the auxologist, best describes the growth process as a whole, is not useful or is even potentially misleading for the clinician.

See Milani (2000) for further problems related to the calculations of raw velocities and Prader *et al.* (1989) for a qualitative illustration of this point.

Growth in the First Zürich Longitudinal Study

Most body dimensions share several features of growth, but also present dissimilarities, which we will summarize using the data of the First Zürich Longitudinal Study. In this study (Prader *et al.*, 1989) the anthropometric variables,

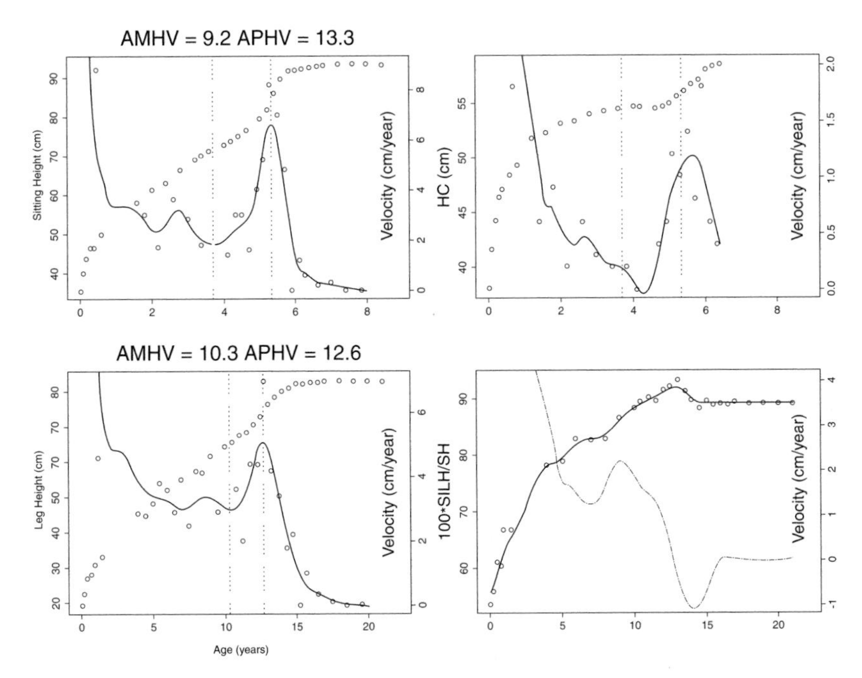

Figure 2.6 Sitting height (SH), head circumference (HC), subischial leg height (SILH) and the quotient SILH/SH for subject 1 in the First Zürich Longitudinal Study. Circles: raw distance measurements. Dots: raw yearly increments (yearly raw velocities). Continuous line: estimated (kernels) velocities for SH, HC and SILH; smoothed distance for the quotient. Dashed line: velocity for the quotient.

not all of interest here, were measured from age 1 month to adulthood (head measurements to 15, occasionally 16, years) in 232 subjects (120 boys, 112 girls) born between 1954 and 1956.

Linear measurements: standing height (H), sitting height (SH), standing subischial leg height (SILH), arm length (AL)

Head: head circumference (HC), biparietal and fronto-occipital diameters

Measures of width: bihumeral diameter, i.e. shoulder width (BHD) and biiliac diameter, i.e. pelvic width (BID)

Muscles and fat: weight (W), upper arm circumference (UAC) and calf circumference; biceps (BS), triceps (TS), subscapular (SSS) and suprailiacal (SIS) skinfold thickness

Derived quantities: body mass index (also called Quetelet's Index, BMI = W/H/H), SH/SILH, AL/SILH and BHD/BID

In Figures 2.6–2.8 observed distance measurements and velocity curves, estimated by kernels, are presented for several parameters for the first subject, a boy, in the First Zürich Longitudinal Study.

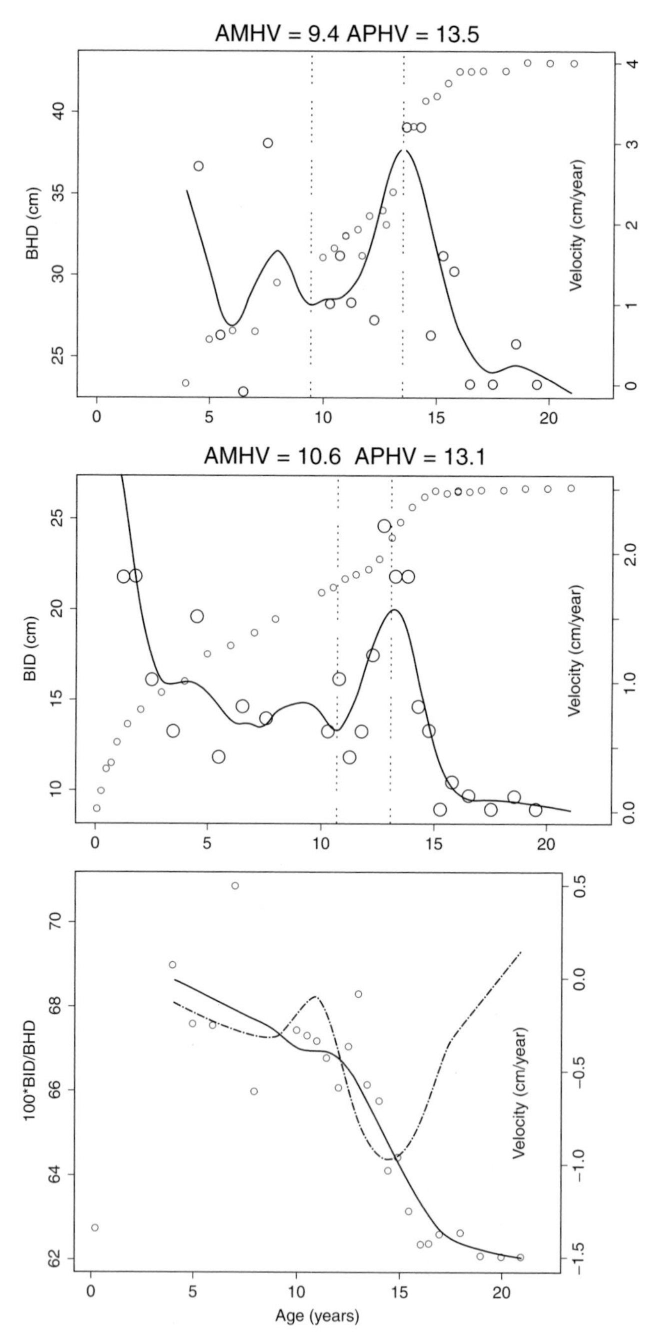

Figure 2.7 Bihumeral diameter (shoulder width, BHD), biiliac diameter (pelvic width, BID) and their quotient for subject 1 in the First Zürich Longitudinal Study. Small circles: raw distance measurements. Large circles: raw yearly increments (yearly raw velocities). Continuous line: estimated (kernels) velocities for BHD and BID; smoothed distance for the quotient. Dashed line: velocity for the quotient. AMHV, age at minimal height velocity; APHV, age at peak height velocity.

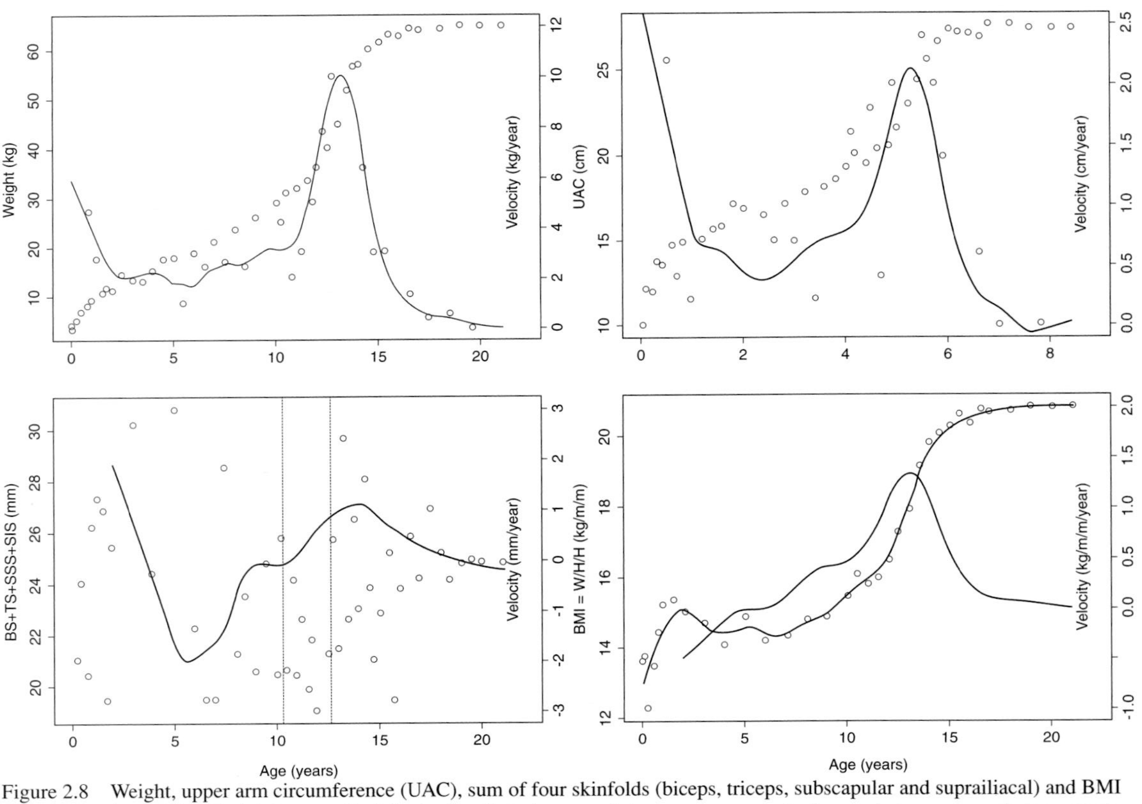

Figure 2.8 Weight, upper arm circumference (UAC), sum of four skinfolds (biceps, triceps, subscapular and suprailiacal) and BMI for subject 1 in the First Zürich Longitudinal Study. Small circles: raw distance measurements. Large circles: raw yearly increments (yearly raw velocities). Continuous line: estimated (kernels) velocities for weight, UAC and sum of skinfolds; smoothed distance for BMI. Dashed line: velocity for BMI.

In Figure 2.6 SH and SILH, HC and the quotient of SILH to SH are presented. Several points are noteworthy:

- The large variability of the yearly and half yearly increments (raw velocities).
- The clear, to some extent unexpected, pubertal spurt of HC.
- The different timing and the different duration of the pubertal spurt for SH and for SILH. This fact is well illustrated by the course of the quotient SILH/SH, which, after increasing from birth to puberty, representing this well-known change of body proportions, decreases again, albeit slightly, when the late pubertal spurt of the trunk occurs.
- We represent the velocity of the quotient SILH/SH by a thin dashed line to indicate that the use of the velocity is not generally recommended for quotients, being related in a complex way to the differential changes of the numerator and denominator.

In Figure 2.7 a similar representation is given for the two parameters of width, BHD and BID; notice the substantial mid-growth spurt in BHD, which is even stronger in other subjects and stronger in some cases, particularly for girls, than the pubertal spurt.

In Figure 2.8, besides weight, variables involving muscle and fat mass are presented. The sum of four skinfolds is used here only to emphasize the non-standard growth course of these measurements. A more in-depth treatment of this topic is given in Gasser *et al.* (1993).

Statistics of the milestones in the First Zürich Longitudinal Study

In Table 2.3 we present a concise description of growth for selected skeletal parameters, in boys and girls of the First Zürich Longitudinal Study, in terms of the milestones introduced above. Generally speaking, the milestones T_6, T_7, T_8 and T_9 as pictured in Figure 2.4 are reasonably well defined in most subjects for all linear measurements. Milestone T_3, the timing of the mid-growth spurt, on the other hand, is ill defined in many subjects, either because the subject is one of those about 15% who do not have a true spurt, or because the subject has multiple spurts before puberty (Butler *et al.*, 1990; R. D. Bock, pers. comm.), which is more often the case for late developers.

There are no such milestones for the skinfolds, although they are still amenable to a similar analysis along these lines (Gasser *et al.* 1993).

The situation is less clear for those body measurements that are related, to a variable extent, to the fat mass, rather than exclusively to the skeleton: primarily weight and upper arm circumference. While for weight, which is determined in most children mainly by the skeleton or muscle mass, reflected by

Table 2.3 *Milestones in the First Zürich Longitudinal Study; means and standard deviations*

Measurement	T_3	T_6	T_7	T_8	T_9
Boys					
Height	7.0 (0.9)	11.1 (1.1)	13.0 (0.9)	14.0 (0.9)	15.2 (0.9)
Sitting height (SH)	6.7 (0.9)	11.2 (1.2)	13.1 (1.0)	14.2 (0.9)	15.5 (1.0)
Standing subischial leg height (SILH)	7.6 (1.2)	11.0 (1.2)	12.6 (1.0)	13.7 (0.9)	15.1 (1.0)
Arm length (AL)	7.2 (1.3)	10.7 (1.1)	12.5 (1.1)	13.8 (0.9)	15.2 (1.0)
Bihumeral diameter (BHD)	6.9 (0.9)	10.4 (1.3)	12.7 (1.2)	14.2 (0.9)	15.7 (1.0)
Biiliac diameter (BID)	7.3 (0.9)	10.8 (1.4)	12.9 (1.2)	14.1 (1.0)	15.4 (1.0)
Girls					
Height	6.9 (0.9)	9.8 (1.0)	11.2 (1.0)	12.2 (1.0)	13.5 (1.0)
Sitting height (SH)	6.6 (1.0)	9.7 (1.0)	11.3 (1.1)	12.4 (1.1)	13.9 (1.1)
Standing subischial leg height (SILH)	7.2 (1.2)	9.4 (1.2)	10.6 (1.2)	11.6 (1.1)	13.3 (1.0)
Arm length (AL)	6.7 (1.1)	9.6 (1.0)	11.0 (1.0)	12.1 (1.0)	13.7 (0.9)
Bihumeral diameter (BHD)	7.0 (1.0)	9.6 (1.0)	11.2 (1.0)	12.6 (1.0)	14.2 (1.0)
Biiliac diameter (BID)	7.2 (1.0)	9.4 (0.8)	11.1 (0.8)	12.4 (1.0)	14.0 (1.0)

Source: Sheehy *et al.* (1999).

a correlation of about 0.5 to 0.7 between weight and height, milestones T_7 and T_8 are still well defined; this is not the case for upper arm circumference. In perhaps 20% of the subjects, depending on the definition, no clear pubertal spurt can be seen.

A brief compilation of interesting findings from the work of Gasser's group follows. The main tool in obtaining these results was structural analysis (Chapter 7). It must be emphasized that several of these findings, rather subtle indeed, could not have been discovered by fitting the individual growth curves using parametric models, which impose fixed relationships, not warranted by the data, between certain aspects of growth (Goldstein (1979: Chapter 4) gives examples for this point).

Growth velocity decreases rapidly (large negative acceleration) during the first year of life. A switch from a rapidly decreasing velocity to a more peaceful slowing of growth, continued over several years, is observed around 1 year of age, in the average, the exact timing depending on the body measurement (Gasser *et al.*, 1991a,b). This phenomenon provides some support to the notion of the infancy phase as a transition from intrauterine to extrauterine growth.

It has long been known that adult height is largely independent of the timing, duration and intensity of the pubertal spurt, for both genders, the best predictor for adult height, besides current height, being the prespurt height velocity VT_6 or the average growth velocity during childhood (Gasser *et al.*, 1985).

The pubertal spurt of legs and trunk is not synchronous. Maximal growth velocity is reached, in the average, about three-quarters of a year later for the trunk than for the legs in girls (half a year later in boys). The pubertal spurt is more intense for boys than for girls, both in absolute and relative terms (i.e. with respect to final size), for all parameters, with the exception of BID (Gasser *et al.*, 1991a,b).

Do the mid-growth spurt and the pubertal spurt correlate? For stature this has been shown not to be the case (Gasser *et al.*, 1985), both for boys and for girls, either for timing, duration or intensity. This again seems to give some support to the postulated childhood and pubertal components of growth. A cursory analysis along the lines of the just-mentioned paper suggests that the same essentially applies for several other measurements, such as sitting and leg height, BHD and BID; with some caveat, however, in the case of girls, where some moderate correlations were observed in the timing of the mid-growth and pubertal spurt for leg height and for BHD (results not shown) (see Gasser *et al.* (1985) for further details).

Distribution of the milestones and of anthropometric measurements

General considerations

Mean and standard deviations of the milestones provide a concise and, for the two genders, easily understandable description of growth for different areas of the body. The next natural question is whether the distribution of these milestones is Gaussian (normal).

Beyond being an intrinsically interesting question (see below), it must be reminded that sophisticated statistical techniques, such as multiple regression and analysis of variance expect Gaussian dependent variables, in order to produce credible standard errors and *P*-values. A complex machinery of transformations is sometimes invoked, when this is not the case (Box and Cox, 1964; Tukey, 1977; Cole, 1988).

It is often assumed in auxology that anthropometric measurements, which are mainly genetically determined, obey a multifactorial additive model and follow, therefore, a Gaussian distribution. This applies to stature, head circumference and other skeletal measurements.

On the other hand, parameters such as weight and skinfold thicknesses, with a moderate to strong exogenous component, in principle not bounded from above, tend to be skewed to the right; their logarithm is often approximately symmetrically distributed, suggesting a lognormal distribution for the original

Table 2.4 *Skewness of milestones in the First Zürich Longitudinal Study*

Measurement[a]	T_3	T_6	T_7	T_8	T_9
Boys					
Height	−0.51	−0.12	0.37	0.68	0.57
SH	−0.52	−1.51	−2.41	0.02	1.14
SILH	−1.63	−0.50	−1.20	0.96	0.38
AL	−1.14	4.84	1.77	1.96	2.39
BHD	0.24	3.72	0.52	0.45	1.58
BID	−0.41	1.05	−0.71	1.27	0.87
Girls					
Height	1.82	2.22	2.30	0.89	1.54
SH	0.43	2.02	1.31	0.99	1.28
SILH	−1.17	−1.06	−1.13	−0.42	0.55
AL	0.34	0.65	−1.70	−0.97	−0.13
BHD	−1.11	−0.43	−2.83	−2.80	0.83
BID	−1.81	0.79	1.90	1.78	0.33

[a] Abbreviations for measurements are as in Table 2.3.

measurement, and, possibly, a multifactorial multiplicative biological mechanism generating it.

In other words, whether the distribution of a body parameter is Gaussian, lognormal, or other, may provide some clues to the mechanisms steering its growth: unifactorial, multifactorial, additive, multiplicative. It should be mentioned at this point that a Gaussian distribution for measurements does not imply a Gaussian distribution for milestones, and vice versa.

As interesting the question of the distribution is, its answer is not, somewhat surprisingly, easily found. The reason for this is that statistical goodness-of-fit tests, such as the Kolmogorov–Smirnov test, or even the Shapiro–Wilk test for normality, have limited power and need large samples, especially when the variable to be tested has a small coefficient of variation (below 5%, say; see Limpert *et al.* (2001) for a discussion of the merits of the Gaussian and lognormal distributions in biology and other fields).

Distributions in the First Zürich Longitudinal Study

Tables 2.4 and 2.5 present coefficients of skewness and kurtosis for the milestones and measurements given in Table 2.3. The coefficients are standardized, i.e. rescaled by their asymptotic standard errors of $\sqrt{(6/n)}$ and $\sqrt{(24/n)}$, respectively, where n is the sample size (Stuart and Keith Ord, 1994). While a number

Table 2.5 *Kurtosis of milestones in the First Zürich Longitudinal Study*

Measurement[a]	T_3	T_6	T_7	T_8	T_9
Boys					
Height	0.31	−0.40	−0.27	−0.40	−0.23
SH	−1.13	−0.47	2.15	0.52	0.67
SILH	−1.10	1.80	3.68	0.47	−0.57
AL	−2.34	1.95	−0.27	0.49	0.01
BHD	−0.89	0.40	−0.66	0.29	−0.85
BID	−1.06	−1.76	0.08	−0.01	−0.93
Girls					
Height	0.63	1.90	0.64	−0.40	0.36
SH	−1.61	1.75	−0.85	−0.68	−0.95
SILH	−1.34	−1.20	−0.77	0.23	−0.16
AL	−2.08	0.23	0.09	−0.04	0.92
BHD	−0.85	0.15	0.73	2.71	1.25
BID	−0.55	0.78	−0.20	−0.18	−0.02

[a] Abbreviations for measurements are as in Table 2.3.

of entries exceed in absolute value the cut-off of 2, indicating significance at the 5% level, and negative values tend to occur in the left portion of the tables, a Shapiro–Wilk test would not reject the null hypothesis that *all* these values come from a standard Gaussian distribution. That is, we would not reject the hypothesis that these milestones have a Gaussian distribution, both in boys and girls.

Turning to the raw anthropometric measurements, one could theoretically expect that Gaussian distributions should not occur very often. Due to the large variability in growth tempo among subjects, mixtures of Gaussian distributions would at best fit such data, and such mixtures do not have, generally, a Gaussian distribution.

Coefficients of skewness (continuous lines) and of kurtosis (dashed lines) versus age, slightly smoothed by splines, are presented for boys (the picture for girls is similar) in Figure 2.9.

Figure 2.9 Skewness (continuous lines) and kurtosis (dashed lines) for sitting height, sitting height velocity, weight and sum of biceps, triceps, subscapular and suprailiacal skinfolds in the boys of the First Zürich Longitudinal Study as functions of age. Thin lines, the original measurements. Thick lines, transformed data: logarithmic transformation for sitting height and weight, $-1/x$-transformation for sum of skinfolds. See text for details.

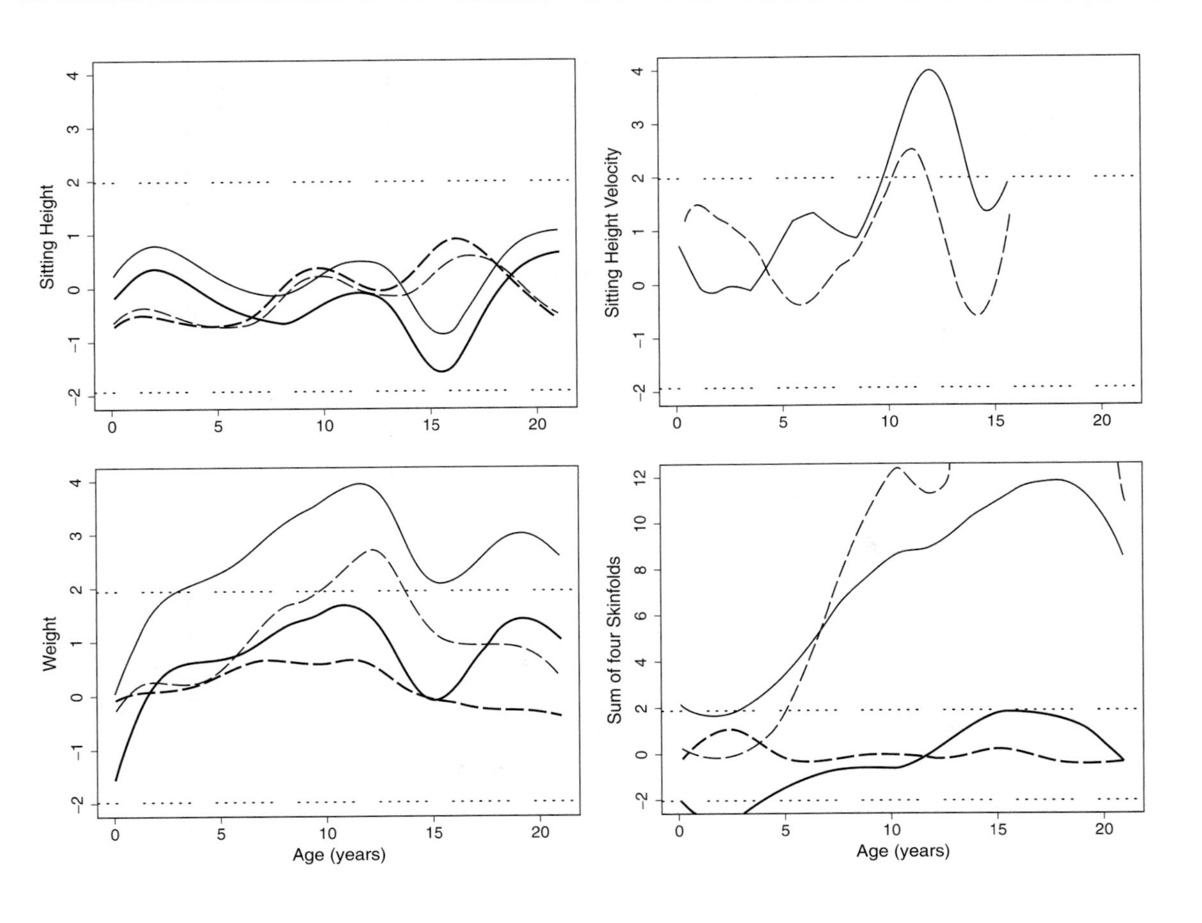
Sitting Height
Sitting Height Velocity
Weight
Sum of four Skinfolds
Age (years)
Age (years)

- For sitting height skewness and kurtosis were calculated for the raw measurements (thin lines) and for their logarithms (thick lines). All lines fall between −2 and 2 indicating that, on this basis, neither the Gaussian distribution nor the lognormal can be rejected at any age.
- For sitting height velocity we have some indications of non-Gaussianity (especially around puberty), which actually contradicts the above statement, because sitting height cannot be Gaussian, if its velocity (a difference) is non-Gaussian.
- Weight untransformed (thin lines) is clearly non-Gaussian for a broad range of ages, but for its logarithm both skewness and kurtosis are not significantly different from zero (thick lines).
- Finally the sum of four skinfolds has massive positive skewness and kurtosis and a strong transformation, such as $-1/x$ (Gasser *et al.*, 1994), is required to reduce both to the range −2 to 2.

Transformations are useful in statistics, but their use may imply subtle considerations and arbitrary criteria, especially with longitudinal data, where to find a common transformation for a range of ages may represent a kind of Procrustean bed and compromises must be accepted. This is further illustrated by the following example: as demonstrated in Figure 2.9 the sum of four skinfolds has a strongly skewed distribution, for example at age 11.5 years for boys, the probability density of which is shown in Figure 2.10 (thin continuous line), together with the fitted Gaussian distribution (thin dashed line), which does not fit at all. The bulk of the distribution, however, between 12 and 30 mm, covering more than 85% of the data, with estimated density given by the thick continuous line, is well fitted by a Gaussian distribution (thick dashed line).

The above questions about distribution and transformation have additional relevance, both auxologically and clinically, for measurements related to obesity. A discussion of this topic goes far beyond the intent of this chapter, but see Johnston and Foster (2001) for an up-to-date treatment of the field.

Stability of growth

It is a well-established axiom in paediatrics that normally developing children follow percentile lines during growth, perhaps apart from the first two years of life, when they must attain their genetically determined centile, while leaving the centile they had been confined to during intrauterine life.

Experience, however, shows that a notable proportion of subjects, leaves, at one time or another, the centile they were following, without any recognizable pathology (Hermanussen *et al.*, 2001b: Figure 2.1). Certainly most of these

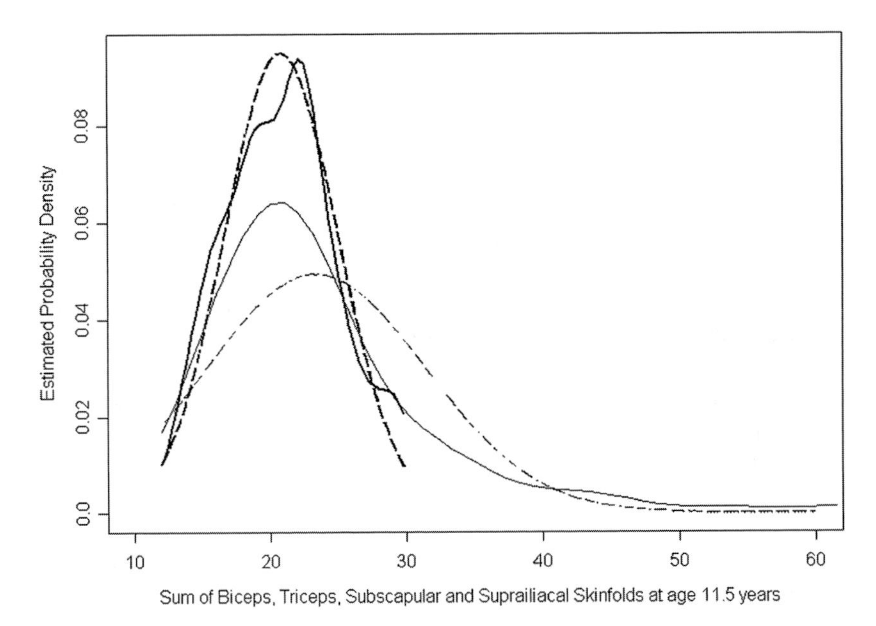

Figure 2.10 Probability density of the sum of four skinfolds in boys of the First Zürich Longitudinal Study at age 11.5. Continuous lines, kernel estimated densities. Dashed lines, fitted Gaussian distributions. Thin lines, all data used. Thick lines, only data below 30 mm (85% of all data). The bulk of the data is well fitted by a Gaussian distribution.

deviations occur during puberty and are due to the different growth tempo of individuals and to the fact that reference curves are cross-sectionally constructed and do not take this into account; even many prepubertal deviations may be due to a deviating tempo of growth, visible, if at all, only later at puberty. (Bone age, or curve registration, i.e. aligning (see Chapter 7 in the present volume) could help here to discriminate between those crossings of centiles due to change in tempo and those due to a true change in growth velocity.)

But whether subjects follow one centile line or cut several of them, they can do it smoothly or with substantial oscillations. By stability of growth, we mean this last aspect, considering an individual growth process stable when the adult value is reached smoothly moving along or across centiles, and unstable otherwise.

A simple way of studying this problem is by looking at the correlation of a measurement at any given age with the corresponding adult measurement. This was done systematically in a descriptive way in Molinari *et al.* (1995). Notable differences were found between the measurements and a curious, not understood, gender difference, girls showing lower correlations throughout.

Stability, as defined here, is an interesting, and puzzling, feature of growth, which has not, to our knowledge, been studied in any depth. Goldstein (1981) defined a coefficient, called *growth constancy index*, in order to compare stability across body parameters.

Sheehy *et al.* (2002) defined two indexes, a long-term and a short-term 'variability' index, to quantify both aspects of the individual growth curve, the crossing of centiles and the smoothness of the centile crossing. These two individually calculated coefficients can be used to compare subjects or groups of subjects (e.g. boys and girls or short and tall subjects) as well as to compare measurements within subject. In that paper possible covariates were also looked for to explain the individual variability of the coefficients.

Some results were obtained: in line with the lower correlation with the adult measurement in girls, a larger long-term variability was found in girls for standing and sitting height, arm length and biiliac width. Also, both the long-term and short-term variability were different for the different measurements, and these differences could not be explained by measurement errors alone. On the other hand no relationship could be established between the tracking behaviour as measured by these indexes and the timing of the pubertal spurt, or with the adult measurements. The sources of between individual differences in tracking behaviour and of differences between anthropometric measurements have not been clarified.

Hermanussen *et al.* (2001a) have introduced a cluster analytic (unsupervised learning) approach, by which growth curves are grouped into classes based on their similarity. The classes are not defined a priori on the basis of previous knowledge, but result exclusively from the data (and from the algorithm and its options). The scientific and clinical relevance of this approach has not yet been demonstrated.

Conclusion

The time for national longitudinal studies covering growth from birth to adulthood may have passed, for economic reasons if for no others. But the preceding considerations are not obsolete, because not all existing studies have yet been fully exploited, and new studies are being started again and again that do not cover the whole age range, or that involve reduced numbers, or subgroups (e.g. diabetic children), of subjects.

Longitudinal studies involving other than the classical anthropometric measurements are now being undertaken; see Largo *et al.* (2002) for studies of neuromotor development. Short-term longitudinal studies, requiring sophisticated mathematical and unprejudiced (non-parametrical) statistical methods, will deepen our understanding of the growth processes, especially if information on correlating endocrinological processes is available.

A number of further growth-related topics has been omitted from this introduction and from the present volume altogether, e.g.:

- Short-term growth, especially aspects of saltation and stasis in short-term growth (Lampl, 1999).
- Prenatal growth.
- Growth in adults.
- Growth and shape.
- Obesity and its measurement during growth (Johnston and Foster, 2001).
- Secular trends.

ACKNOWLEDGEMENT

This work was supported by the Swiss National Science Foundation (Project No 32.45829.95).

REFERENCES

Bertino, E., Di Battista, E., Bossi, A., *et al.* (1996). Fetal growth velocity: kinetic, clinical, and biological aspects. *Archives of Disease in Childhood*, **74**, F10–F15.

Bock, R. D. and Thissen, D. M. (1976). Fitting multi-component models for growth in stature. *Proceedings of the 9th International Biometric Conference*, vol. 1, pp. 431–42. Boston, MA: The Biometric Society.

Bogin, B. (1999). *Patterns of Human Growth*, 2nd edn. Cambridge, UK: Cambridge University Press.

Box, G. E. P. and Cox, D. R. (1964). An analysis of transformations. *Journal of the Royal Statistical Society B*, **26**, 211–52.

Butler, G. E., McKie, M. and Ratcliffe, S. G. (1990). The cyclical nature of pre-pubertal growth. *Annals of Human Biology*, **17**, 177–98.

Cole, T. J. (1988). Fitting smoothed centile curves to reference data. *Journal of the Royal Statistical Society A*, **151**, 385–418.

Count, E. W. (1943). Growth patterns of the human physique: an approach to kinetic anthropometry. *Human Biology*, **15**, 1–32.

Efron, B. and Tibshirani, R. J. (1993). *An Introduction to the Bootstrap*. New York: Chapman and Hall.

Eiholzer, U., Bodmer, P., Bühler, M., *et al.* (1998). Longitudinal monthly body measurements from 1 to 12 months of age: a study by practitioners for practitioners. *European Journal of Pediatrics*, **157**, 547–52.

Eveleth, P. B. and Tanner, J. M. (1991). *Worldwide Variation in Human Growth.* Cambridge, UK: Cambridge University Press.

Falkner, F. and Tanner, J. M. (1986). *Human Growth: A Comprehensive Treatise*, 2nd edn. New York: Plenum Press.

Gasser, T., Köhler, W., Müller, H.-G., *et al.* (1985). Human height growth: correlational and multivariate structure of velocity and acceleration. *Annals of Human Biology*, **12**, 501–15.

Gasser, T., Kneip, A., Ziegler, P., *et al.* (1990). A method for determining the dynamics and intensity of average growth. *Annals of Human Biology*, **17**, 459–74.

Gasser, T., Kneip, A., Binding, A., Prader, A. and Molinari, L. (1991a). The dynamics of linear growth in distance, velocity and acceleration. *Annals of Human Biology*, **18**, 187–205.

Gasser, T., Kneip, A., Ziegler, P., *et al.* (1991b). The dynamics of growth of width in distance, velocity and acceleration. *Annals of Human Biology*, **18**, 449–62.

Gasser, T., Ziegler, P., Kneip, A., *et al.* (1993). The dynamics of growth of weight, circumferences and skinfolds in distance, velocity and acceleration. *Annals of Human Biology*, **20**, 239–59.

Gasser, T., Ziegler, P., Seifert, B., *et al.* (1994). Measures of body mass and of obesity from infancy to adulthood and their appropriate transformation. *Annals of Human Biology*, **21**, 111–25.

Goldstein, H. (1979). *The Design and Analysis of Longitudinal Studies*. London: Academic Press.

(1981). Measuring the stability of individual growth patterns. *Annals of Human Biology*, **8**, 549–57.

Hermanussen, M., Lange, S. and Grasedyck, L. (2001a). Growth tracks in early childhood. *Acta Paediatrica*, **90**, 381–6.

Hermanussen, M., Largo, R. H. and Molinari, L. (2001b). Canalization in human growth: a widely accepted concept reconsidered. *European Journal of Pediatrics*, **160**, 163–7.

Johnston, F. E. and Foster, G. D. (2001). *Obesity, Growth and Development*. London: Smith-Gordon.

Jolicoeur, P., Pontier, J., Pernin, M. O. and Sempé, M. (1988). A lifetime asymptotic growth curve for human height. *Biometrics*, **44**, 995–1003.

Karlberg, J., Jalil, F., Lam, B., Low, L. and Yeung, C. Y. (1994). Linear growth retardation in relation to the three phases of growth. *European Journal of Clinical Nutrition*, **48** (Suppl. 1), S25–S44.

Lampl, M. (1999). *Saltation and Stasis in Human Growth and Development: Evidence, Methods and Theory*. London: Smith-Gordon.

Largo, R. H. and Prader, A. (1983a). Pubertal development in Swiss boys. *Helvetica Paediatrica Acta*, **38**, 211–28.

(1983b). Pubertal development in Swiss girls. *Helvetica Paediatrica Acta*, **38**, 229–43.

Limpert, E., Stahel, W. and Abbt, M. (2001). Lognormal distributions across the sciences: keys and clues. *BioScience*, **51**, 341–52.

Largo, R. H., Fischer, J. E. and Caflisch, J. A. (2002). *Zürcher Neuromotorik*. Zürich, Switzerland: AWE-Verlag.

Milani, S. (2000). Kinetic models for normal and impaired growth. *Annals of Human Biology*, **27**, 1–18.

Molinari, L., Largo, R. H. and Prader, A. (1980). Analysis of the growth spurt at age seven (mid-growth spurt). *Helvetica Paediatrica Acta*, **35**, 325–34.

Molinari, L., Gasser, T., Largo, R. H. and Prader, A. (1995). Child–adult correlations for anthropometric measurements. In *Essays on Auxology*, eds. R. Hauspie, G. Lindgren and F. Falkner, pp. 164–77. Welwyn Garden City, UK: Castlemead.

Prader, A., Tanner, J. M. and Harnack, G. A. V. (1963). Catch-up growth following illness or starvation: an example of developmental canalization in man. *Journal of Pediatrics*, **62**, 646–59.

Prader, A., Largo, R. H., Molinari, L. and Issler, C. (1989). Physical growth of Swiss children from birth to 20 years of age. *Helvetica Paediatrica Acta*, Suppl. 52.

Sandland, R. L. and McGilchrist, C. A. (1979). Stochastic growth curve analysis. *Biometrics*, **35**, 255–71.

Scammon, R. E. (1927). The first seriatim study of human growth. *American Journal of Physical Anthropology*, **10**, 329–36.

(1930). The measurement of the body in childhood. In *The Measurement of Man*, eds. J. A. Harris, C. M. Jackson, D. G. Patterson and R. E. Scammon, pp. 171–215. Minneapolis, MN: University of Minnesota Press.

Sheehy, A., Gasser, T., Molinari, L. and Largo, R. H. (1999). An analysis of variance of the pubertal and mid-growth spurts for length and width. *Annals of Human Biology*, **26**, 309–31.

Sheehy, A., Gasser, T., Largo, R. H. and Molinari, L. (2002). Short-term and long-term variability of standard deviation scores for size in children. *Annals of Human Biology*, **29**, 202–18.

Smith, D. W., Truog, W., Rogers, J. E., *et al.* (1976). Shifting linear growth during infancy: illustration of genetic factors in growth from fetal life through infancy. *Journal of Pediatrics*, **89**, 225–30.

Stuart, A. and Keith Ord, J. (1994). *Kendall's Advanced Theory of Statistics*, vol. 1. London: Charles Griffin.

Tanner, J. M. (1955). *Growth at Adolescence*. Oxford: Blackwell Scientific Publications.

(1981). *A History of the Study of Human Growth.* Cambridge, UK: Cambridge University Press.

(1986). Growth as a target-seeking function: catch-up and catch-down growth in man. In *Human Growth: A Comprehensive Treatise,* vol. 1, *Developmental Biology: Prenatal Growth*, 2nd edn, eds. F. Falkner and J. M. Tanner, pp. 167–79. New York: Plenum Press.

Tanner, J. M. and Cameron, N. (1980). Investigation of the mid-growth spurt in height, weight and limb circumferences in single-year velocity data from the London 1966–67 growth survey. *Annals of Human Biology*, **7**, 565–77.

Tanner, J. M., Whitehouse, R. H. and Takaishi, M. (1966). Standards from birth to maturity for height, weight, height velocity, and weight velocity: British children, 1965. *Archives of Disease in Childhood*, **41**, 454–71, 613–35.

Thompson, W. D'Arcy (1942). *On Growth and Form*, abridged edn. Cambridge, UK: Cambridge University Press.

Tukey, J. (1977). *Exploratory Data Analysis*. Boston, MA: Addison-Wesley.

Ulijaszek, S. J., Johnston, F. E. and Preece, M. A. (1998). *The Cambridge Encyclopedia of Human Growth and Development*. Cambridge, UK: Cambridge University Press.

Waterlow, J. C. and Schürch, B. (1994). Causes and mechanisms of linear growth retardation. *European Journal of Clinical Nutrition*, **48** (Suppl. 1), S1–S216.

Zankl, A., Eberle, L., Molinari, L. and Schinzel, A. (2002). Growth charts for nose length, nasal protrusion, and philtrum length from birth to 97 years. *American Journal of Medical Genetics*, **111**, 388–91.

3 *Sampling for growth studies and using growth data to assess, monitor and survey disease in epidemiological settings*

Edward A. Frongillo
Cornell University

Sampling for growth studies

The purpose of sampling is to take measurements on a representative portion of the population so that the whole population does not have to be measured. Each observation in the sample can be thought of as representing a certain number of population members. The reciprocal of the number in the population represented by an observation is the sampling proportion.

Sampling schemes

Most schemes for obtaining samples fall into one of three categories. Sampling can be done: (1) from a complete or nearly complete list; (2) from a set of people who go somewhere or do something; or (3) in two or more stages (Fowler, 1984). The first category implies that there is a list of the population available in advance from which then to take a sample. Examples of the second category are sampling children who get services from a hospital or who shop at a store with a parent. The third category includes schemes such as sampling, in turn, states or provinces, counties or departments, neighbourhoods or villages, households, and then individuals.

Methods in Human Growth Research, eds. R. C. Hauspie, N. Cameron and L. Molinari.

Characteristics to evaluate sampling

Three characteristics important for evaluating a sampling scheme are: (1) comprehensiveness; (2) known probability of (or equal) selection; and (3) efficiency (Fowler, 1984). Comprehensiveness refers to whether everyone in the population of interest from which the sample is to be drawn had a chance to be selected into the sample. That is, a sampling scheme is not comprehensive if some people in the population are excluded from being possibly sampled. For example, if one is trying to sample the population of people living in a city, a sampling scheme might be based on randomly choosing telephone numbers and addresses from the telephone directory. But such a scheme is not comprehensive because people living in the city without telephones have no chance at being selected into the sample.

A good sampling scheme will either give each individual an equal chance of being selected into the sample, or, if not, will know what is the probability of being selected for each individual selected. If this information is not known, then it is not possible to estimate accurately the relation between statistics calculated from the sample and the corresponding population parameters.

Efficiency refers to the costs that are inherent in selecting each individual into the sample on average. Sampling schemes that require screening many individuals or travelling long distances to identify an individual for possible selection into the sample may not be efficient.

Types of samples

Samples can be divided into three types: (1) simple; (2) stratified with equal probability of selection; and (3) complex. Simple samples are ones for which each individual has equal probability of being selected, and individuals are selected directly in one step or stage. That is, each individual in the sample represents the same number of population members. Table 3.1 shows several of the ways one might obtain a simple sample.

A stratified sample is obtained through a structured process in which individuals are sampled from within non-overlapping subpopulations (i.e. strata) that are defined according to one or a few known characteristics of the individuals (Cochran, 1977; Levy and Lemeshow, 1999). The objective of stratification is to reduce the usual sampling variability and to produce a sample that is more likely to represent well the underlying population. For example, one might divide a village into four geographic areas, and then sample individuals from within each of the areas. This guarantees that some individuals are sampled from each of the four areas and avoids the problem that, by chance, no individual is sampled

Table 3.1 *Ways of obtaining a simple sample*

Name of method	Description
Random	Members of a population are selected independently of each other
Systematic	The first member of the population might be chosen at random, but successive members are selected according to some fixed rule such as taking every 18th person
Snowball	One sampled individual is used to identify the next individual to be sampled
Accidental	The sample is obtained from those who happen to be at hand
Quota	A predetermined number of individuals are identified by certain characteristics
Spinning bottle	A mechanism is employed to indicate in which direction to go to find the next individual to sample

Source: Persson and Wall (2000).

from one of the areas. If the number of individuals sampled from each area is proportional to the size of the population in each area, then the probability of selection for each individual is equal.

Surveys that use either a simple random sample or a stratified sample in which each person has an equal probability of being chosen have two important disadvantages. First, they may not include enough observations from particular subpopulations to allow for accurate estimates for these subpopulations. 'Oversampling' is needed to make sure that a minimum number are available for particular subpopulations. Second, they may not be practical to implement on a large scale. It is usually much cheaper to conduct a large survey in stages. A complex survey overcomes these two disadvantages because it samples observations disproportionately within strata and has multiple stages of sampling (Levy and Lemeshow, 1999). That is, complex surveys differ from simple or equal-probability stratified surveys in two fundamental ways.

First, a complex survey samples individuals within strata typically with disproportionate sampling proportions. For example, in many national surveys, strata are defined based upon geographic region. Individuals sampled from one stratum may represent more or fewer population members than individuals sampled from another stratum. That is, the sampling proportions across strata differ. Often adjustments to the sampling proportions are made after the data are collected; this is called post-stratification adjustment. To have the sample represent the population, information on the sampling proportions – usually called sample weights – is required.

Second, a complex survey has multiple stages of sampling. At every stage except the lowest stage, clusters of observations are sampled. At the lowest stage, individual observations are sampled. For example, a survey of schoolchildren may be done by first sampling schools, then classrooms within schools, and finally children within classrooms. This type of sampling is often required because it is logistically impossible, difficult or expensive to sample observations directly.

Implications of type of sampling for analysis

Over-sampling, i.e. sampling disproportionately more members of particular subpopulations, ensures that the sampling is adequate for these subpopulations, but then the estimate for the total population must be adjusted for the fact that some subpopulations are overrepresented in the sample. The consequence of disproportionate sampling within strata is that estimates of means, population totals, regression coefficients and other statistics should be made using the sampling weights if the estimates are to reflect the population accurately. Also, the variances (or standard errors) of those estimates should be calculated using the sampling weights. If the sampling weights cover a broad range of probabilities, then use of the sampling weights will result in larger variances of estimates than if a simple survey had been done. Software is needed that will allow the incorporation of sampling weights. General statistical procedures (e.g. for obtaining descriptive statistics and doing linear regression) in software such as SAS and SPSS can incorporate the sampling weights to obtain weighted estimates of the parameters of interest (e.g. means, regression coefficients). These general procedures may not, however, obtain correct weighted estimates of the standard errors of those parameters of interest. The statistical software STATA and SAS present options and explanations for doing this correctly.

The use of multistage cluster sampling means that observations cannot be assumed to be independent as is commonly done for a simple survey; observations that are from the same cluster will likely be more similar to each other than to observations from a different cluster. The consequence of this is that variances of estimates will typically be larger than in a simple survey. We usually assume that the cluster sampling does not affect estimates themselves, only their variances. The effect on variances of complex sampling is quantified as the design effect, which is the ratio of the variance under complex sampling divided by the variance that would have been obtained if a simple survey of the same size could have been done. Design effects can differ within the same survey markedly, depending upon the variable of interest, the subpopulation and, in regression analyses, the variables in the regression model. For example, across the means of four anthropometric variables and six subpopulations in the second National Health and Nutrition Examination Survey from the United States, design effects ranged from 1.0 to 5.7.

When surveys are complex, it is best to use statistical procedures specifically designed to handle disproportionate sampling and multiple stages of sampling. These procedures either implement a linearization method, replication methods, or both (Lee *et al.*, 1989; Skinner *et al.*, 1989; Lehtonen and Pahkinen, 1995; Levy and Lemeshow, 1999). Linearization methods approximate the complicated formulae for the variances of estimates by using Taylor series expansion. Replication methods re-sample the sampling units at the highest stage (i.e. the primary sampling units) to estimate the variability contributed by that stage using one of three operations: balanced repeated replications, jackknife, or bootstrap. Typically, both sets of methods only use directly information about the highest and lowest sampling stage, even if multiple stages were used for sampling. Depending on the survey to be analysed and how the design information for the survey has been constructed and provided, one method may be more suitable than the other.

Several statistical packages perform correct analyses of data from complex surveys by including weights and information about clustering. Some examples are SUDAAN, STATA, SAS and WesVarPC (for information about and reviews of these and other software packages for analysing complex surveys see Harvard University (2004)).

Another, more recent approach to analysing data from complex surveys is to use software for multilevel modelling. Levels in the model are specified that correspond to the stages of sampling; this accounts for the cluster sampling. Sample weights must be used to account for unequal sampling probabilities. Procedures such as SAS PROC MIXED and MLn can incorporate sample weights in a multilevel analysis. The procedures, however, use weights differently; SAS produces different results depending upon how the model is specified, variances will be severely inflated if the weights are not normalized properly, and both programs are documented poorly with regard to the use of weights. Therefore, this approach should be used with caution.

Sample size considerations

Sample size should be large enough so that precision is adequate for the intended uses of the data. Estimating adequate sample size for means of groups or differences among groups is straightforward when variables are normally distributed (e.g. body length), but it requires a good estimate of the expected variability and specifications of the power and the smallest meaningful difference. The power should be specified as 90% for costly and logistically challenging field studies that would be difficult to repeat (Frongillo and Rowe, 1999). The smallest meaningful difference is the smallest difference that would be substantively important, and not the difference expected or the difference that others have found. If variables follow distributions other than the normal (e.g. whether or not a child is underweight), similar calculations can be done. To illustrate the process of estimating sample size, two examples are given. Both involve estimating differences in growth velocity in infants, from 4 to 6 months and from birth to 4 months.

In the Honduras study of Cohen *et al.* (1994), the intent was to compare growth for infants who were exclusively breast-fed versus those complemented with solid foods during 4 to 6 months. Standard deviations were 326 g for weight gain and 1.1 cm for length gain from 4 to 6 months, values similar to those reported by the World Health Organization (World Health Organization Working Group on Infant Growth, 1994, 1995). The smallest meaningful difference in growth increment over 4 to 6 months is debatable. A smallest meaningful difference of 0.5 standard deviations corresponds to about 10% of the range of the distribution of increments; this value is consistent with criteria for judging growth faltering of infants (Frongillo *et al.*, 1990; Frongillo and Habicht, 1997; Frongillo and Rowe, 1999). Applying this difference translates to a difference of 163 g for weight gain and 0.55 cm for length gain for 2-month increments from 4 to 6 months. With these smallest meaningful differences, the required sample size per group is 85, assuming a power of 90%, $\alpha = 0.05$, and a two-tailed test. Using smaller differences of 120 g for weight gain and 0.4 cm for length gain, the required sample size per group is 160 (Frongillo and Habicht, 1997).

Another use of growth data is to judge the adequacy of a new infant formula. Given that a new infant formula will probably be used by a large number of infants, a population rather than individual perspective is the most salient. In the United States, it has previously been recommended that the smallest meaningful difference in growth increment is 3 g/day, or 318 g for the period of 14 to 120 days (American Academy of Pediatrics, 1988). If two different infant formulas resulted in a difference of 318 g of weight at 120 days of age, that would mean that the entire distribution of weight would be shifted by that amount. To gauge whether 318 g is a small or large difference, it is helpful to reflect both on the distribution of infant growth as captured in a reference and on differences due to other factors that we know about and accept as being important. The difference of 3 g/day is about the difference between the 25th and 50th (and 50th and 75th) percentiles of the increments in the Iowa and Iowa/Fels data (Nelson *et al.*, 1989; Guo *et al.*, 1991). The difference of 318 g is nearly as big as the difference in birthweight between low and very high altitude (Haas *et al.*, 1982) and is about 50% larger than the effect of smoking during gestation on birthweight of about 200 g (Institute of Medicine, 1990). From these perspectives, a difference of 3 g/day is meaningful.

Are differences smaller than 3 g/day also meaningful? For weight gain during the first 120 days, a clinically meaningful difference of about 0.5 standard deviations (Frongillo *et al.*, 1990; Frongillo and Habicht, 1997) would correspond to a difference of about 2.5 g/day. But these results may not be applicable to the evaluation of differences in weight gain of a new infant formula because of a mismatch on the basis of age of the infants, type of data (cross-sectional versus longitudinal) and use (clinical versus population). Another perspective comes from examining results that have been found previously when comparing different formulas. For example, Roche *et al.* (1993) investigated differences among three formulas in weight gain from birth to 120 days for about 263 infants. The differences in weight gains that were observed for males of 210, 270 and 480 g seem important, but the largest difference for females of 120 does not.

The choice of the smallest meaningful difference has implications for the sample size needed. If the smallest meaningful difference is 3 g/day, the power is set at 90% and the test is two-tailed, then the sample size needed per group is 67, assuming a standard deviation of 5.3 g/day (Nelson *et al.*, 1989; Guo *et al.*, 1991). If the smallest meaningful difference is instead 2 g/day (or 212 g from 14 to 120 days), then the sample size needed per group is 149.

When sampling occurs in stages, the usual formula for calculating sample size for a simple sample does not apply. For example, for two-stage sampling, sampling variability will occur at each of the stages. The usual formula for sample size can easily be adapted to cover this situation. To have adequate power (90%) for a two-tailed test at $\alpha = 0.05$ to detect a smallest meaningful difference D between two conditions across the clusters:

$$D = \sqrt{2\left(z_{\alpha/2} + z_{\beta}\right)^2 \left(s_2^2/k + s_1^2/mk\right)}$$

where k is the number of clusters at the second stage, m is the number of individuals per cluster, s_2^2 is the variance expected at the second stage and s_1^2 is the variance expected at the first stage. For a two-tailed test, with a Type I error equal to 0.05, we have $z_{\alpha/2} = 1.96$, and with a Type II error rate equal to 0.10 (power $= 90\%$), we have $z_{\beta} = 1.28$. These two variances must be estimated from prior research in similar settings. From this formula, it can be seen that the total sampling variability is a function of the variances and the sizes of the sample at each stage. Minimizing the sampling variability depends upon the relative size of the two variances. Determining the most efficient sampling from both statistical and logistical perspectives depends upon the variances and also the relative costs of obtaining more clusters versus more individuals per cluster.

This approach for estimating sample size with multistage sampling does not take into account the increase in sampling variability due to disproportionate sampling proportions. When complex sampling (i.e. both multistage and disproportionate sampling) is to be done, an alternate approach is to use an estimate of the design effect from prior research in similar settings. The usual formula can again easily be adapted to incorporate the design effect. To have adequate power (90%) for a two-tailed test at $\alpha = 0.05$ to detect a smallest meaningful difference D between two different conditions:

$$D = \sqrt{2\left(z_{\alpha/2} + z_{\beta}\right)^2 (s^2/n)(DEFF)}$$

where n is the number of individuals, s^2 is the variance among individuals and $DEFF$ is the design effect.

Examples of sampling strategies

To illustrate how sampling might proceed in practice, imagine that a study is to take a cross-sectional sample of children under 5 years of age in a city to be able to collect interview information and measure anthropometry. Several strategies might be employed.

A general household survey is useful when well-defined neighbourhoods exist and household interviews are acceptable. The survey might proceed by setting boundaries and numbering city blocks, taking a simple random sample of city blocks, and then visiting all (or a random sample of) households in selected blocks.

When well-defined neighbourhoods do not exist, a general household survey may be inefficient. If it is possible to identify and locate some children under 5 years of age, these children can be used as index children to construct a sample. The procedure might be to move around the neighbourhood near an index child using a set rule, and visit every household until a certain number of children under 5 years of age are found.

If most families own telephones, random-digit dialling can be used to conduct screening interviews by telephone. When a child under 5 years of age is identified, then a full interview can be scheduled and carried out in the home.

When birth records are available from hospitals or government registries, then a list of the population of interest can be built from these records, and then a random or representative sample taken. This is useful when family addresses are given in the records, mobility is low, and most children identified meet whatever selection criteria have been set by the study.

Another sampling strategy is to take advantage of institutions such as day-care centres, kindergartens and paediatrician offices that are frequented by children under 5 years old. This strategy is useful when a high proportion of children under the age of 5 attend, officials in the institutions are co-operative, and perhaps it is advantageous to obtain measurements from children at these institutions rather than in the home. A general procedure would be to gather listings from the institutions, check for duplicates since children can attend more than one institution, prepare a list from which to sample, and then take a random or representative sample. For this strategy to be successful, it is important to establish that the sampling list provides good coverage of the population of children under 5 years old.

Using growth data in epidemiological settings

Because anthropometric data are relatively easy to collect, especially data on weight and height, information on body size is frequently obtained in epidemiological studies. In elders, weight loss is indicative of failing health. In adults, relative weight measured as body mass index (BMI, calculated as weight in kilograms divided by the square of height in metres) is indicative of poor health, with low BMI (less than 18.5) indicative of undernutrition and high BMI (greater than 30) indicative of obesity (World Health Organization, 1995). Low and high BMI are associated with increased mortality (Troiano *et al.*, 1996). In children, small size and low rates of growth are indicative of exposure to infections, severe disease, and poor nutrition. To illustrate how growth data are used in epidemiological settings to assess, monitor and survey disease, the remainder of this chapter discusses briefly the relation of morbidity and mortality in young children and the assessment and monitoring of child health and nutrition.

Indicating morbidity and mortality

In the 1940s, clinicians and researchers began to understand that malnutrition and infection are synergistic (Scrimshaw, 2003). Infections worsen nutritional status, and malnutrition lowers immunological resistance, resulting in more severe infections. This synergistic relation is illustrated in Figure 3.1. By the late 1960s, there were ample data documenting this synergism, and also the mechanisms through which infections worsen nutritional status were largely understood. The understanding of how malnutrition reduces resistance to infections in human populations has mostly come since then (Scrimshaw, 2003).

The knowledge of this synergism has strongly influenced public health programmes aiming to reduce illness and malnutrition (Scrimshaw, 2003). Nevertheless, despite the strength of the research documenting the synergism, the influence of this knowledge in national and international health policy was limited (Booth, 1994) until the 1990s when Pelletier *et al.* (1993, 1994, 1995)

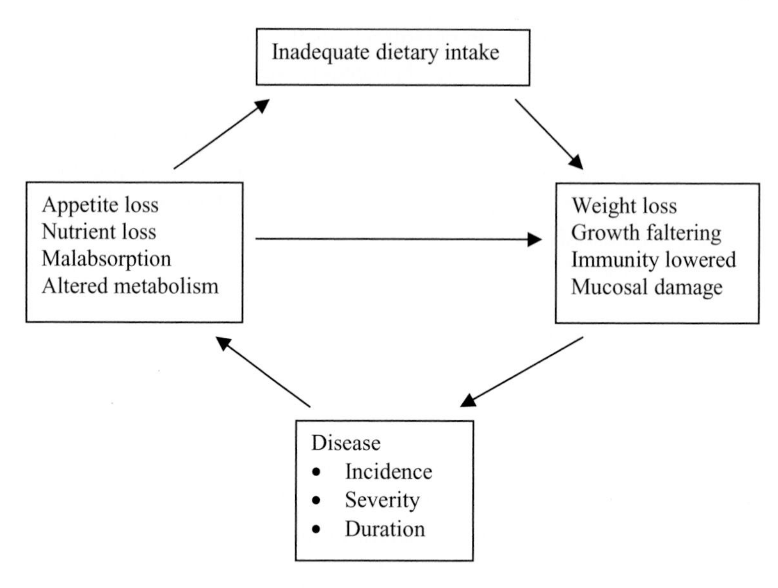

Figure 3.1 Malnutrition and infection cycle (Tompkins and Watson, 1989).

demonstrated epidemiologic evidence for this synergism in relation to child mortality. The meta-analysis carried out by Pelletier *et al.* was made possible by the availability of data on attained growth status measured by weight and on subsequent mortality from multiple cohort studies of infants and young children. This analysis resulted in the estimate that more than one-half of all child deaths in the world are due, in part, to malnutrition, particularly mild-to-moderate malnutrition. It also reinforced the understanding that child survival depends on multiple causes, and that simply partitioning causes of death into mutually exclusive categories of disease paints a misleading picture. These insights have provided a foundation that has led to the determination that underweight in children is the number one risk factor for the global burden of disease (World Health Organization, 2002).

Assessing and monitoring child health and nutrition

Growth data on children play an important role in the assessment and monitoring of child health and nutrition. For example, one of the five indicators for evaluating progress toward the Millennium Development Goal to eradicate extreme poverty and hunger is the prevalence of underweight children under 5 years of age (United Nations, 2000). Information on child growth status is the best global indicator of well-being in children, and the prevalence of child underweight (i.e. weight less than −2 standard deviations compared to the international reference) and stunting (i.e. height less than −2 standard deviations compared to the international reference) is used to monitor child health and nutrition globally. The prevalence of stunting in developing countries fell from 47% to 33% in the 20 years from 1980 to 2000 (Administrative Committee on Coordination/Subcommittee on Nutrition, 2000; de Onis *et al.*, 2000). Regional analyses showed that stunting increased in Eastern Africa, but decreased in southeastern Asia, south central Asia and South America. Modest improvement was seen in northern Africa and the Caribbean, while little progress was seen in western Africa and Central America. Data on child weight and height are also used within countries to monitor child health, malnutrition and obesity (for example, Ogden *et al.*, 1997).

Child growth data are also used in epidemiological investigations aimed at understanding the determinants of poor child health and nutrition at household and higher levels. For example, a few studies have examined what are the characteristics of developing countries that are related to why the well-being of children is better in some countries than in other countries with similar resources (Frongillo and Hanson, 1995; Frongillo *et al.*, 1997; Smith and Haddad, 2001). Recently, longitudinal data on child underweight at the national and subnational levels were used to test the effects of improvements in general child malnutrition that have occurred in the past 20 years on changes in child survival over that period (Pelletier and Frongillo, 2003).

ACKNOWLEDGEMENTS

Some information about complex sampling was adapted from three newsletters on this topic from the Office of Statistical Consulting, Cornell University and authored by Edward Frongillo, Cara Olsen and Karen Grace-Martin (Cornell University, 2004). Some information about sample size calculations was first developed by the author in an unpublished report prepared for the United States Food and Drug Administration. The examples for sampling strategies were adapted from an unpublished report prepared for the World Health Organization Multicentre Growth Reference Study by Cesar Victora. Karen Grace-Martin provided comments on an earlier draft.

REFERENCES

Administrative Committee on Coordination/Subcommittee on Nutrition (2000). *Fourth Report of the World Nutrition Situation*. Geneva, Switzerland: International Food Policy Research Institute.

American Academy of Pediatrics (1988). *Clinical Testing of Infant Formulas with respect to Nutritional Suitability for Term Infants*. Elk Grove Village, IL: Committee on Nutrition.

Booth, I. W. (1994). Costing malnutrition: add or multiply? *Lancet*, **343**, 554–5.

Cochran, W. G. (1977). *Sampling Techniques*, 3rd edn. New York: John Wiley.

Cohen, R. J., Brown, K. H., Canahuati, J., Landa Rivera, L. and Dewey, K. G. (1994). Effects of age of introduction of complementary foods on infant breast milk intake, total energy intake, and growth: a randomized intervention study in Honduras. *Lancet*, **344**, 288–93.

Cornell University (2004). http://www.human.cornell.edu/admin/statcons/

de Onis, M., Frongillo, E. A. and Blössner, M. (2000). Is malnutrition declining? An analysis of changes in levels of childhood malnutrition since 1980. *Bulletin of the World Health Organization*, **78**, 1222–33.

Fowler, F. J. (1984). *Survey Research Methods*. Beverly Hills, CA: Sage Publications.

Frongillo, E. A., Rothe, G. E. and Lambert, J. K. J. (1990). Determining growth faltering with a tracking score. *American Journal of Human Biology*, **2**, 491–501.

Frongillo, E. A. and Habicht, J. P. (1997). Investigating the weanling's dilemma: lessons from Honduras. *Nutrition Reviews*, **55**, 390–5.

Frongillo, E. A. and Hanson, K. M. P. (1995). Determinants of variability among nations in child growth. *Annals of Human Biology*, **22**, 395–411.

Frongillo, E. A. and Rowe, E. M. (1999). Challenges and solutions in using and analysing longitudinal growth data. In *Human Growth in Context*, eds. F. E. Johnston, P. Eveleth and B. Zemel, pp. 51–64. London: Smith-Gordon.

Frongillo, E. A., de Onis, M. and Hanson, K. M. P. (1997). Socioeconomic and demographic factors are associated with worldwide patterns of stunting and wasting. *Journal of Nutrition*, **127**, 2302–9.

Guo, S., Roche, A. F., Fomon, S. J., *et al.* (1991). Reference data on gains in weight and length during the first two years of life. *Journal of Pediatrics*, **119**, 355–62.

Haas, J. D., Moreno-Black, G., Frongillo, E. A., *et al.* (1982). Altitude and infant growth in Bolivia: a longitudinal study. *American Journal of Physical Anthropology*, **59**, 251–62.

Harvard University (2004). http://www.fas.harvard.edu/~stats/survey-soft/survey-soft.html

Institute of Medicine (1990). *Nutrition during Pregnancy*. Washington, DC: Subcommittee on Nutritional Status and Weight Gain during Pregnancy, Institute of Medicine, National Academy of Sciences.

Lee, E. S., Forthofer, R. N. and Lorimor, R. J. (1989). *Analyzing Complex Survey Data*, Sage University Paper no. 71. Newbury Park, CA: Sage Publications.

Lehtonen, R. and Pahkinen, E. J. (1995). *Practical Methods for Design and Analysis of Complex Surveys*. New York: John Wiley.

Levy, P. S. and Lemeshow, S. (1999). *Sampling of Populations*, 3rd edn. New York: John Wiley.

Nelson, S. E., Rogers, R. R., Ziegler, E. E. and Fomon, S. J. (1989). Gain in weight and length during early infancy. *Early Human Development*, **19**, 223–39.

Ogden, C. L., Troiano, R. P., Briefel, R. R., *et al.* (1997). Prevalence of overweight among preschool children in the United States, 1971 through 1994. http://www.pediatrics.org/cgi/content/full/99/4/e1

Pelletier, D. L. and Frongillo, E. A. (2003). Changes in child survival are strongly associated with changes in malnutrition in developing countries. *Journal of Nutrition*, **133**, 107–19.

Pelletier, D. L., Frongillo, E. A. and Habicht, J. P. (1993). Epidemiologic evidence for a potentiating effect of malnutrition on child mortality. *American Journal of Public Health*, **83**, 1130–3.

Pelletier, D. L., Frongillo, E. A., Schroeder, D. G. and Habicht, J. P. (1994). A methodology for estimating the contribution of malnutrition to child mortality in developing countries. *Journal of Nutrition*, **124**, 2106S–2122S.

(1995). The effects of malnutrition on child mortality in developing countries. *Bulletin of the World Health Organization*, **73**, 443–8.

Persson, L. A. and Wall, S. (2000). *Epidemiology for Public Health*. Umeå, Sweden: Umeå University.

Roche, A. F., Guo, S., Siervogel, R. M., Khamis, H. J. and Chandra, R. K. (1993). Growth comparison of breast-fed and formula-fed infants. *Canadian Journal of Public Health*, **84**, 132–5.

Scrimshaw, N. S. (2003). Historical concepts of interactions, synergism and antagonism between nutrition and infection. *Journal of Nutrition*, **133**, 316S–321S.

Skinner, C. J., Holt, D. and Smith, T. M. F. (eds). (1989). *Analysis of Complex Surveys*. New York: John Wiley.

Smith, L. C. and Haddad, L. (2001). How important is improving food availability for reducing child malnutrition in developing countries? *Agricultural Economics*, **26**, 191–204.

Tompkins, A. and Watson, F. (1989). *Malnutrition and Infection: A Review*, Nutrition Policy Discussion Paper no. 5. Geneva, Switzerland: Administrative Committee on Coordination/Subcommittee on Nutrition.

Troiano, R. P., Frongillo, E. A., Sobal, J. and Levitsky, D. A. (1996). The relationship between body weight and mortality: a quantitative analysis of combined information from existing studies. *International Journal of Obesity*, **20**, 63–75.

United Nations (2000). Resolution adopted by the General Assembly 55/2. United Nations Millennium Declaration. 18 September, A/RES/55/2. http://www.un.millenniumgoals/

World Health Organization (1995). *Physical Status: The Use and Interpretation of Anthropometry*, WHO Technical Report Series no. 854. Geneva, Switzerland: World Health Organization.

(2002). *World Health Report 2002*. Geneva, Switzerland: World Health Organization.

World Health Organization Working Group on Infant Growth (1994). *An Evaluation of Infant Growth*. Geneva, Switzerland: World Health Organization.

(1995). An evaluation of infant growth: the use and interpretation of anthropometry in infants. *Bulletin of the World Health Organization*, **73**, 165–74.

4 *Measuring growth*

Noël Cameron
Loughborough University

Introduction

This chapter deals with measuring growth using primarily anthropometric techniques and external dimensions of the body such as height, weight and measures of subcutaneous fat. Of course, there are situations in which changes in various components of body composition are also valid measures of growth and in these situations non-invasive anthropometry will not be the method of choice. Chapter 6 discusses these techniques in detail.

Our knowledge of the process of human growth and development is directly dependent on the methods employed to measure that process and the scientific and/or clinical context within which those methods are employed. The methods concern not only the instrumentation and measurement technique but also the frequency of assessments. Daily assessments will lead to a different model of the growth process than weekly, monthly, quarterly or yearly assessments (Lampl *et al.*, 1992). Clearly the accuracy and reliability with which we can measure will play a major role in determining the frequency because too short a period of time between assessments may mean that any apparent change is due to error rather than to growth.

The context will usually be one of three types: screening, surveillance or monitoring. Screening is concerned with the identification of a particular subset of the population with certain prescribed characteristics. Most usually they will be outside (above or below) a certain cut-off point for height, weight, or a combination of the two such as weight-for-height or body mass index (BMI). Such cut-off points are replete within the literature pertaining to growth. Stunting, for example, is defined as a height-for-age Z-score of less than -2 (Ong *et al.*, 2000). Overweight, using the UK reference charts (Cole *et al.*, 2000), is defined as a BMI-for-age greater than the 91st centile and obesity by a BMI-for-age greater than the 98th centile. These values correspond to BMIs

Methods in Human Growth Research, eds. R. C. Hauspie, N. Cameron and L. Molinari.
Published by Cambridge University Press.

of 25 and 30 at 18 years of age but to lesser values at younger ages. In the USA a 'danger of overweight' is defined by a BMI greater than the 85th centile and 'obesity' by a BMI greater than the 95th centile of 2002 CDC Growth Charts (Barlow and Dietz, 1998). To identify children outside all of these cut-offs it is necessary to measure the dimension of interest with known accuracy and reliability. Without such knowledge the possibility of falsely classifying a child as normal when they are not (false negative), or unwell when they are not (false positive), is unknown. Screening is a cross-sectional, once-off activity and usually involves large samples of children. Once identification is complete it will lead on to further assessment or intervention. In this context of large sample sizes and the need for high levels of measurement efficiency, growth status may be characterized by only a few dimensions such as height and weight. Instrumentation tends to be basic and portable and assessment is undertaken by more than one observer with each one working independently.

Surveillance is concerned with assessing the process of growth on more than one occasion on a sample of children and may well follow a screening exercise. For instance, the screening process may have identified a subsample of stunted children below -2 Z-scores for height-for-age. It is decided that such children are to take part in a nutritional supplementation programme and that they will be reassessed at some future date to see if their height-for-age Z-scores have improved. This situation is now termed surveillance and may involve reassessment over a number of months and/or years. Clearly this context requires rather more attention to detail than the first. The samples tend to be smaller, the need for a more comprehensive description of growth requires more dimensions to be assessed, and the number of observers will be fewer with perhaps greater expertise. In addition, and because assessment on two or more occasions is required, accuracy and reliability of measurement will become more of an issue.

It may be that this process of surveillance leads to the identification of a child with an unusual pattern of growth and the decision is made to reassess the child regularly in a clinical situation with access to treatment regimes if a treatable abnormality is diagnosed. This final context is termed monitoring and involves the long-term high-frequency assessment of a single child or small sample of children. This context requires a specialist setting with dedicated space, high-quality instrumentation, and a broad range of dimensions to assess properly all aspects of the growth pattern.

Screening provides only cross-sectional data but in the latter two scenarios the outcome data includes both cross-sectional and longitudinal components. Knowledge is gained both of the current growth status and of the rate of change of growth with time, or 'growth velocity'. The usefulness of the velocity data will be dependent on the period of time between assessments; too short and the outcome is marred by diurnal variation and short-term fluctuations characteristic of saltatory growth, too long and any meaningful shorter-term fluctuations will not be noticed.

Within these contexts it is important to understand the terms accuracy, precision, reliability and validity because an understanding of these concepts is fundamental to obtaining anthropometric estimations of growth that can be used to provide a true picture of the growth process. Although a dictionary definition of accuracy may be 'careful, precise, in exact conformity with truth' (*Oxford English Dictionary*), within an anthropometric context we do not know what 'truth' is. For example, when we measure a child's height we only have the estimation of that height, we do not know what the actual, true height is. We can improve the accuracy and decrease error by ensuring that we use an appropriate, specific, purpose-made, valid, calibrated instrument to measure the child and by using a properly trained observer. Accuracy may be determined from a test–retest experiment in which the standard error of measurement (SE_{meas}) or technical error of measurement (TEM) is calculated (Cameron, 1984; Lohman *et al.*, 1988). Precision is the proximity with which an instrument can measure a particular dimension. Note that this is different from accuracy in that it relates to the smallest unit of measurement possible with a particular instrument. So a stadiometer that measures height to the nearest centimetre is not as precise as one that measures to the nearest millimetre. Clearly accuracy and precision are related in that the proximity to the true measurement can only be within the limits imposed by the precision of the instrument. Thus two observers may appear to have equal accuracy when they use a stadiometer to measure height to the last completed centimetre but a more precise instrument, that measures in millimetres, might demonstrate that one of the observers is more accurate than the other. Reliability is the extent to which an observer or instrument consistently and accurately measures whatever is being measured on a particular subject. Note that reliability involves three sources of error: the observer, the instrument and the subject. Also, reliability involves two characteristics: consistency and accuracy. Reliability can be assessed in a test–retest experimental design by calculating both the SE_{meas} and the standard deviation of differences (SD_d) (Cameron, 1984; Lohman *et al.*, 1988). Finally, validity is the extent to which a measurement procedure measures or assesses the variable of interest. For instance, the measurement of total body fat may be measured by dual X-ray absorptiometry or by body density via hydrodensitometry. The former is less indirect than the latter and is thus more valid. Similarly, growth in stature is most validly assessed by measuring height rather than by measuring leg length and predicting stature.

Within the measurement of human growth it is important to know one's accuracy and reliability both for the absolute (cross-sectional) determination of a child's height and for the determination of growth velocity. Obviously when taking two measurements of height to calculate growth velocity the estimation is affected by two sources of error, one for each height estimation. The interval between measurement occasions will be determined to some extent by the reliability of the observer. If, for instance, the observer has a reliability (SD_d or TEM) of 0.3 cm and the child is growing at a rate of 4 cm/year then growth will not be certain to have occurred (with 95% confidence limits) until the difference in heights between two measurement occasions is greater than 1.96×0.3 cm, or 0.59 cm. It will take this child 54 days to grow 0.6 cm, thus measurements taken on a monthly or even 2-monthly basis will be subject to observer error. The minimal time between measurement occasions for this child should be at least 3 months to ensure that false growth due to observer error is minimal.

Generally linear dimensions, e.g. height, are more accurately assessed than those involving soft tissue, e.g. skinfolds. Linear dimensions are most often taken between bony landmarks with little or no compressible tissues to interfere with accurate measurement. Soft tissue measurements, such as arm circumference, by definition involve compressible subcutaneous fat and muscle that will reduce accuracy and reliability. It is thus of some importance for the observer to be aware of exactly what is being measured within a particular dimension. For instance an observer measuring triceps skinfold without an understanding of the arrangement of subcutaneous fat and muscle is unlikely to appreciate the importance of the need to separate the subcutaneous fat layer from the muscle layer during measurement. Similarly, an observer ignorant of skeletal anatomy will not appreciate the need for straightening the vertebral column or of applying pressure to the mastoids prior to measurement of height or sitting height.

This chapter on measuring growth is necessarily prescriptive providing the most appropriate measurement techniques and instrumentation to get the most accurate assessments of the process of human growth. Lengthy discussions of instrumentation and measuring techniques may be found in Cameron (1978, 1983, 1984) and Lohman *et al.* (1988). The organization of this chapter is such that a measurement technique is described with the instrumentation recommended for the most accurate and reliable results. Thus the measurement of height will be accompanied by descriptions of stadiometers that allow accurate and reliable measurement; skinfold measurements will be accompanied by descriptions of skinfold callipers, and so on. The measurements chosen for description are those that are recommended either as baseline measurements (e.g. height, weight, skinfolds) or as examples of measurement techniques that may be applied to other similar dimensions. Thus where appropriate, such as for skinfolds and girths or circumferences, a generic technique is described that can be applied to any assessment site as well as specific techniques for the recommended sites. For example, a generic technique for the assessment of subcutaneous fat is described and specific techniques for the triceps, biceps, subscapular and suprailiac sites. These four sites are the most commonly measured but the researcher or clinician may wish to increase the number of sites because they are undertaking a more extensive investigation of fat patterning or have a particular clinical interest in fat distribution. In that case the generic technique can be applied to any skinfold site as long as the site is accurately identified. The following measurements will be described followed by the recommended instruments to obtain accurate values:

Linear dimensions: stature, sitting height, supine length, crown–rump length, tibial length, biiliac diameter, biepicondylar humerus
Circumferences/girths: head, arm, waist and hip
Skinfold thickness: triceps, biceps, subscapular and suprailiac
Weight.

Finally a glossary of anatomical surface landmarks is provided.

Measuring procedures

The accuracy with which the following measurements may be obtained can be maintained at a high level by following a few simple rules of procedure:

1. Ensure that the subject is in the minimum of clothing or at least in clothing that in no way interferes with the identification of surface landmarks.
2. Familiarize the subject with the instrumentation, which may appear frightening to the very young subject, and ensure that he/she is relaxed and happy. If necessary involve the parents or carers to help in this procedure by conversing with the child.
3. Organize the laboratory so that the minimum of movement is necessary and so that the ambient temperature is comfortable and the room well lit.
4. When possible use another person to act as a recorder who will complete a measurement form. Place the recorder in such a position that he or she can clearly hear the measurements and is seated comfortably at a desk with enough room to hold recording forms, charts and so on.
5. Measure the left-hand side of the body unless the particular research project dictates that the right-hand side should be used or unless the comparative projects have used the right-hand side.
6. Mark the surface landmarks with a water-soluble felt-tip pen prior to starting measurements.
7. Apply the instruments gently but firmly. The subject will tend to pull away from the tentative approach but will respond well to a confident approach.

8. Call out the results in whole numbers – for example, a height of 112.1 cm should be called out as 'one, one, two, one' not as 'one hundred and twelve point one' nor as 'eleven, twenty-one'. Inclusion of the decimal point may lead to recording errors and combinations of numbers may sound similar, for example, 'eleven' may sound similar to 'seven'.
9. Establish the reliability of measurement (accuracy and repeatability) prior to the study using subjects with similar characteristics (age, sex, etc.) to the participants in the study proper.
10. During the course of the study create a quality-control procedure by repeating measurements on randomly selected subjects. At the end of each session these can be compared to the a priori reliability figures obtained in the pilot study.
11. When measurement are repeated, e.g. skinfolds, the recorder should check that the repeat value is within the known reliability of the observer. If it is not then a third measurement is indicated. For a final value average the two that fall within the limits.
12. Do not try to measure too many subjects in any one session. Fatigue will detract from reliable measurement for which concentration is vital.

Measurements

Stature (stadiometer)

The subject presents for the measurement of stature dressed in the minimum of clothing, preferably just underclothes, but if social custom or environmental conditions do not permit this then at the very least without shoes and socks. The wearing of socks will not, of course, greatly affect height, but socks may conceal a slight raising of the heels that the observer from his/her upright position may not notice.

The subject is instructed to stand upright against the stadiometer such that his/her heels, buttocks and scapulae are in contact with the backboard, and the heels are together (Figure 4.1). If the subject suffers from 'knock-knees' then the heels are slightly spread so that the knees touch, but do not overlap. As positioning is of the greatest importance the observer should always check that the subject is in the correct position by starting with the feet and checking each point of contact with the backboard as he/she moves up the body. Having got to the shoulders he/she then checks that they are relaxed, by running his/her hands over them and feeling the relaxed trapezius muscle. The observer then checks that the arms are relaxed and hanging loosely at the sides. The head should be positioned in the 'Frankfurt plane' (see p. 102), and the headboard of the instrument then moved down to make contact with the vertex of the skull.

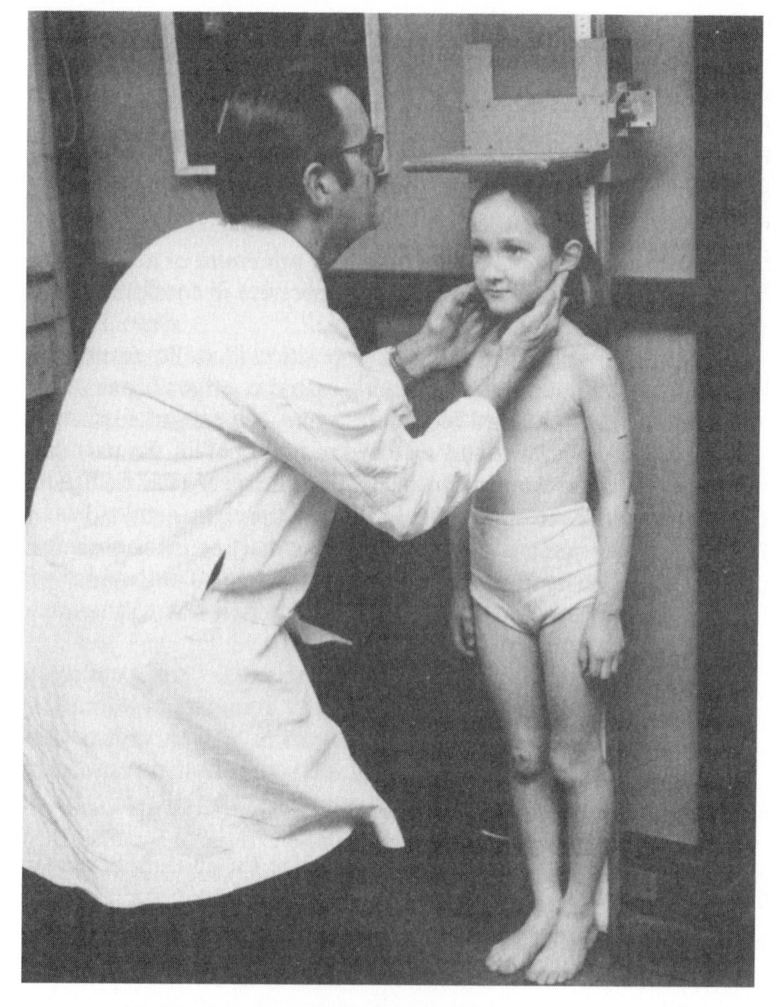

Figure 4.1 The measurement of stature.

To ensure that the Frankfurt plane has been achieved the observer may find it helpful to use either of the following techniques. The observer bends down so that his/her eyes are level with the plane and notes that the lower orbits of the eyes and the external auditory meatuses are in a horizontal line. Alternatively, when using instruments with a counter, he may grip the head with his open hands and pivot it backwards and forwards, in a nodding motion, and at the same time observe the counter. The counter should register the greatest height when the head is tilted not too far forward or backwards. It is thus a relatively easy matter to ensure correct positioning.

It is advisable to place a weight, of about 0.5 kg, on the headboard. This weight presses down on the hair thus flattening any hairstyle and overcomes the natural friction of the machine so that any upward or downward movement during the measurement is recorded on the counter.

With the subject in the correct position he/she is instructed: 'Take a deep breath and stand tall.' This is done to straighten out any kyphosis or lordosis and produce the greatest unaided height. It is at this point that the observer applies pressure to the mastoid processes – not to physically raise the head but to hold it in the position that the subject has lifted it to by breathing deeply. The subject is then instructed to 'Relax' or to 'Let the air out' and 'Drop the shoulders'. The shoulders are naturally raised when the subject takes a deep breath and thus tension is increased in the spinal muscles and prevents total elongation of the spine. Relaxing or breathing out releases this tension and commonly produces an increase of about 0.5 cm in absolute height. The effect of this pressure or traction technique is to counteract the effect of diurnal variation that works to reduce stature during the normal course of a day. This reduction may be as much as 20 mm (Strickland and Shearin, 1972) but the pressure technique reduces that to less than 4.6 mm over the whole day (Whitehouse *et al.*, 1974). Stature is read to the last completed unit whether from a counter or graduated scale. Height is not rounded up to the nearest unit as this will produce statistical bias and almost certainly invalidate estimates of height velocity.

Various physical defects will prevent rigorous adherence to this procedure, but simple rules may be applied to obtain accurate estimates of stature. If there is asymmetry of the legs then a suitable block may be put under the shorter leg and stature measured on the longer or the subject may be instructed to stand on the shorter leg for the measurement. The recorder would then note two heights for the individual and the technique used to acquire them. Scoliosis is another common problem; it may be countered to some extent by ensuring that the subject is achieving his or her maximum height and noting the degree of scoliosis on that and each subsequent occasion. This may then be taken into account when assessing the growth velocity and comparing it to an increase in subischial leg length, which should be roughly half the total increase in stature prior to puberty. Foot deformities may be allowed for by noting the unique posture of the subject.

The correctness of applying gentle pressure to the subject in order to decrease postural and diurnal effects must ultimately be decided by the investigator. Certainly many studies have not used this technique but when measuring children the application of upward pressure gives more reliable readings from day to day than other techniques.

Stadiometers

The stadiometer is composed of a horizontal headboard and vertical backboard. The backboard must be so designed that it maintains a vertical position whether free-standing or wall-mounted and the headboard must always move freely over the surface of the backboard. The method of displaying the value of the dimension must be clear and, if necessary, capable of easy calibration. Usually a counter mechanism or graduated rule is used for this purpose. It has been found that counter mechanisms lead to fewer reading errors but, in an uncontrolled environment, they may be broken by abuse. Reading from graduated rules requires the observer to be at the same height as the cursor to avoid parallax errors. It is very important to appreciate that in order to obtain accurate and reliable measures of growth a purpose-built stadiometer incorporating the design principles of the type described below is essential. It is not acceptable to use less suitable equipment such as the wooden rule attached to the traditional weighing scale to be found in doctor's surgeries throughout the world, which is usually extremely unstable and imprecise. It is better in situations when purpose-built machines are not available to use a vertical wall and a method of determining the level of the vertex of the skull. A book held at right angles to touch both a child's head and a vertical wall would, for instance, suffice. With the child properly positioned the book can be placed on the child's head so that one side touches the wall and the other the child's head. A mark is then made on the wall at the inferior surface of the book. The distance from the floor to the book can be measured with a tape measure after the child has moved away.

Harpenden Stadiometer

This stadiometer is a counter-recording instrument (Holtain Ltd, 2003) in which the counter gives a reading in millimetres over a range of 600 mm to 2100 mm. It is wall-mounted and made of light alloy with a wooden headboard fixed to a metal carriage that moves freely on ball-bearing rollers. The face of the stadiometer is finished in plastic for easy cleaning. The complete instrument is 232 cm tall and weighs 12.7 kg.

This stadiometer has been in use in many clinics, hospital outpatient departments and growth centres for a number of years. When treated properly it gives consistently accurate results, but the counter will break if the headboard is 'raced' up or down the backboard. For this reason it is recommended that the headboard is always locked or moved to its topmost position when not in use to prevent children or inexperienced adults from breaking the counter.

Calibration of this instrument is straightforward and takes very little time. A metal rod of known length is placed between the headboard and the floor so that it stands vertically. If the counter does not record the correct length of the rod then it may be loosened by undoing the two metal retaining screws, and pulled away from the main fibre cog of the carriage. In this position the small metal cog of the counter may be turned until the counter records the true length of the metal rod. The counter is then pressed against the backplate so that the teeth of the counter cog and carriage cog engage and the retaining screws are tightened. The headboard is then moved up and down the backboard a number of times to ensure that the counter continues to give an accurate reading. If not, the counter must be replaced. It is recommended that the instrument be calibrated prior to every measuring session, particularly if the stadiometer is left in a situation that allows public access.

Portable stadiometers for screening and surveillance

Portable stadiometers are a requirement for the majority of studies in the screening and surveillance contexts. A number of excellent portable alternatives to the wall-mounted stadiometer are available depending on considerations of budget and degree of portability. Anthropometers may be used for measuring height as long as the problems of instrument stability can be adequately solved. The instrument must be held vertically and the simple answer of an assistant holding the main bar is not really adequate. It is possible for the anthropometer to be inserted into a block or tripod, which might also allow the inclusion of a footboard for the subjects.

Sitting height (sitting-height table)

Accurate measurement of sitting height requires impeccable positioning from the subject and, therefore, great attention to detail from the observer. Whether on a specially constructed sitting-height table, or a suitable flat surface, the subject is positioned so that the head is in the Frankfurt plane, the shoulders relaxed, the back straight, the upper surface of the thighs horizontal and the feet supported so that a right angle is formed with the thighs and the tendons of the biceps femoris, at the back of the knee, are just clear of the table. The arms are loose at the sides and the hands rest in the subject's lap (Figure 4.2).

Each part of this position may be checked, as for stature, by starting at the feet and moving up to the head. It is essential to have the knees raised away from the table but only to the point where a right angle is formed with the thighs. Too acute an angle tends to cause the subject to roll backwards, and too obtuse an angle or the thighs touching the table tends to roll the subject forwards. Both situations reduce sitting height, and prevent the subject from maintaining a straight back. If the subject's hands are not resting in the lap or along the thighs then he/she may use them to push upwards from the surface

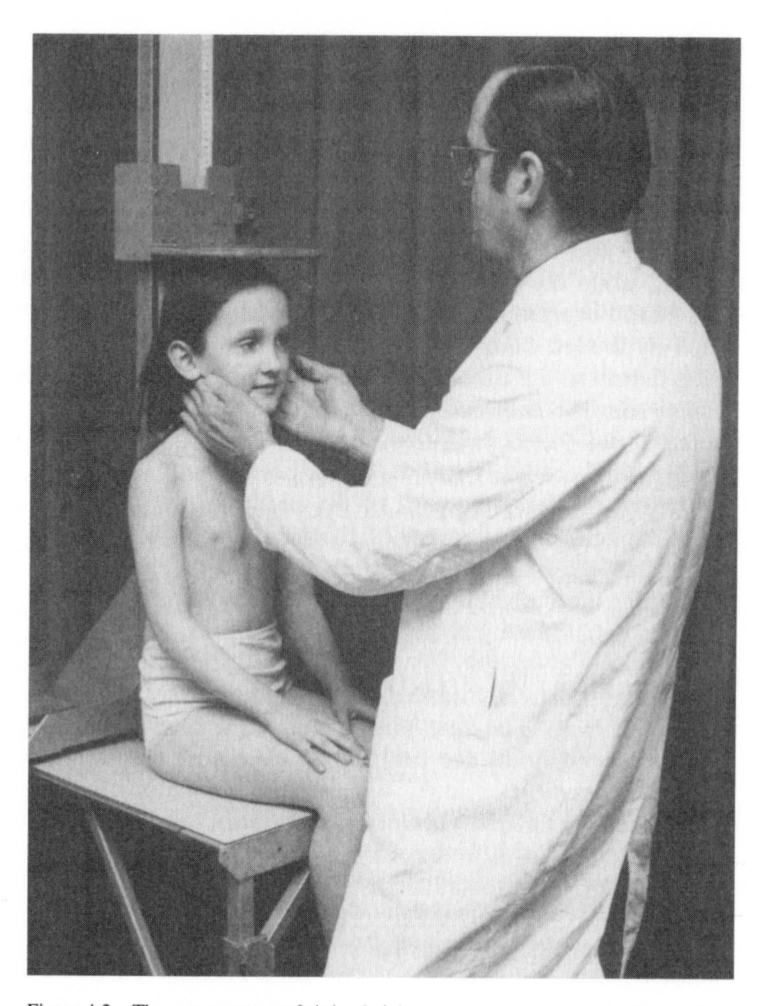

Figure 4.2 The measurement of sitting height.

of the table and falsely increase sitting height. Similarly, contraction of the thighs or buttocks will raise the subject away from the table. The straightness of the back is the most difficult part to accomplish but may be achieved by giving the subject clear instructions to, for instance, 'Sit up straight' or 'Sit tall' and at the same time running the fingers up the spine from the sacrum to the thoracic vertebrae causing the subject to move involuntarily away from the observer's hand and thus straighten any slouching or rounding of the spine common to most subjects when asked to sit. The shoulders may be checked, as with stature, by running the hands along them, and the head position may be checked by either observing the counter or by visually inspecting the imaginary line between meatus and lower orbit. Having lowered the headboard, again with a small weight to compress the hair, the subject is instructed to 'Take a deep breath and relax.' Pressure is applied to the mastoids to prevent the head from falling on relaxation and the measurement read to the last completed unit.

Harpenden Sitting-Height Table

To date, the Harpenden Sitting-Height Table (Holtain Ltd, 2003) is the only instrument manufactured specifically to measure sitting height. It is designed as a free-standing table with a horizontally moving carriage that holds the vertical backboard and the horizontal headboard. This carriage is necessary to accommodate different thigh lengths in the subjects and there is a movable footboard to accommodate different calf lengths and so ensure correct positioning of the subject. The headboard carriage is similar to that of the stadiometer and recording is by a counter mechanism over the range 320 mm to 1090 mm. The cost and size of this instrument make it suitable only for a permanent measuring situation but in such a situation its ease of use and design qualities, that ensure good subject positioning, make it a valuable tool. Calibration and cleaning requirements are similar to those of the stadiometer.

Sitting-height tables for screening and surveillance

Assuming that the anthropometric kit includes a portable stadiometer then a stool or box of fixed height may be placed against it on which the subject sits. After recording sitting height in the normal way the height of the stool or box is subtracted from the stadiometer reading to obtain actual sitting height. It is obviously simpler and will involve less chance of error if the box is of a standard height – for example, 50 cm. With a little imagination the box may be used to carry other equipment and so be useful when not in use for measuring sitting height.

Supine or recumbent length (Supine-length/recumbent-length table; neonatometer)

The measurement of supine length requires two observers, one to hold the head and the other to hold the feet and move the cursor. The subject lies on the supine-length table or in the neonatometer such that the head is positioned in a supinated Frankfurt plane and the vertex of the head touches the fixed end of the apparatus. The head is held in this position throughout the measurement by an observer who constantly checks that the correct position is being maintained and that contact between head and headboard is constant. The second observer checks that the rest of the body is relaxed and that the child is not arching the spine or bending the knees. This observer holds the feet such that the ankles are at right angles and the toes not bending over to interfere with the cursor. The cursor is then moved into contact with the feet and slight pressure is applied to the ankles to straighten the legs. This normally causes the head to be moved away from the headboard so that the other observer must gently pull the head back into contact with it (Figure 4.3). This dual pulling of the child has the same effect as deep breathing in the measurement of stature – to overcome diurnal variation in posture.

Depending on the age of the child, various problems arise during this measurement. Very young children will automatically bend the knees and the observer must apply downward pressure on them with his/her forearm or elbow. The shoulders will also be lifted off the board and the observer holding the head must use the index fingers to press them gently back into contact. It is sometimes necessary to release one of the feet if the child fights so strongly that accurate measurement is compromised and indeed in the very young this is often easier than trying to struggle with both feet. It should be emphasized that these problems will arise more often if steps are not always taken to relax the subject and familiarize him or her with the apparatus. Cuddly toys suspended above the table or pictures on the ceiling are good methods of attracting the attention of slightly older subjects but on the very young it is a great help to allow the parent or carer to lean over the child and talk to it to reassure it.

Crown–rump length (supine-length/recumbent-length table; neonatometer)

Crown–rump length is the equivalent measurement to sitting height when applied to children who are too young to stand or present with standing difficulties. The subject lies on a supine-length table or similar device with one observer holding the head as for supine length and the other observer holding the feet such that the knees are above the child and the buttocks are presented to (the face of) the cursor. It is quite usual for the child to object strongly to this measurement and attempt to straighten the legs. In this situation the observer can gain more control by quickly pushing the back of the knees thus causing them to bend and then pushing downwards so that the legs stay bent. The observer's other hand is used to move the cursor up to the buttocks to make firm contact. It is extremely difficult to describe the amount of pressure to be applied to the cursor. Modern instruments have a pressure switch that locks the cursor when

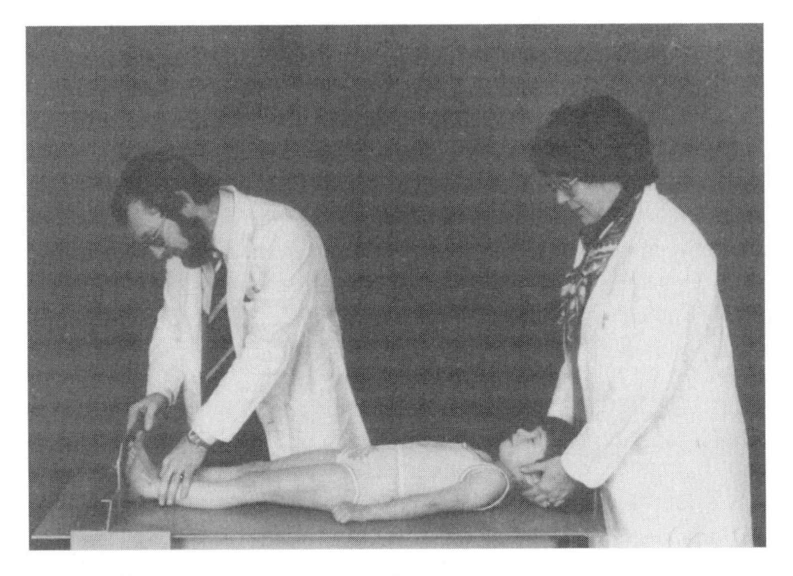

Figure 4.3 The measurement of crown–rump length.

a pressure of 0.5 kg resists its movement and it is about that degree of pressure that is required. The observer should not be pushing so hard that the child's spine is being compressed nor so lightly that the soft tissue of the buttocks is not being compressed. As with supine length, the observer holding the head must take great care that the supinated Frankfurt plane is being maintained and that the shoulders are in contact with the table. Once again, the use of visual aids to maintain the interest of the child is an advantage.

Harpenden Neonatometer

The Harpenden Neonatometer (Holtain Ltd, 2003) is constructed as a rectangular light-alloy frame with a curved metal headrest at one end and a cursor carriage at the other. In common with the other Harpenden instruments, the recording is by a counter mechanism. The important addition to this instrument is the locking mechanism attached to the cursor that locks the footboard when a pressure of 0.5 kg is exerted against it. Such a mechanism prevents the observer from having to fight with the unruly baby to maintain the leg in a straight position for longer than a few seconds. The highly portable nature of the instrument allows it to be placed over the recumbent baby rather than disturbing the baby by placing the subject inside the instrument. A short version is made to fit inside most incubators. This measures over the range 180 mm to 600 mm compared with 188 mm to 750 mm for the standard model. This instrument is necessary for any neonatal clinic but the general growth clinic, dealing with all ages of subjects, can perfectly well measure supine length accurately with a longer, all-age instrument.

Harpenden Infantometer

The Harpenden Infantometer (Holtain Ltd, 2003) was designed to fill the instrument gap between neonates and school-age children, measuring over a range of 300 mm to 940 mm. Bearing in mind the fact that many studies on this age range of subjects are set in the home rather than the growth centre, it is designed as a portable instrument weighing about 6.75 kg. It is constructed of light-alloy with a flat headboard and footboard fixed to a movable cursor and counter recording mechanism. As with the Neonatometer a locking device is fitted to aid measurement when the subjects are active.

Harpenden Infant Measuring Table

The Harpenden Infant Measuring Table (Holtain Ltd, 2003) is the non-portable version of the Infantometer. The base is constructed of light-alloy with a fixed wooden headboard and footboard fixed to a carriage and counter mechanism as with the other instruments. There is no locking mechanism. The measuring range is 230 mm to 1200 mm so it is suitable for postneonates up to preschool children. The lack of a locking mechanism reflects the fact that the subject should be more co-operative and that the measuring position can be maintained for longer.

Harpenden Supine Measuring Table

This is the full-length supine table (Holtain Ltd, 2003) similar in construction to the stadiometer. It is recommended that this instrument is mounted on permanent wall-brackets, but adjustable legs may be supplied at an additional cost. As in the Infant Measuring Table, the head and footboards are made of wood and the latter is fixed to a cursor and counter mechanism. The measuring range is 300 mm to 2100 mm and so accommodates all age ranges of children and adults.

Calibration

Calibration of all these instruments is similar to that employed for calibrating the stadiometers. A metal rod of known length is used to ensure that the counter is reading correctly. If not, then suitable adjustment can be made by loosening the retaining screws of the counter and turning its metal cog to the correct

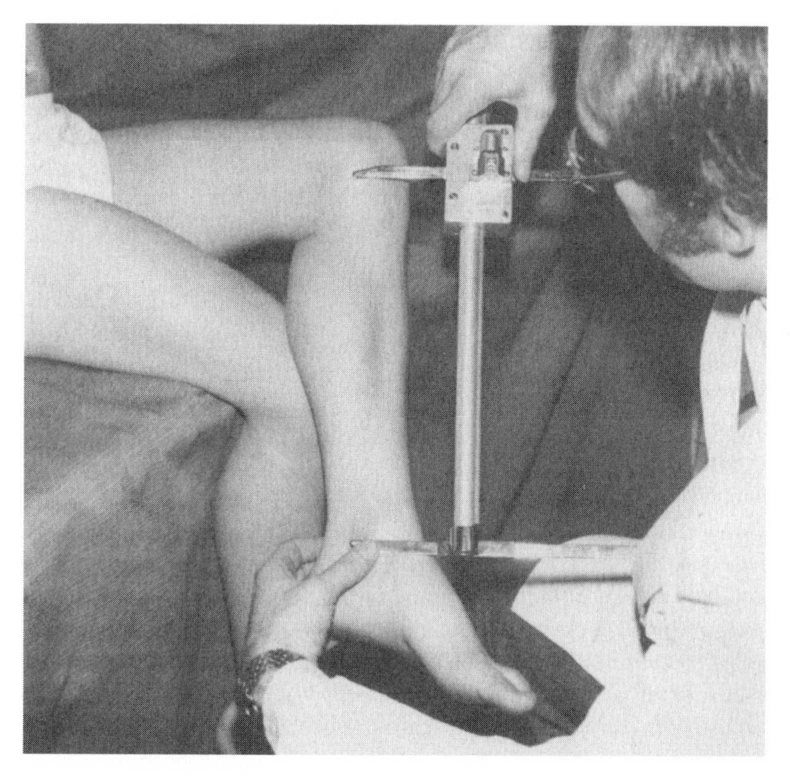

Figure 4.4 The measurement of tibial length.

measurement. The footboard is moved forwards and backwards a few times to check the reliability, and if this is suspect a new counter is fitted.

Tibial length (anthropometer)

The subject crosses the legs so that the left ankle proximal to the malleolus rests on the right knee (Figure 4.4). In this position it is possible to feel the proximal–medial border of the tibia, which should be marked, and the distal border of the medial malleolus, which should also be marked. The anthropometer is applied so that the fixed end is distal to the malleolus and the counter end is brought into contact with the marked proximal end of the tibia. Care should be taken that the beam of the anthropometer is parallel to the tibia. This is achieved by altering the lengths of the blades so that the fixed blade is longer and crosses the full width of the ankle and the free blade is shorter and goes only part way across the knee.

Anthropometers

The anthropometer is one of the most versatile anthropometric instruments and as such will be used more than any other instrument. It is suitable for almost any linear dimension from height to foot length and therefore the instrument should be chosen with care. Two basic designs of anthropometer are available but one has significant advantages over the other for measuring growth.

The Martin anthropometer

The Martin anthropometer is used universally in physical anthropology. It is manufactured by GPM (SiberHegner, 2003) and is thus sometimes referred to as the GPM anthropometer. The original version is composed of four metal rods with graduations in millimetres and centimetres engraved on them. A sliding cursor runs the length of the rods when they are joined together to give a maximum reading of two metres. Curved or straight blades may be inserted into the cursor housing and fixed end so that the distance between them may be read from the graduated main beam. The major disadvantage becomes apparent when this anthropometer is used to measure children. In these situations it is important to be able to feel the landmarks of the body as the blades of the anthropometer are applied to the marked positions. It must be possible, therefore, to move the cursor housing whilst holding the tips of the blades. The frictional forces involved in the Martin anthropometer make this operation virtually impossible. A more recent version of the Martin instrument features a beam of square cross-section but the problem with frictional forces still remains.

Harpenden anthropometer

The Harpenden instrument (Holtain Ltd, 2003) was designed to overcome the problem of frictional forces when holding the instrument by the tips of the blades. The cursor runs on miniature ball-bearing rollers allowing a free movement that is without crossplay. As with the other Harpenden instruments this is a counter display instrument giving readings over the range 50 mm to 570 mm. As with the Martin anthropometer beam extensions may be added to the main bar but because of the counter display, constants, equivalent to the length of the beams, must be added on to the counter reading.

Biiliac diameter (anthropometer)

The subject stands with his/her back to the observer, feet together and hands away from his/her sides to ensure a clear view of the iliac crests (Figure 4.5). For measurements depending on the identification of any bony landmark

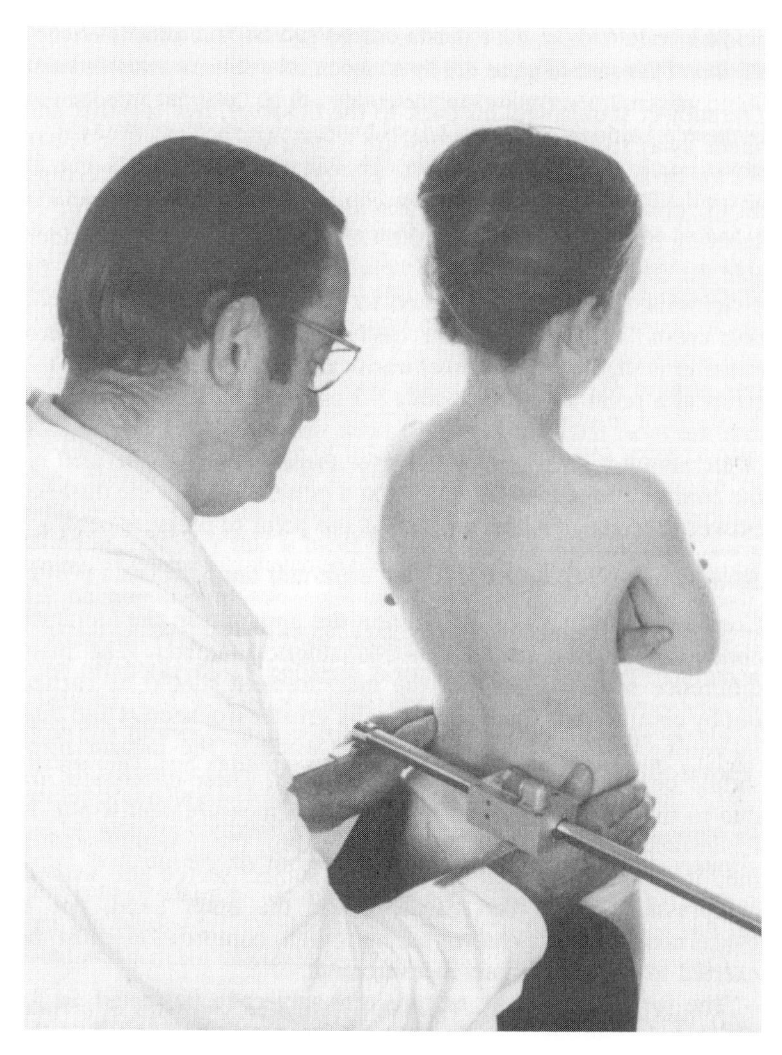

Figure 4.5 The measurement of biiliac diameter.

(e.g. biacromial diameter) it is a good procedure for the observer to feel the position and shape of the landmark prior to measurement. Thus in this case the iliac crests should be palpated prior to applying the instrument, especially when the subject has considerable fat deposits in that region. The anthropometer is held so that the blades rest medially to the index fingers and over the angle formed by the thumb and index finger. The index fingers rest on top of the blades, to counteract the weight of the bar and counter mechanism, and the middle fingers of each hand are free to palpate the measurement points immediately prior to measurement. In this position the observer can quite easily move the blades of the anthropometer so long as it is of the counter type. Other anthropometers have too great a frictional force opposing such movement and must be held by the main bar so that the blades are remotely applied to the most lateral points of the iliac crests. This will be more easily accomplished if the anthropometer is slightly angled downwards and the blades applied to the crests at a point about 2–3 cm from the tips. To ensure that the most lateral points have been obtained it is a useful point of technique to 'roll' the blades over the crests. It will be seen on the counter of the instrument that at a particular point the distance between the crests is greatest; this is the point of measurement.

Biepicondylar humerus (bicondylar vernier, anthropometer)

The subject sits facing the observer with the left arm raised to the level of the shoulder and the elbow bent to a right angle. The bicondylar vernier is applied to the medial and lateral epicondyles of the humerus. This will normally be at an angle to the horizontal because the medial epicondyle is lower than the lateral. Pressure is exerted to compress the soft tissue.

Biepicondylar vernier

Biepicondylar measurements may be assessed using either a specific instrument such as the bicondylar vernier manufactured by Holtain Ltd (2003), or by using an anthropometer. The bicondylar vernier is a small version of the anthropometer and thus allows for more convenient measurement.

Circumferences/girths (tape measure)

Head circumference

Head circumference used to be described as a 'fronto-occipital' circumference or a Frankfurt plane circumference, but both techniques have largely been replaced by simply measuring the maximum head circumference.

The subject stands sideways to the observer such that the left-hand side is closer. The head is held straight or in the Frankfurt plane but the plane is of little consequence as long as the head is straight and the eyes looking forward (Figure 4.6). The subject's arms are relaxed. It is easier for the observer if he/she is positioned so that his/her eyes are level with the subject's. The tape is opened and passed around the head from left to right. The free and fixed ends are then transferred to the opposite hands so that the tape now passes completely around the head and crosses in front of the observer. Using the middle finger of the left

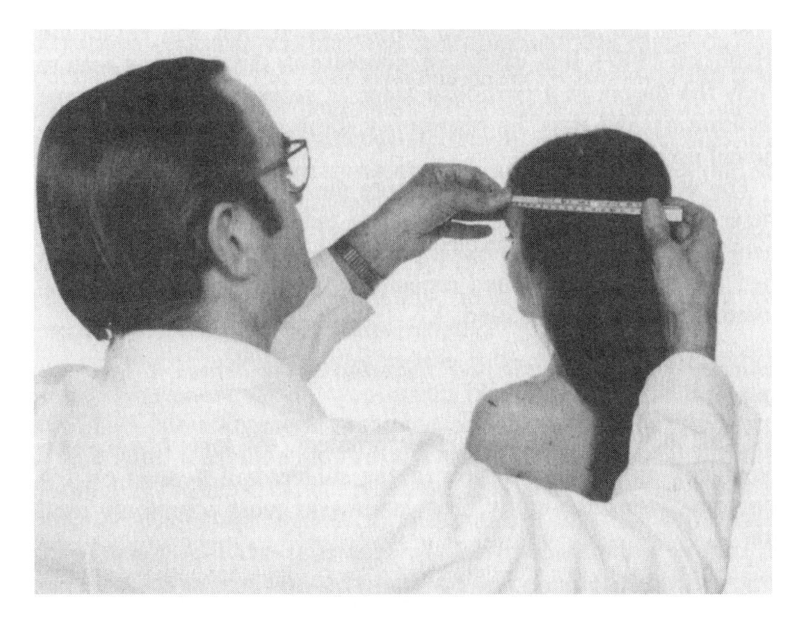

Figure 4.6 The measurement of head circumference.

hand the observer presses the loose tape to the forehead of the child and moving the finger up and down determines the most anterior part of the head. Having done this he pulls the tape tighter and repeats this procedure with the middle finger of the right hand to determine the most posterior part of the occiput. Once determined, the tape is pulled tight to compress the hair and the measurement read to the last completed unit. Head circumference is the only circumference in which the tape is pulled tight and even then should cause the child no undue discomfort.

Arm circumference

In a good deal of the anthropometric literature arm circumference is described as 'upper arm circumference' when, in fact, no such anatomical area exists. The anatomical arm or brachium lies between the shoulder and the elbow and is proximal to the forearm or antebrachium. The circumference or girth that is measured is thus properly described as the arm circumference. The subject stands in the same position as for head circumference measurement – sideways with the left arm hanging loose at the side (Figure 4.7). The mid-arm level is determined as described below (see p. 104). The tape is passed around the arm from left to right and the free and fixed ends transferred as for head circumference. Ensuring that the tape is at the same level as the midpoint of the arm mark, it is tightened so that it touches the skin all round the circumference but

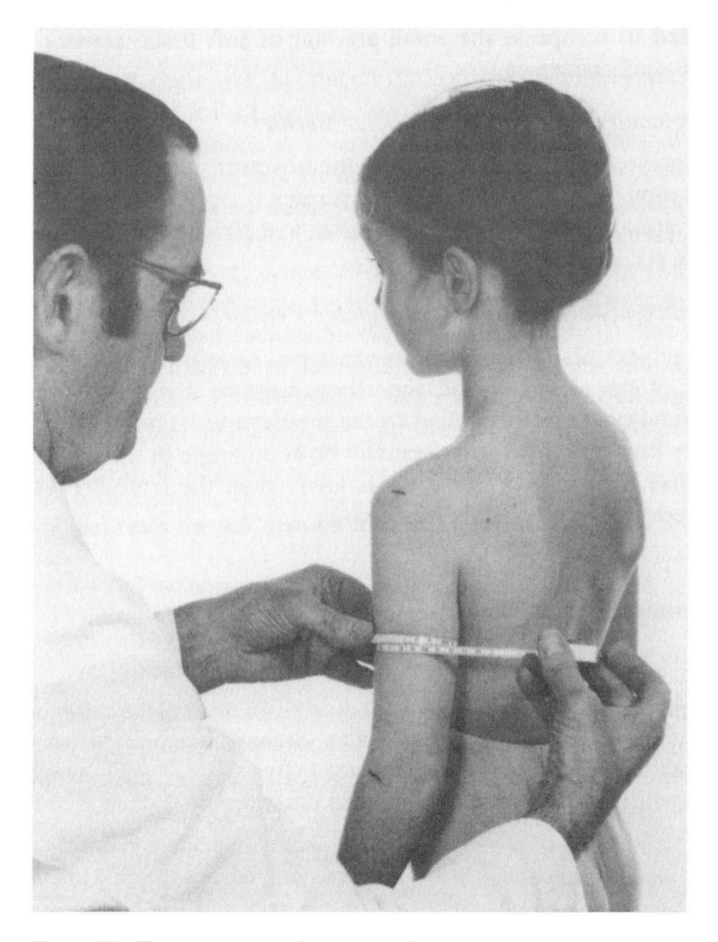

Figure 4.7 The measurement of arm circumference.

does not compress the tissue to alter the contour of the arm. The circumference is then read to the last completed unit. Because the arm in cross-section is not an exact circle but rather oval, some difficulty is usually met in ensuring that the tape actually touches the skin on the medial side of the arm. To ensure that this is done the middle finger of the left hand can be used to press the tape gently to the skin.

Waist circumference (abdominal circumference)

Some confusion exists in the literature with regard to the level of measurement. However, the minimum circumference between the iliac crests and lower ribs would appear to be the most reliable to determine. The general technique is for the subject to stand erect facing the observer with the arms away from the body.

The tape is passed around the body and tightened at the required level ensuring that it is horizontal and not compressing the soft tissue.

Hip circumference (hip girth)
Hip circumference should be measured at the level of the greatest protrusion of the buttocks when the subject is standing erect with the feet together. The subject stands sideways to the observer with the feet together and arms folded. The observer passes the tape around the body at the level of the most prominent protrusion of the buttocks so that it lightly touches but does not compress the skin. In most cases the subject will be dressed in underclothes, which will obviously affect the measurement, and so the observer should either provide standard thin undergarments or other appropriate clothing.

Tape measures

Many tape measures are available that are suitable for anthropometric use. Suitability depends on fulfilling five criteria:

1. Flat cross-section. Tapes with a curved cross-section are difficult to bend maintaining a smooth outline.
2. Millimetre graduations. The graduations must be in centimetres and millimetres and preferably marked on both edges of the tape. Thus at the crossover position it makes little difference whether the lead of the tape is above or below the reading.
3. Blank leader strip. The tapes should contain a blank leader strip prior to the graduations commencing. This enables the observer to hold the leading part without obscuring the zero value.
4. Metal or fibreglass construction. It has always been recommended that only steel tapes should be used so that they did not stretch or deteriorate with use. Fibreglass tapes are now being manufactured that are guaranteed not to stretch.
5. Minimum length of 1 metre.

I have not described any single tape in detail because various types are available that fulfil most or all of these criteria.

Skinfolds

The technique of picking up the fold of subcutaneous tissue measured by the skinfold calliper is often referred to as a ‘pinch’ but the action to obtain the fold is to sweep the index or middle finger and thumb together over the surface of the skin from about 6 to 8 cm apart. This action may be simulated by taking a piece of paper and drawing a, say, 10 cm line on its surface. If the middle finger and thumb are placed at either end of this line and moved together such that they do not slide over the surface of the paper but form a fold of paper between them then that is the action required to pick up a skinfold. To 'pinch' suggests a small and painful pincer movement of the fingers, and this is not the movement made. The measurement of skinfolds should not cause undue pain to the subject, who may be apprehensive from the appearance of the callipers and will tend to pull away from the observer, and, in addition, a pinching action will not collect the quantity of subcutaneous tissue required for the measurement.

The measurement of skinfolds is prone to many sources of error. Location of the correct site is critical (Ruiz *et al.*, 1971), but greater errors may arise from the consistency of subcutaneous tissue and the individual way in which each observer collects the fold of tissue. The observer should practice to obtain an awareness of how a correct skinfold should 'feel' and thus be aware of those occasions when a true skinfold is not being obtained, i.e. when a full fold of fat has not been obtained.

Triceps skinfold

The level for the triceps skinfold is the same as that for the arm circumference – midway between the acromion and the olecranon when the arm is bent at a right angle. It is important that the skinfold is picked up both at a midpoint on the vertical axis of the arm and a midpoint between the lateral and medial surfaces of the arm. If the subject stands with his back to the observer and bends the left arm the observer can palpate the medial and lateral epicondyles of the humerus. This is most easily done with the middle finger and thumb of the left hand, which will eventually grip the skinfold. The thumb and middle finger are then moved upwards, in contact with the skin, along the vertical axis of the upper arm until they are at a level about 1.0 cm above the marked midpoint. The skinfold is then lifted away from the underlying muscle fascia with a sweeping motion of the fingers to the point at which the observer is gripping the 'neck' of the fold between middle finger and thumb (Figure 4.8). The skinfold calliper, which is held in the right hand with the dial upwards, is then applied to the neck of the skinfold just below the middle finger and thumb at the same level as the marked midpoint of the upper arm. The observer maintains his/her grip with the left hand and releases the trigger of the skinfold calliper with his right to allow the calliper to exert its full pressure on the skinfold. In almost every case the dial of the calliper will continue to move but should come to a halt within a few seconds at which time the reading is taken to the last completed 0.1 mm. In larger skinfolds the calliper may take longer to reach

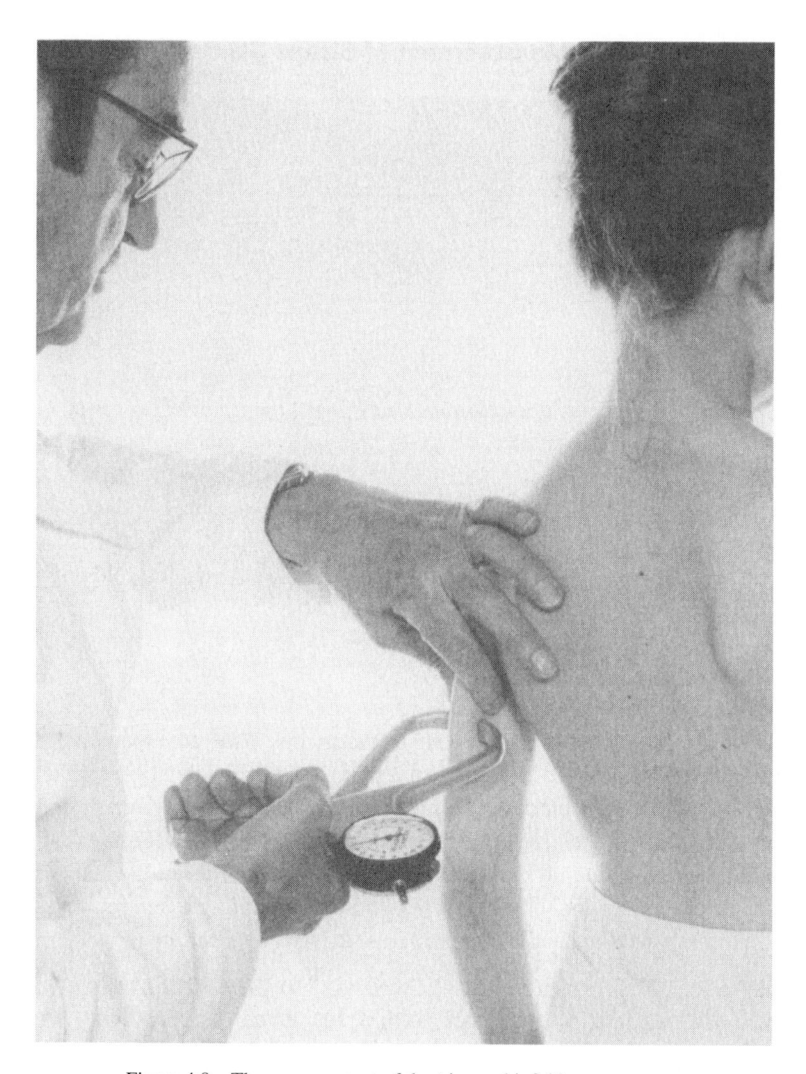

Figure 4.8 The measurement of the triceps skinfold.

a steady state but it is unusual for this to be longer than 7 seconds. Indeed, if the calliper is still moving rapidly it is doubtful that a true skinfold has been obtained and the observer must either try again or admit defeat. This situation is only likely to occur in the more obese subject with skinfolds greater than 20 to 25 mm – that is, above the 97th centile of British charts. Within the 97th and 3rd centiles skinfolds are relatively easy to obtain but they do require a great deal of practice.

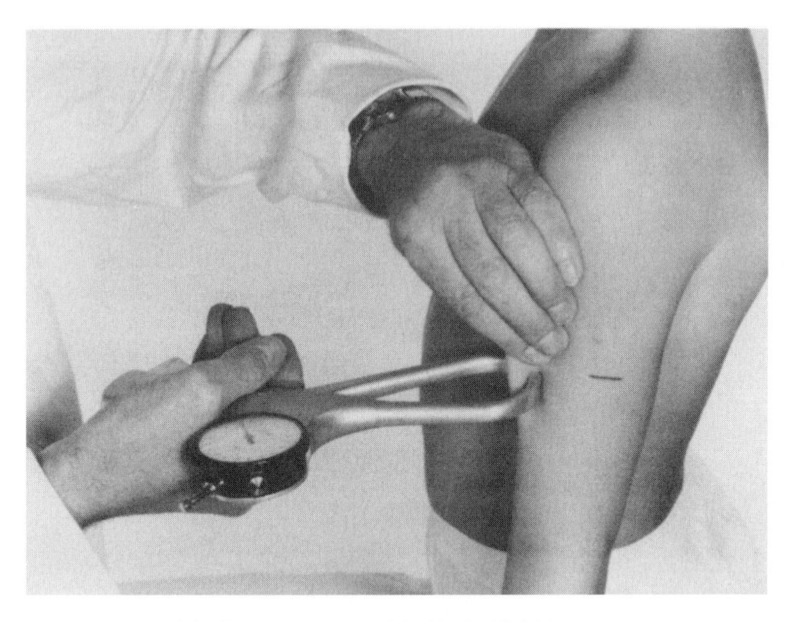

Figure 4.9 The measurement of the biceps skinfold.

Biceps skinfold

The biceps skinfold is the exact opposite of the triceps skinfold, being on the anterior aspect of the arm and at the same midpoint level as previously described for triceps skinfold. It is picked up with the subject facing the observer and his/her left arm hanging relaxed but with the palm facing forwards (Figure 4.9). The middle finger and thumb sweep together at a point 1 cm above the marked midpoint level coming together at the vertical axis joining the centre of the antecubital fossa and the head of the humerus. It is unusual for the movement of the dial to present any problem with this skinfold measurement, as it is not a site for major fat deposits.

Subscapular skinfold

The point of measurement is located immediately below the inferior angle of the scapula. The subject stands with his/her back to the observer and the shoulders relaxed and arms hanging loosely at the sides of the body. This posture is most important to prevent movement of the scapulae; if the subject folded his/her arms, for instance, the inferior angle of the scapula would move laterally and upwards and therefore no longer be in the same position relative to the layer of fat. The skinfold is picked up, as for the triceps skinfold, by a sweeping motion of the middle finger and thumb, and the calliper applied to the neck of the fold immediately below the fingers (Figure 4.10). The fold will naturally be at an

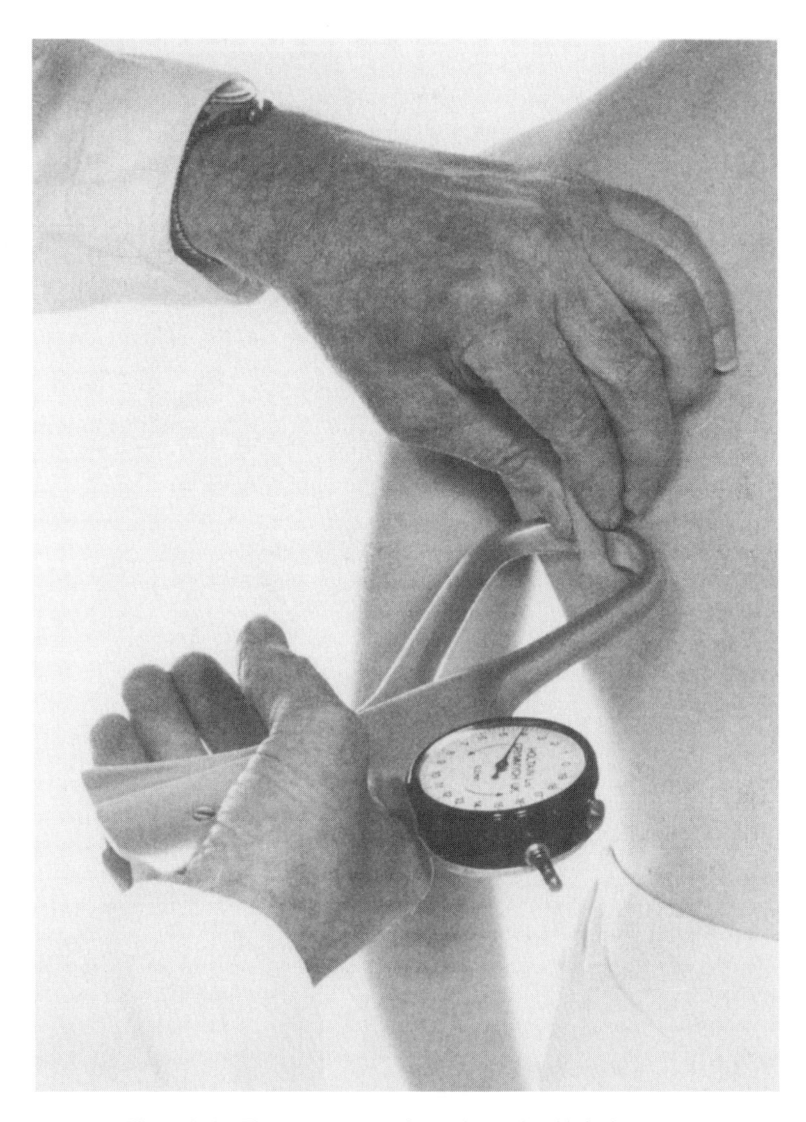

Figure 4.10 The measurement of the subscapular skinfold.

angle laterally and downwards and will not be vertical. Once again, the dial of the calliper will show some movement that should soon cease.

Suprailiac skinfold

The point of measurement for the suprailiac skinfold is 1 cm superior and 2 cm medial to the anterior superior iliac spine. This is best palpated with the subject

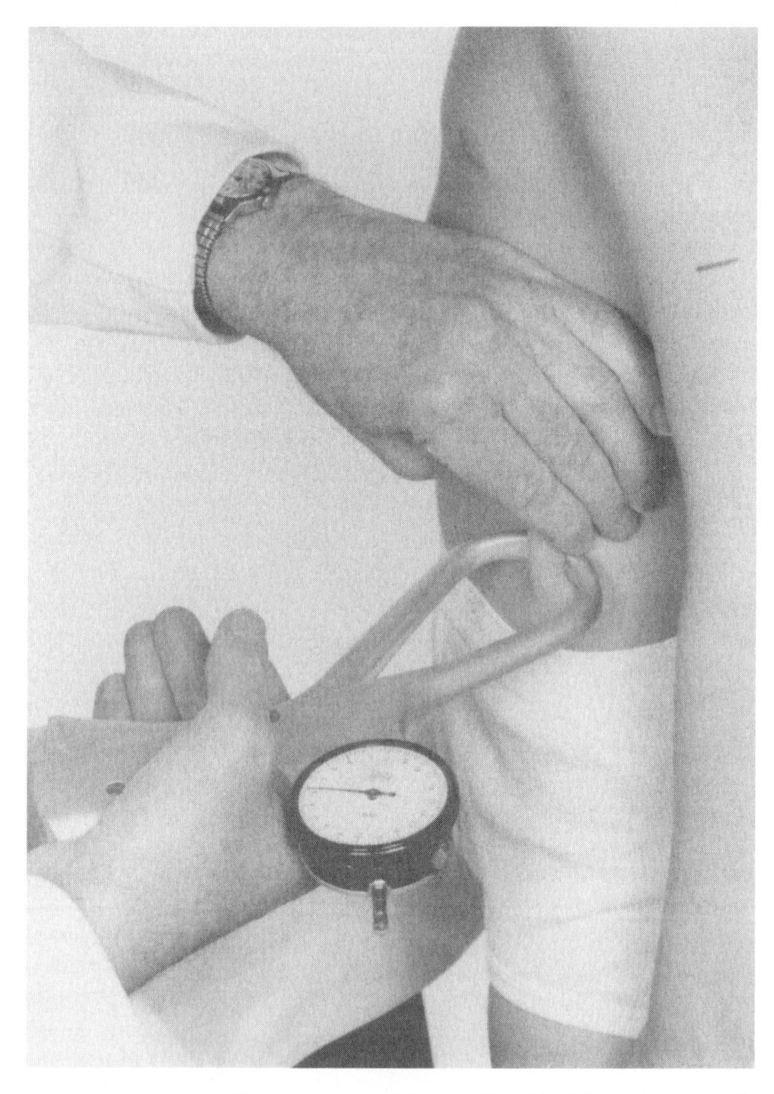

Figure 4.11 The measurement of the suprailiac skinfold, superior and medial to the anterior superior iliac spine.

standing facing the observer. The skinfold is picked up with a sweep of the middle finger and thumb and is a vertical skinfold (Figure 4.11). Once again the calliper is applied below the fingers and, after the dial has stopped moving, the measurement is read to the last completed 0.1 mm. It should be noted that Durnin and Rahaman (1967) and Durnin and Womersley (1974), who developed many of the linear regression equations to predict total body fat from skinfolds, do not use this conventional suprailiac site for their measurement to derive total body fat. Instead they use a mid-axillary suprailiac skinfold just above the iliac crest but still vertical. Harrison *et al.* (1988) recommend a skinfold that is aligned inferomedially at 45 degrees to the horizontal. However, they also make the point that skinfolds at various locations in the suprailiac area appear highly correlated both with each other and with body density, so that no single position appears to offer unique information.

Skinfold callipers

Skinfold callipers are designed to measure the thickness of a fold of subcutaneous fat that has been picked up at a specific landmark on the body by the anthropometrist. The compressible nature of these skinfolds mean that the callipers have had to be designed to exert a constant pressure at all settings of the calliper jaws. Constant-pressure callipers have been developed in America and England that exert a pressure of 10 g/mm^2 of the jaw face area at all settings.

The global obesity epidemic has resulted in a burgeoning interest in measures of body fat. Indeed an industry has developed around the production of callipers of diverse designs and usefulness (see for example Bodytrends (2003)). For research purposes it is fundamentally important to obtain a calliper that will accurately and precisely record the skinfold of interest and be accepted by comparative researchers as appropriate for the purpose of assessing subcutaneous fat. Of the numerous models available three could be described as universally acceptable within human growth research; the Lange, the Harpenden, and the Holtain callipers. I describe these three below although other callipers are available e.g. the Lafayette skinfold calliper, but they have not, as yet been extensively tested in clinical and research situations (see for example Schmidt and Carter (1990) for a comparison of Harpenden, Lange and Lafayette callipers).

Lange calliper

The Lange calliper was introduced by Lange and Brozek (1961) to provide for 'persistent demands for a light compact skinfold calliper' (Lange, 2003). It is composed of a slender handle opposed by a thumb lever. Pressure on this lever opens the jaws uniformly to a maximum of 6 cm with a reading accuracy to ± 1 mm. The lever is released to clamp these jaws on to the skinfold. The jaws have an area of approximately 30 mm^2 and a constant pressure of 10 g/mm^2 irrespective of the size of the skinfold, and they are pivoted to adjust automatically for parallel measurement of the skinfold. The reading is displayed by a fine pointer on a semicircular dial 6 cm in diameter, with an almost linear scale and divisions 150% of natural size.

Harpenden skinfold callipers

The Harpenden calliper (Tanner and Whitehouse, 1955) resulted from the investigation of Edwards *et al.* (1955), who recommended a design that included jaw faces of size 6 × 15 mm, with well-rounded edges and corners, pressure at the faces of 10 g/mm^2 that does not vary by more than 2.0 g/mm^2 over the range of openings 2–40 mm, and a scale such that readings can be taken to the nearest 0.1 mm.

These callipers may be calibrated by fixing them in a bench clamp so that the jaws are parallel to the floor. When a weight equivalent to 10 times jaw face area is applied (90 g if the face is 6 × 15 mm as recommended) the jaws should stay closed but if another 1 g is added they should open to their fullest extent. There is always some leeway around this figure but it has been demonstrated that pressures between 9 g/mm^2 and 13 g/mm^2 make little difference to the skinfold reading.

Holtain (Tanner–Whitehouse) skinfold callipers

The Holtain/Tanner–Whitehouse skinfold calliper (Holtain Ltd, 2003) is the improved version of the Harpenden calliper. The design principle regarding jaw pressure, jaw face area and readability are maintained but the Holtain calliper is lighter than the Harpenden model and easier to hold, thus making repeated measurements less tiring and perhaps creating greater accuracy.

Calibration may be checked in the same way as for the Harpenden model. Both models require little or no servicing beyond cleaning and care. There has been some criticism of these callipers due to inconsistency of spring pressure but if they are calibrated when first received and checked regularly thereafter, as with all other instruments, no problems should arise.

Weight

The measurement of weight should be the simplest and most accurate of the anthropometric measurements. Assuming that the scales are regularly calibrated, the observer ensures that the subject is either dressed in the minimum of clothing or a garment of known weight that is supplied by the observer. The subject stands straight, but not rigid or in a 'military position', and is instructed to 'Stand still'. If the instrument is a beam balance then the observer moves the greater of the two counterweights until the nearest 10 kg point below the child's weight is determined. The smaller counterweight is then moved down the scale until the nearest 100 g mark below the point of overbalance is reached and this is recorded as the true weight. This procedure is necessary to determine weight to the *last completed unit*. If the weight is taken as the nearest 100 g above true weight then that 100 g is greater than actual weight and the last unit has not been completed.

Determining the weight of neonates can be a noisy and tearful procedure but need not be if the help of the parent or carer is solicited. The observer simply weighs adult and child together and then transfers the baby to his/her assistant's arms and weighs the adult alone. The baby's weight can thus be determined by difference – (weight of adult + baby) – (weight of adult) – and the child is left relatively undisturbed.

Weighing scales

The measurement of weight should be one of the easiest of anthropometric measurements and yet results are often inadequate due to inappropriate instrumentation. Accuracy to 0.1 kg (100 g) is acceptable as long as regular calibration ensures the minimum chance of error and appropriate steps are taken to tare for any clothing the subjects are required to wear.

The mechanical instrument best suited to repeatedly accurate weight measurements is that designed on the balance-arm principle with two balance arms. One major balance arm measures to greater than 110 kg in steps of 10 kg and the other, minor arm, to 10 kg in steps of 100 g. Such an instrument is capable of measuring individuals from birth through to adulthood with appropriate modification to include a baby-pan and seat for subjects unable to stand and/or too large for the baby-pan. Electronic balances are also available at less than the cost of mechanical machines and appear well suited to growth clinic use as long as their power source (usually batteries) is regularly checked.

A catalogue of instruments is available from Chasmors (CMS) Weighing Equipment (Chasmors, 2003).

Surface landmarks

Various measurements in anthropometry are taken between points of the axial skeleton. Its landmarks are shown in Figure 4.12.

Acromion process (lateral border of the acromion)

The acromion projects forwards from the lateral end of the spine of the scapula with which it is continuous (Figure 4.13). The lower border of the crest of the spine and the lateral border of the acromion meet at the acromial angle, which may be the most lateral point of the acromion. Great diversity in the shape of the acromion between individuals means that sometimes the acromial angle

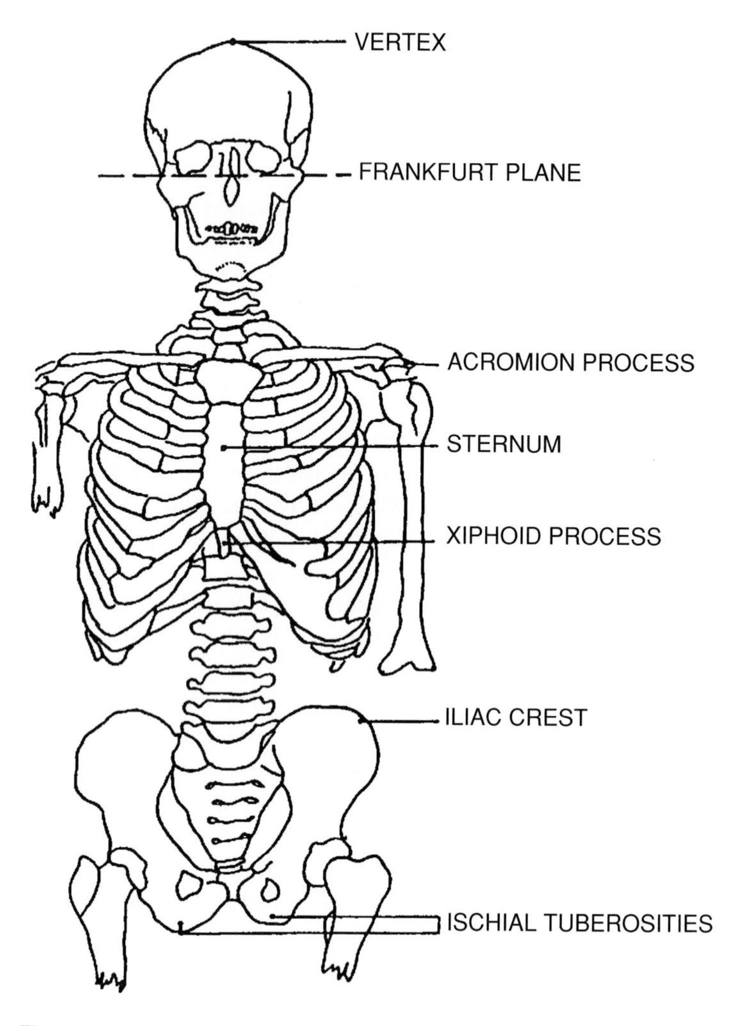

Figure 4.12 Skeletal landmarks of the axial skeleton.

is not the most lateral point. Palpation of the most lateral part may best be performed by running the anthropometer blades laterally along the shoulders until they drop below the acromia. If the blades are then pushed medially the most lateral part of the acromia must be closest to the blades and may be felt below the surface marks left by the blades. There is the possibility of the inexperienced anthropometrists confusing the acromioclavicular joint with the lateral end of the acromion. Great care must be taken to distinguish between these two landmarks prior to measurement.

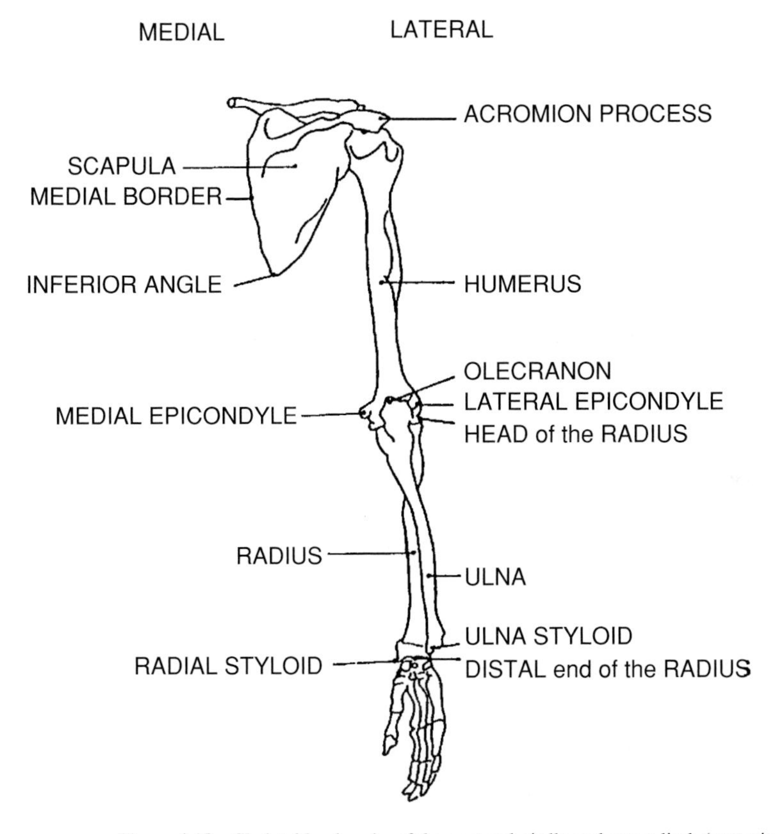

Figure 4.13 Skeletal landmarks of the pectoral girdle and upper limb (posterior view).

Anterior superior iliac spine

This is the anterior extremity of the ilium which projects beyond the main portion of the bone and may be palpated at the lateral end of the fold of the groin (Figure 4.14). It is important to distinguish the iliac crest from the anterior spine when measuring biiliac diameter (see p. 84).

Biceps brachii

The biceps brachii is the muscle of the anterior aspect of the upper arm. Its two heads, the short and the long, arise from the coracoid process and the supraglenoid tubercle of the scapula respectively and are succeeded by the muscle bellies before they end in a flattened tendon that is attached to

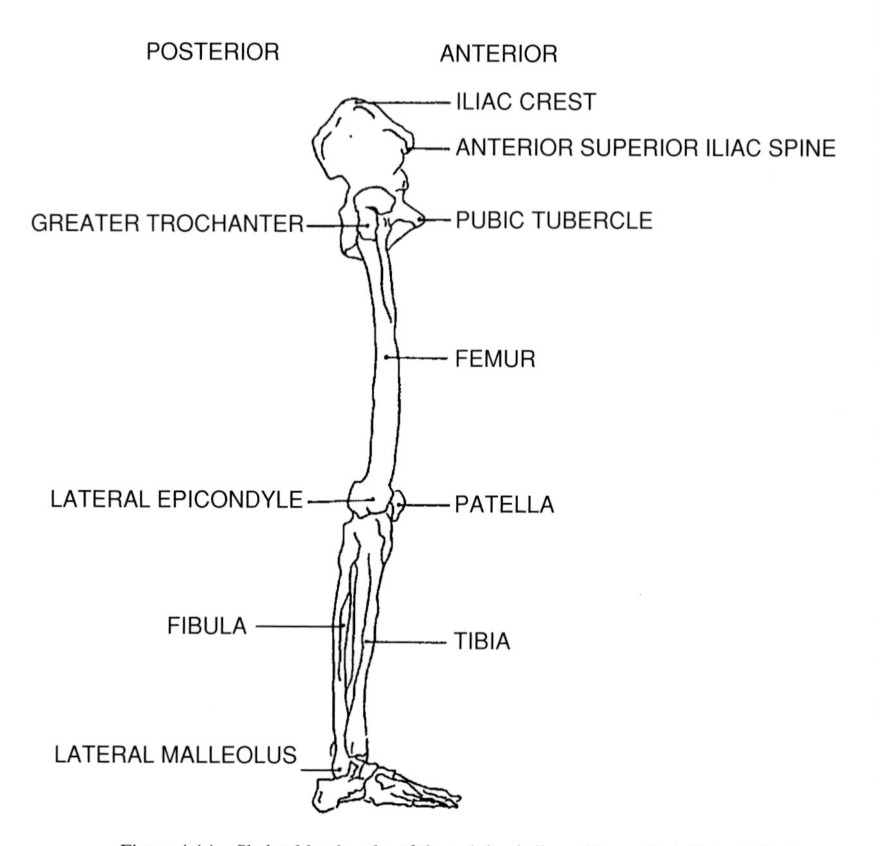

Figure 4.14 Skeletal landmarks of the pelvic girdle and lower limb (lateral view).

the posterior part of the radial tuberosity. When relaxed the muscle belly has its greatest bulge towards the radius but when contracted with the arm flexed the belly rises to a point nearer the shoulder. Thus relaxed and contracted arm circumferences, taken at the maximum bulge of the muscle, are not at exactly the same level.

Distal end of the radius

This is the border of the radius proximal to the distal–superior borders of the lunate and scaphoid and medial to the radial styloid (Figure 4.13). It may be palpated by moving the fingers medially and proximally from the radial styloid (see 'radial styloid', below).

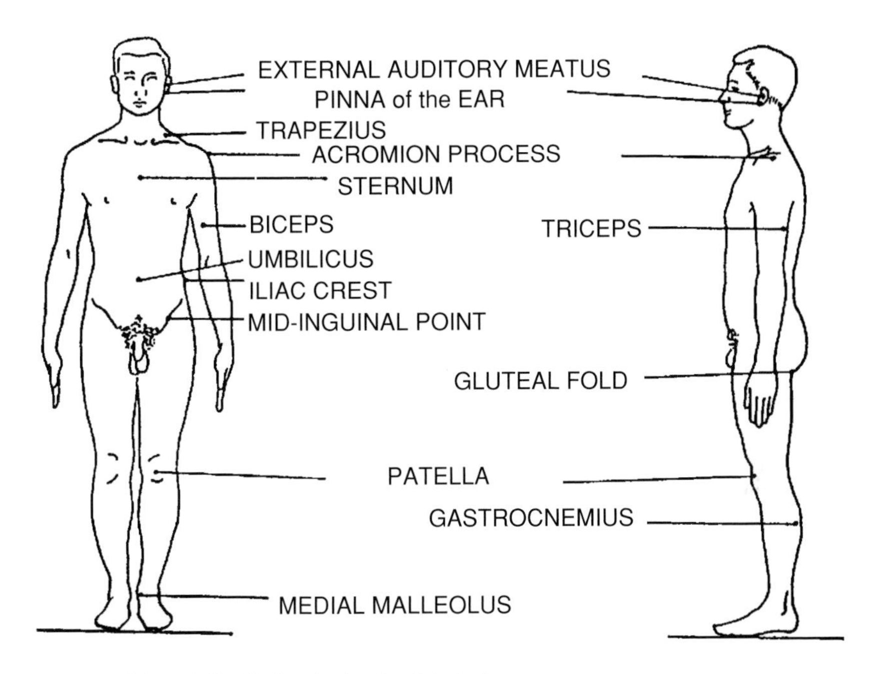

Figure 4.15 Surface landmarks of the body.

External auditory meatus

This landmark, used to obtain the Frankfurt plane, is also called the external acoustic meatus and leads to the middle ear from the external auricle (Figure 4.15). In terms of a surface landmark it is therefore simply present as a hole in the external ear and may therefore be easily seen. The tragus, the small curved flap that extends posteriorly from the front of the external ear, overlaps the orifice of the meatus and may be used to gauge the level of the orifice.

Femur epicondyles

The lower end of the femur consists of two prominent masses of bone called the condyles, which are covered by large articular surfaces for articulation with the tibia (Figure 4.14). The most prominent lateral and medial aspects of the condyles are the lateral and medial epicondyles. These may be easily felt through the overlying tissues when the knee is bent at a right angle, as in the sitting position. If the observer's fingers are then placed on the medial and lateral aspects of the joint the epicondyles are the bony protuberances immediately above the joint space.

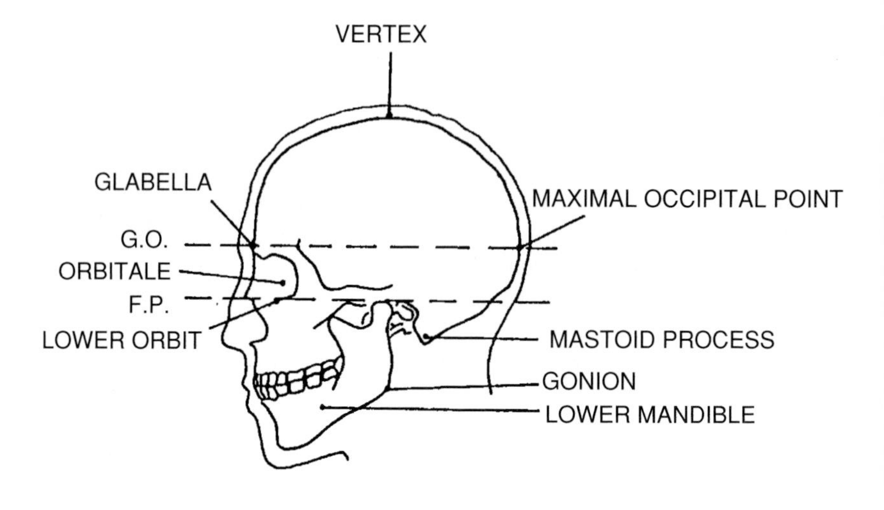

Figure 4.16 Skeletal and surface landmarks of the head.

Frankfurt plane

This plane, used extensively in anthropometric measurement, is obtained when the lower margins of the orbital openings and the upper margins of the external acoustic (auditory) meatuses lie in the same horizontal plane (Figure 4.16). The supinated Frankfurt plane, used in the measurement of recumbent and crown–rump length, is vertical rather than horizontal.

Gastrocnemius

This is the most superficial of the group of muscles at the rear of the lower leg and forms the belly of the calf.

Glabella

This landmark is in the midline of the forehead between the brow ridges and may be used as the most anterior point of the head (Figure 4.16).

Gluteal fold

This fold or furrow is formed by the crossing of the gluteus maximus and the long head of the biceps femoris and semitendinosus. It may therefore be viewed from the lateral aspect or the posterior aspect as the crease beneath the buttock.

In some subjects, perhaps because of a lack of gluteal development, a crease may not be present. In this case the level of the gluteal fold is judged from the lateral profile of the buttocks and posterior thigh.

Head of the radius

This may be palpated as the inverted, U-shaped bony protuberance immediately distal to the lateral epicondyle of the humerus when the arm is relaxed with the palm of the hand facing forwards (Figure 4.13).

Humeral epicondyles

These are the non-articular aspects of the condyles on the lower surface of the humerus (Figure 4.13). The medial epicondyle forms a conspicuous blunt projection on the medial aspect of the elbow when the arm is held at the side of the body with the palm facing forward. The lateral epicondyle may be palpated opposite and a little above the medial epicondyle.

Iliac crest

This may be palpated as the most superior edge of the ilium (Figure 4.14) and may be easily felt through the overlying soft tissue. Greater difficulty will be experienced with the more obese subject but it is quite possible with the anthropometer blades to compress the tissue and feel the crest.

Malleoli

The medial malleolus is the bony protuberance on the medial side of the ankle (Figure 4.15). It is the inferior border of this malleolus that is palpated and used as a landmark for the measurement of tibial length.

Mastoid process

This is the conical projection below the mastoid portion of the temporal bone (Figure 4.16). It may be palpated immediately behind the lobule of the ear and is larger in the male than in the female.

Midaxillary line

The axilla is the pyramidal region situated between the upper parts of the chest wall and the medial side of the upper arm. The midaxillary line is normally taken as the line running vertically from the middle of this region to the iliac crest.

Midinguinal point (inguinal crease)

The inguinal ligament runs from the anterior superior iliac spine to the pubic tubercle at an angle of 35 to 40 degrees and is easily observed in all individuals.

The midpoint between the anterior spine and the pubic tubercle on the line of the inguinal ligament is taken as the midinguinal point.

Midpoint of the arm

The midpoint of the arm, used for arm circumference, is taken as the point on the lateral side of the arm midway between the lateral border of the acromion and the olecranon when the arm is flexed at 90 degrees. This may be most easily determined by marking the lateral border of the acromion and applying a tape measure to this point. If the tape is allowed to lie over the surface of the arm, the midpoint may easily be calculated and marked. Alternatively, tape measures do exist with a zero midpoint that are specifically designed to determine this landmark. (It has been common to refer to this point, and the circumference or girth at this level, as the 'mid upper arm' landmark/circumference. This terminology is specifically not used here because it is anatomically incorrect to describe this area of the brachium or arm as the 'upper arm'.)

Occiput

The occipital bone is situated at the posterior and base of the cranium (Figure 4.16). The occiput is the most posterior part of this bone and may be clearly seen from the side view of the subject.

Olecranon

The olecranon is the most proximal process of the ulna (Figure 4.13) and may be easily observed when the arm is bent as the point of the elbow.

Patella

The patella is the sesamoid bone anterior to the knee joint embedded in the tendon of the quadriceps muscle (Figure 4.14). It is flat, triangular below and curved above. When the subject is standing erect its lower limit lies above the line of the knee joint and its upper border may be palpated at the distal end of the quadriceps muscle.

Pinna of the ear

The pinna of the ear is more correctly called the lobule and is the soft part of the auricle that forms the ear lobe.

Radial styloid

The radial styloid is the distal projection of the lateral surface of the radius (Figure 4.13). It extends towards the first metacarpal and may be palpated as a bony projection on the lateral surface of the wrist when the hand is relaxed.

Scapula

The scapula is the large, triangular flattened bone on the posterolateral aspect of the chest, and is commonly known as the shoulder blade (Figure 4.13). Its medial border slopes downwards and laterally to the inferior angle that may be easily palpated, and lies over the seventh rib or seventh intercostal space when the arm is relaxed.

Sternum

The sternum or breastbone is the plate of bone inclined downwards and a little forwards at the front of the chest (Figure 4.12). It is composed of three parts: the manubrium at the top, the body or mesosternum at the centre and the xiphoid process at the lower end. The mesosternum and xiphoid process are important landmarks in anthropometry. The mesosternum is marked by three transverse ridges or sternebrae and the junction between the third and fourth sternebrae form a landmark in chest measurement. The fourth sternebra may not be easily palpated but the junction lies below the more easily palpated third sternebra. The xiphoid process may be palpated by following the line of the sternum to its end. The sternum is considerably larger in males than in females.

Trapezius

The trapezius is a flat, triangular muscle extending over the back of the neck and the upper thorax.

Triceps

The triceps muscle is the large muscle on the posterior side of the arm. When the arm is actively extended two of the three triceps heads may be seen as medial and lateral bulges.

Trochanters

The greater and lesser trochanters are projections at the proximal end of the femur (Figure 4.14). The lesser trochanter cannot be palpated on the living subject because it lies on the posterior surface of the femur and is covered by the large gluteal muscles. The greater trochanter, however, is palpable as the bony projection on the lateral surface of the upper thigh approximately a hand's breadth below the iliac crest.

Ulna styloid

The styloid process of the ulna is present as a short, rounded projection at the distal end of the bone (Figure 4.13). It may be easily palpated on the posterior–medial aspect of the wrist opposite and about 1 cm proximal to the styloid process of the radius.

Umbilicus

The umbilicus, or naval, is clearly observable in the centre of the abdomen. It is variable in position, lying lower in the young child due to lack of abdominal development.

Vertex of the skull

This is the topmost point of the skull (Figure 4.16) and theoretically comes into contact with the stadiometer headboard when height is being properly measured. With the head in the Frankfurt plane the vertex is slightly posterior to the vertical plane through the external auditory meatus and may be easily palpated.

REFERENCES

Barlow, S. E. and Dietz, W. H. (1998). Obesity evaluation and treatment: expert committee recommendations. The Maternal and Child Health Bureau, Health resources and Services Administration, and the Department of Health and Human Services. *Paediatrics*, **102**, E29.

Bodytrends (2003). Catalogue of calipers. http://www.bodytrends.com/buycalip.htm

Cameron, N. (1978). The methods of auxological anthropometry. In *Human Growth: A Comprehensive Treatise*, eds. F. Falkner and J. M. Tanner, pp. 35–90. New York: Plenum Press.

(1983). The methods of auxological anthropometry. In *Human Growth: A Comprehensive Treatise*, 2nd edn, eds. F. Falkner and J. M. Tanner, pp. 3–46. New York: Plenum Press.

(1984). *The Measurement of Human Growth.* London: Croom-Helm.

Chasmors (2003). Catalogue of weighing instruments. Chasmors (CMS) Weighing Equipment, 18 Camden High Street, London NW1 0JH, UK. http://www.chasmors.com

Cole, T. J., Bellizzi, M. C., Flegal, K. M. and Dietz, W. H. (2000). Establishing a standard definition for child overweight and obesity worldwide: international survey. *British Medical Journal*, **320**, 1240–3.

Durnin, J. V. G. A. and Rahaman, M. M. (1967). The assessment of the amount of fat in the human body from measurements of skinfold thickness. *British Journal of Nutrition*, **21**, 681–9.

Durnin, J. V. G. A. and Womersley, J. (1974). Body fat assessed from total body density and its estimation from skinfold thickness: measurements on 481 males and females aged from 16 to 72 years. *British Journal of Nutrition*, **32**, 77–97.

Edwards, D. A. W., Hammond, W. H., Healy, M. J. R., Tanner, J. M. and Whitehouse, R. H. (1955). Design and accuracy of callipers for measuring subcutaneous tissue thickness. *British Journal of Nutrition*, **9**, 133–43.

Harrison, G. G., Buskirk, E. R., Carter, J. E. L., *et al.* (1988). Skinfold thicknesses and measurement technique. In *Anthropometric Standardization Reference Manual*, eds., T. G. Lohman, A. F. Roche and R. Martorell, pp. 55–70. Champaign, IL: Human Kinetics Books.

Holtain Ltd (2003). Catalogue of anthropometric equipment. Holtain Ltd, Crosswell, Crymych, SA41 3UF, UK. http://www.fullbore.co.uk/holtain/medical/welcome.html

Lampl, M., Veldhuis, J. D. and Johnson, M. L. (1992). Saltation and stasis: a model of human growth. *Science*, **258**, 801–3.

Lange, Inc. (2003). The Lange calliper. http://www.langecaliper.com

Lange, K. O. and Brozek, J. (1961). A new model of skinfold calliper. *American Journal of Physical Anthropology*, **19**, 98–9.

Lohman, T. G., Roche, A. F. and Martorell, R. (eds.) (1988). *Anthropometric Standardization Reference Manual*. Champaign, IL: Human Kinetics Books.

Ong, K. K. L., Ahmed, M. L., Emmett, P. M., Preece, M. A. and Dunger, D. B. (2000). Association between postnatal catch-up growth and obesity in childhood: prospective cohort study. *British Medical Journal*, **320**, 967–71.

Ruiz, L., Colley, J. R. T. and Hamilton, P. J. S. (1971). Measurement of triceps skinfold thickness. *British Journal of Preventive and Social Medicine*, **25**, 165–7.

Schmidt, P. K. and Carter, J. E. (1990). Static and dynamic differences among five types of skinfold callipers. *Human Biology*, **62**, 369–88.

SiberHegner (2003). The Martin anthropometer. http://www.siberhegner.com

Strickland, A. L. and Shearin, R. B. (1972). Diurnal height variation in children. *Pediatrics*, **80**, 1023–5.

Tanner, J. M. and Whitehouse, R. H. (1955). The Harpenden skinfold calliper. *American Journal of Physical Anthropology*, **13**, 743–6.

Whitehouse, R. H., Tanner, J. M. and Healy, M. J. R. (1974). Diurnal variation in stature and sitting height in 12–14-year-old boys. *Annals of Human Biology*, **1**, 103–6.

5 *Measuring maturity*

Noël Cameron
Loughborough University

Introduction

This chapter concerns the way in which maturity may be assessed. It will concentrate on the assessment of the process of maturation from birth through childhood and adolescence, i.e. the period of time in which maturation interacts with growth. It is important, therefore, to understand the difference between 'growth' and 'maturation'. Bogin (1999) defines the former as 'a quantitative increase in size or mass' such as increases in height or weight. Development or maturation, on the other hand, is defined as 'a progression of changes, either quantitative or qualitative, that lead from an undifferentiated or immature state to a highly organized, specialized, and mature state'. The end point of maturation, within the context of the growth, is the attainment of adulthood, which may be defined as a 'functionally mature individual'. Functional maturation, in a biological context, implies the ability to successfully procreate and raise offspring who themselves will successfully procreate. We know that in addition to the obvious functional necessities of sperm and ova production, reproductive success within any mammalian society is also dependent on a variety of morphological characteristics such as size and shape. The too short or too tall, the too fat or too thin are unlikely to achieve the same reproductive success as those within an 'acceptable' range of height and weight values that are themselves dependent on the norms in a particular society. Thus in the broadest context, maturation *and* growth are intimately related and both must reach functional and structural end points that provide the opportunity for successful procreation.

Like the previous chapter on measurement techniques this chapter will be prescriptive in that it will concentrate on the techniques that have been shown to be most appropriate, i.e. accurate and reliable, within the context of growth assessments in both research and clinical environments.

Methods in Human Growth Research, eds. R. C. Hauspie, N. Cameron and L. Molinari.
Published by Cambridge University Press.

Initial considerations

It is inevitable that those reading this book may wish to develop their own methods to assess maturation. In so doing the rationale that has been developed over the last 50 years by which we identify and manage a possible maturity indicator is of importance. This identification and management is based on six considerations. It is important first to appreciate that maturation is not linked to time in a chronological sense. In other words, one year of chronological time is not equivalent to one year of maturational 'time'. This is perhaps best illustrated in Figure 5.1 in which three boys and three girls of precisely the same chronological ages demonstrate dramatically different degrees of maturity as evidenced by the appearance of secondary sexual characteristics. In addition they exhibit changes in the proportion and distribution of subcutaneous fat, and the development of the skeleton and musculature, that result in sexually dimorphic body shapes in adulthood.

Whilst each individual has passed through the same chronological time span they have done so at very different rates of maturation. Thus an age scale to represent maturity fails because no particular age can be associated with full maturity and stages prior to full maturity because of the lack of a constant relationship between maturity and time both between and within the sexes. The early 'atlas' techniques to assess skeletal maturity (e.g. Greulich and Pyle, 1959) did not overcome this problem and represented full maturity as a chronological age of 18 years. Later techniques by Acheson (1954, 1957) and Tanner *et al.* (1962, 1975, 1983, 2001) moved away from an age-based method and developed bone-specific scoring techniques in which scores rather than ages were assigned to any particular stage of maturation (appearance of a maturity indicator). The basic principle is that the appearance of maturity indicators within a particular bone represents the maturation of that single bone. Because of the process of uneven maturation and the discrete nature of maturity indicators (see below) bones within any one area may not be at the same stage of maturation and the scoring process needs to minimize the overall disagreement between different bones. Thus the sum of squares of the deviations of the bone scores about the mean score is minimized. The mathematical details are provided by Tanner *et al.* (1975, 1983, 2001) and are of general importance in the development of methods to assess maturity. In the Fels method of Roche *et al.* (1988) the grades of maturity indicators are submitted to a computer program that determines the mean and standard error of estimate of skeletal age. The program is based on a statistical procedure that combines the grades into a single continuous measure of skeletal maturity scaled to years by using latent trait analysis. The chronological ages, determined by logit regression, at which 50% of children exhibit the maturity indicators form the parameters used in the construction of

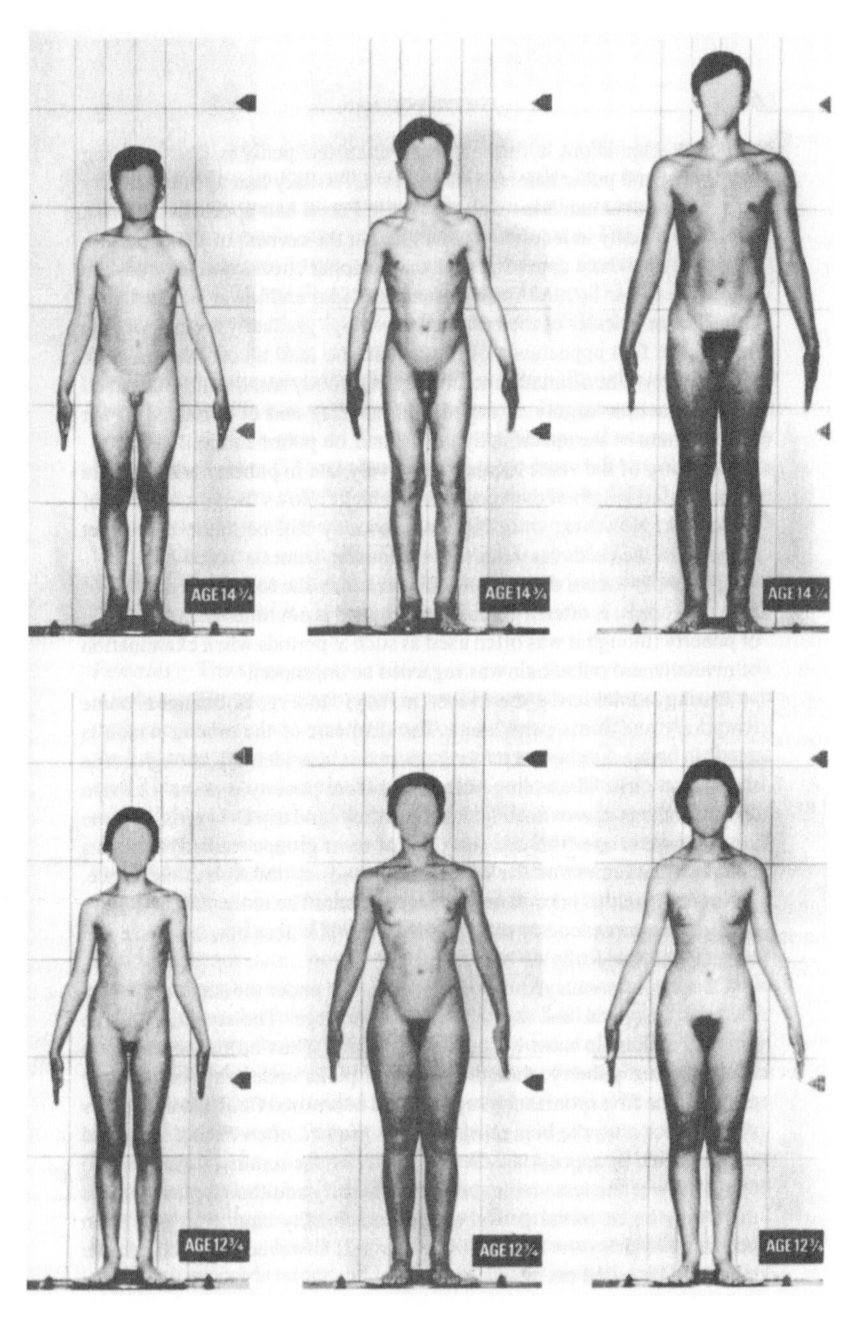

Figure 5.1 Three boys and three girls photographed at the same chronological ages within sex; 12.75 years for girls and 14.75 years for boys. (From Tanner, J. M. (1980). Growth and endocrinology of the adolescent. In: *Endocrine and Genetic Diseases of Childhood*, 2nd edn, ed. L. Gardner. Philadelphia, PA: W. B. Saunders.)

the scale. Thus Fels combines estimates of age of appearance of each indicator from each bone into the skeletal maturity scale. Thus whilst it is not a scoring technique, in that its output is an age rather than a maturity score, it is rather more sophisticated than the atlas techniques and forms an intermediate stage between direct age comparison and bone-scoring.

Second, maturation is most often assessed by the identification of 'maturity indicators'. Such indicators are discrete events or stages recognizable within the continuous changes that occur during the process of maturation. Thus the indicators that identify changes in the radiographic appearance of the radial epiphysis or changes in breast or pubic hair development divide the *continuous* changes that occur during skeletal and sexual maturation into *discrete* stages. Maturity indicators must conform to certain prerequisites if they are to be useful. They must possess the quality of *universality* in that each grade of each indicator must be present in all normal children of both sexes and those grades must appear *sequentially*, and in the *same* sequence, in all children. Maturity indicators should have the ability to distinguish between children of the same chronological age or *discrimination* as well as being *reliable* with good inter- and intra-observer reliability and *valid* with the ability to reflect genuine maturational change, i.e. they ought to reflect a continuous process of maturation rather than a discontinuous process. Finally the quality of *completeness* is desirable. Completeness requires the indicator to increase in prevalence from zero to 100% within a relatively short period of time. Roche *et al.* (1988), for instance, use the example of the hook of hamate, which was virtually absent in their source sample of boys at 10 years but 'almost universal' after 13 years of age. This rapid change in prevalence makes such an indicator very informative.

Third, there is variability of maturation *within* the individual *between* maturational processes. For instance, whilst skeletal and secondary sexual maturation are associated they are not correlated so significantly that one can categorically associate a particular stage of sexual maturation with a particular skeletal 'age' (Marshall and Tanner, 1969, 1970). In the closest association, of skeletal age to menarcheal age, it is possible to state that a girl with a skeletal age less than 12 years is unlikely to have experienced menarche and that one with a skeletal age of 15 years is likely to be postmenarcheal. However, we cannot state with any real degree of confidence that the association of these two maturational processes is closer than that.

Fourth, there is variation *within* maturational processes. For instance within sexual maturation, it is apparent that different structures, e.g. genitalia and pubic hair, will not necessarily be at precisely the same level of maturity. Similarly, within a particular anatomical region, such as the hand and wrist, all bones will not be at precisely the same stage of maturation. Thus we have a process of 'uneven maturation'.

Fifth, there is clear sexual dimorphism within human growth and maturation such that females tend to be advanced relative to males at any particular chronological age. In Figure 5.1 for instance, the females are aged exactly 12.75 years and the males 14.75 years yet their levels of secondary sexual maturity are similar. In addition, this sexual dimorphism extends to the functional importance of particular maturational events. Spermarche and menarche, for instance, are often thought to be equivalent stages of maturation in males and females yet their positions within the sequence of maturational events during puberty are quite different and thus their associations with other aspects of maturation also differ. Menarche, for instance, occurs following peak height velocity and towards the latter part of secondary sexual development, i.e. in breast stage 3, 4, or 5 (Marshall and Tanner, 1969). Relatively sparse data on spermarche identifies its occurrence at approximately 14 years in boys, which would be in the early or middle part of the adolescent growth spurt and thus indicative of an earlier stage of pubertal maturation.

Sixth, maturation is not related to size except in very general terms; a small human is likely to be a child and thus less mature than a large human who is more likely to be an adult. As the ages of the two individuals approach each other so the distinction between size and maturity narrows and disappears such that, within a group of similar maturity there will be a range of sizes and within a group of similar size there will be a range of maturity levels. Thus when maturation is assessed size must be controlled for or excluded from the assessment method. The only exception to this rule is in the assessment of testicular volume; 4 ml represents the initiation of pubertal development and 12 ml mid-puberty. This is not to say that there is no variation in testicular volume. Like all aspects of growth and development, variability is an inherent aspect of testicular growth. Clinicians, however, use the above measures as indicators of normal testicular growth and of the initial and middle stages of pubertal development.

These six considerations, the relationship of maturity to time, the quantification of the continuous process of maturation by using discrete events, the relative independence of different processes of maturation within the individual, the appreciation of uneven maturation, sexual dimorphism, and the lack of a relationship between maturity and size, have governed the development of techniques for the assessment of maturation.

Methods of assessment

Maturation is assessed using a combination of processes and events. Maturational 'processes' include secondary sexual development, dental development and skeletal development. Maturational 'events' include those aspects of maturation that occur once and provide an unambiguous signal that the individual has reached a particular level of maturity. Examples are the exact age at which menarche (the first menstrual period) is experienced in girls or the exact age of peak height velocity during the adolescent growth spurt.

The following sections describe the various techniques involved in the assessment of maturity using skeletal maturity, secondary sexual development, dental maturity, and landmarks on the growth curve. Where necessary the conceptual rationale that underlies the methods have been described because there is the possibility that the reader will wish to develop their own methods and thus to learn from previous attempts.

Skeletal development

Radiation dosage

The assessment of skeletal maturation is a relatively uncommon occurrence in the current ethical climate which prevents the child being exposed to any and all potential risks. Such protection is a laudable aim but the exposure of a child's hand and wrist to radiation for the purposes of obtaining a hand–wrist radiograph requires the minimum of exposure levels. The typical dosage is 0.01 milliSieverts (mSv) per radiograph. Background radiation dosage in the UK, i.e. the dosage to which everyone is exposed, is 2.2 mSv per annum. Thus a hand–wrist radiograph exposes the child to the radiation dosage that he/she would naturally receive in 1.7 days in the UK. The World Health Organization in 1977 recommended a maximum dosage to children involved in research projects of 0.5 mSv. Their justification was that this level of radiation lies within the expected variations in normal background radiation that an individual could well experience in changing geographical locations. Thus in either clinical or research settings the risk of hand–wrist radiographs is minimal and should not prevent a well-designed research project from obtaining ethical approval.

Skeletal maturity methods

Whilst a number of techniques exist to assess skeletal maturity assessment procedures have been dominated by two different approaches to the problem; the 'atlas' technique of Greulich and Pyle (1959) and the 'bone-specific scoring' technique of both Tanner and his colleagues (Tanner *et al.*, 1962, 1975, 1983, 2001). The relatively recent appearance of the Fels technique of Roche and his colleagues (Roche *et al.*, 1988; Chumlea *et al.*, 1989) means that whilst it is used extensively in American clinical work it has yet to make a broad impact in research. All of these methods use the left hand and wrist to estimate a skeletal age or bone age yet they are different both in concept and in method. Greulich–Pyle bone ages are most commonly assessed by comparing a radiograph to a series of standard radiographs photographically reproduced in the atlas. The chronological age assigned to the standard most closely approximating the radiograph is the bone age of the subject. In practice a more precise estimate of bone age may be obtained by assessing each bone in the hand and wrist separately, but this is rarely done. Thus, there are errors in most Greulich–Pyle estimations because the uneven maturation present in the hand and wrist is not acknowledged. The system is based on subjects from Cleveland, Ohio, who were assessed during the 1920s and 1930s. The Tanner–Whitehouse system requires 20 bones of the hand and wrist to be assessed individually and a score to be assigned to each. The summation of the scores results in a bone maturity score, which is equivalent to a particular bone age. This technique used subjects from a variety of studies conducted in the south of England during the 1950s and 1960s. Although the latter is more recent, the effect of positive secular trends and population differences in average maturity status means that estimates of skeletal age based on either technique must be viewed with some caution. However, the statistical rationale of the bone-specific scoring technique can be applied to any series of radiographs from a representative sample of a population. In contrast to the atlas technique it is thus possible to develop specific national references for the assessment of skeletal maturity using a bone-specific approach, which would result in a more sensitive clinical appraisal. This process would be similar with the Fels hand–wrist technique requiring logit analyses of the appearance of each grade of maturity indicator *within* the population of interest and the construction of a new skeletal maturity scale.

The atlas technique of Greulich and Pyle (1959) is now almost obsolete to be replaced by either the Tanner–Whitehouse or the Fels methods. However, in some research settings it is necessary for comparative purposes to use the atlas technique. Its use will thus be described below along with both the Tanner–Whitehouse and Fels methods.

Practical approaches to these techniques will be described below. In all cases it is fundamentally important to obtain a radiograph of the left hand and wrist that clearly portrays the bones in their standard relationship to each other. Thus the posing of the hand–wrist is of great importance. The upper limb should be abducted so that it is level with the shoulder and the arm and forearm are in the same horizontal plane. The palm faces downwards in contact with the radiographic cassette. The fingers are slightly spread and the thumb is placed in a natural degree of rotation such that its axis forms an angle of about 30 degrees with the index finger. Tanner *et al.* (2001) recommend a tube–film distance of 76 cm with the tube being centred directly above the head of the third metacarpal. Naturally suitable protection to the child and observers should be ensured by using lead-lined protective sheeting.

Atlas techniques

The atlas technique has its origins in the pioneer work of Dr T. Wingate Todd, who published an *Atlas of Skeletal Maturity* in 1937 (Todd, 1937). Todd based his atlas on the hand–wrist radiographs of 1000 children from the Brush Foundation Study of Human Growth and Development, which started in 1929 in Cleveland, Ohio. The children were only admitted to the study on the application of a paediatrician and were thus, in the Midwest of the 1930s, a socially advantaged group later described by Greulich and Pyle (1950) as 'above average in economic and educational status'.

From each chronological age group the films were examined bone by bone and the film exhibiting the modal maturity of the bone for that group was selected and the maturity indicators of that particular bone described. The appearance of these indicators was taken as typical for a healthy child of that age and sex. Having described these indicators for each bone and each age, the series was re-examined to identify radiographs showing, for every bone, the modal maturity for that age and sex. Each of these standards was assigned a 'skeletal age' determined by the age of the children on whom the standard was based and it is those standards that appeared in the atlas. Continuing Todd's work, Drs William Walter Greulich, Idell Pyle and Normand Hoerr published a variety of atlases between 1950 and 1969 to describe the skeletal maturation of the hand and wrist, knee, and foot and ankle (Greulich and Pyle, 1950, 1959; Pyle and Hoerr, 1955). That for the hand and wrist is the best known and is referred to universally as the 'Greulich–Pyle Atlas'.

Whilst other techniques of assessing skeletal maturity have a prescribed system for rating each bone Greulich and Pyle (1959) made no specific recommendation as to the technique to be used in the atlas method. However, they did provide a five-stage procedure which, they say, 'may prove helpful to the reader' but expect the observer to replace it by a method of his/her own devising that he/she finds better adapted to their needs or preferences. This lack of prescription has resulted in the majority of raters simply comparing their radiographs with standards and assigning the skeletal age of the standards to their radiographs or making a very subjective interpolation between those standards and the adjacent ones. The method suggested by Greulich and Pyle is as follows:

1. Compare the film with a standard of the same sex and the nearest chronological age.
2. Compare with the adjacent standards.
3. Select for detailed comparison the standard, which superficially appears to resemble it most closely.
4. For each individual bone, in an orderly sequence, compare to the same bone on the standard. If it appears to be the same assign that skeletal age to it, if not compare to the adjacent standards until a similar bone is found and an age assigned. If none is found then a skeletal age should be estimated from those that it most closely resembles.

If the whole hand corresponds to the standard then that is the skeletal age. If it is intermediate between standards then interpolate between the adjacent skeletal ages. The technique that has been evolved and recommended by Roche (1970) is much more rigorous. He suggests that the comparison to a standard should be done but only to orientate the assessor as to the likely development of the child. Each bone is then compared to the maturity indicators, at the rear of the book, and a standard number assigned to each bone. The median of these 30 numbers is used to determine the skeletal age. Say the median for a boy's bones is 17.5; this falls midway between the skeletal ages of 8 and 9 'years' and provides interpolated skeletal age of 8.5 'years'. Using the median as opposed to the mean is important, as this statistic does not unfairly bias the assessment when a centre is late in appearing and therefore scores very low. The advantage of using Roche's technique is that the assessor is prevented from relying on his own subjective assessment of a few bones and is forced to consider all the bones of the hand and wrist using clearly defined, objective criteria. It seems unlikely that Greulich and Pyle expected raters to use just a few of the many bones available and this practice violates the need to avoid errors due to ignorance of uneven maturation.

Bone-specific scoring techniques

Bone-specific techniques were developed in an attempt to overcome the two main disadvantages of the atlas techniques. These were the concept of the 'evenly maturing skeleton' and the difficulty of using 'age' in a system measuring maturity. Acceptance of the evenly maturing skeleton was compulsory if one used the atlas method of comparing the radiograph with standard plates. This acceptance decreased the significance of individual variation within the bones of the hand–wrist. Similarly the acceptance of 'age' from the standards implied the acceptance of a chronological time series. The Tanner–Whitehouse methods evolved from the Oxford method of Acheson (1954, 1957) in which bone-specific scoring was used for the first time.

The Tanner–Whitehouse method

In 1959 and 1962 J. M. Tanner and R. H. Whitehouse published their first attempt at a bone-specific-scoring system. This was known as TW1 but was later revised and published as TW2 (Tanner *et al.*, 1962, 1975, 1983) and most recently as TW3 (Tanner *et al.*, 2001). The basic rationale was that the development of each single bone reflected a single process that they defined as maturation. 'Scores' could be assigned to the presence of particular maturity indicators within the developing bones. Ideally each of the *n* scores from each of the bones in a particular individual ought to be the same. This common score, with suitable standardization, would be the individual's maturity. To arrive at a practical technique a variety of modifications to this rationale had to be made. In addition Tanner and his colleagues were highly critical of the method and how it operated in practice – in how well it served the paediatric and research communities for whom it was intended. Their monitoring of the system promoted the various modifications that resulted in TW2 and most recently TW3.

The underlying rationale of the Tanner–Whitehouse techniques was based on dissatisfaction with a maturity system based on chronological age and thus the need to define a maturity scale that does not refer directly to age. The result of such a system would be that in any particular population the relationship between maturity and age could be studied and 'maturity standards', similar to height or weight standards, could be produced. Concentrating on the bones of the hand and wrist, they defined series of eight maturity indicators for each bone and nine for the radius. (As with the Oxford method, the sesamoid bones were ignored.) These maturity indicators were then evaluated, not in relation to chronological age, but in relation to their appearance within the full passage of each specific bone from immaturity to maturity. Thus, for example, it was possible to say that a particular indicator on the lunate first appeared at 13% maturity and that a process of fusion in the first metacarpal started at 85% maturity. In addition Tanner and his colleagues were of the opinion that the metacarpals and phalanges, being greater in number than the carpal bones, would weight the final scores in favour of the 'long' bones; they therefore omitted rays 2 and 4 from the final calculations. Further they weighted the scores so that half of the mature score derived from the carpal bones and half from the long and short bones. The scores were so proportioned that the final mature score totalled 1000 points. Five thousand radiographs of normal British children were then rated, using this technique, to arrive at population 'standards' that related bone maturity scores to chronological ages. The resulting curve of bone maturity score against age was sigmoid, demonstrating a non-linear relationship between skeletal maturity and chronological age.

There were three objections to TW1. First, some of the maturity indicators involved the assessment of size relations between bones, which may be altered by pathological conditions; this violates the requirement of universality in the selection of maturity indicators. Second, by constraining the number of maturity indicators to eight, Tanner *et al.* were weakening their system by of necessity ignoring the fact that some bones may exhibit more or fewer maturity indicators than the eight required by the TW1 system. Third, the contribution of the carpus to 50% of total maturity presents a problem in terms of the repeatability of assessing maturity indicators (i.e. the carpus is less reliable) and because the carpus is known not to play a major role either in growth in height or in epiphyseal fusion.

Tanner and his colleagues took cognizance of these criticisms in their development of the TW2 system, which was in general use in Europe for 20 years. They did not change the maturity indicators but they changed the scores assigned to the individual bones to allow the calculation of a bone maturity score based on the radius, ulna and short bones (RUS) only or the carpal bones (CARPAL) only in addition to the full 20-bone score (TW2 (20)).

The mathematical basis of the system is complex and not particularly germane to this chapter but may be studied in the second and third editions of the Tanner–Whitehouse technique (Tanner *et al.*, 1983, 2001) or in Healy and Goldstein (1976).

Most recently Tanner and his colleagues have published an updated method now known as TW3 (Tanner *et al.*, 2001). It had been almost 20 years since the second edition of the Tanner–Whitehouse book and, as with all systems within growth research that rely on source samples from a particular historical time, Tanner and his colleagues were well aware of the secular trend. That trend is almost universal and has been a recognized aspect of generational differences in human growth for many years. The secular trend affects both growth in overall size and in maturity such that size gets larger and maturational events occur earlier with each succeeding generation. Thus rate of skeletal maturation will also have advanced and bone-specific scoring techniques ought to reflect or allow for that advancement.

In addition, some important conceptual advances had occurred in the intervening 20 years. One of these is that it is now widely recognized that 'standards' and 'references' are similar but not the same. Standards are now viewed as being *prescriptive* and are based on desirable growth of groups of healthy children living in optimal environments, i.e. disease-free and environmentally ideal. References are *descriptive* and are based on the growth of children living in normal environments in which they experience normal levels of infectious diseases and are not protected from environmental insult, i.e. it is growth 'as is'. The source samples from which the reference charts within TW3 are constructed are not composed of children with optimal growth living in optimal environments. They therefore reflect a process of normal growth and should be called 'references'.

There are four major differences between TW2 and TW3. The most important thing, however, is that the descriptions and manual ratings of the stages of the bones have *not* been altered. They remain the same so that previous ratings and calculations of bone maturity scores in TW2 are still valid for TW3. However, the TW2 (20) bone score has been abolished. This is because it was felt that the mixture of the carpal maturity scores with the RUS maturity scores was not of major value. Skeletal maturity of the carpal bones in isolation is problematical in most situations. They appear to give different information about the process of maturity. The RUS data are certainly more useful both in terms of reflecting general skeletal maturity and in the prediction of adult height. Second, the source samples for the reference charts have been updated so that they now reflect the norms for more recent samples of children from Europe and North America. Thus the conversion to bone age also changes particularly from about 10 years onwards.

The third and fourth changes relate to the height prediction technique rather than the assessment of skeletal maturity per se. RUS bone score is now used rather than bone age in the prediction equations and the source sample has been improved by using more appropriate data from the First Zürich Longitudinal Study.

The new maturity score, now called EA90 (to reflect the European and American sources) or TW3, are based on data from samples of children in Europe, North America and Japan assessed in the 1970s, 1980s and 1990s. These included Belgian data (21 174 boys and 10 000 girls) from the Leuven growth study, Spanish children (2000, with over 5000 radiographs) from the Bilbao study, Japanese children (1000) from Tokyo, Italian children (950 boys and 880 girls) from Genoa, Argentinian data from the early 1970s and data from Project Heartbeat of about 1000 normal European-American children in Texas (for details of these samples see Tanner *et al.*, 2001). The new EA90 bone age values were chosen to match this scoring system but mainly concentrated on the Belgian, Spanish and American samples.

The differences in RUS scores between TW2 and TW3 for both sexes are shown in Figure 5.2. At preadolescent ages the scores for boys vary very little between the systems. After 9 years of age in boys and from about 5 years in girls the differences increase quite dramatically so that, for instance, a boy scoring 405 would have had a TW2 bone age of 13 years and a TW3 bone age

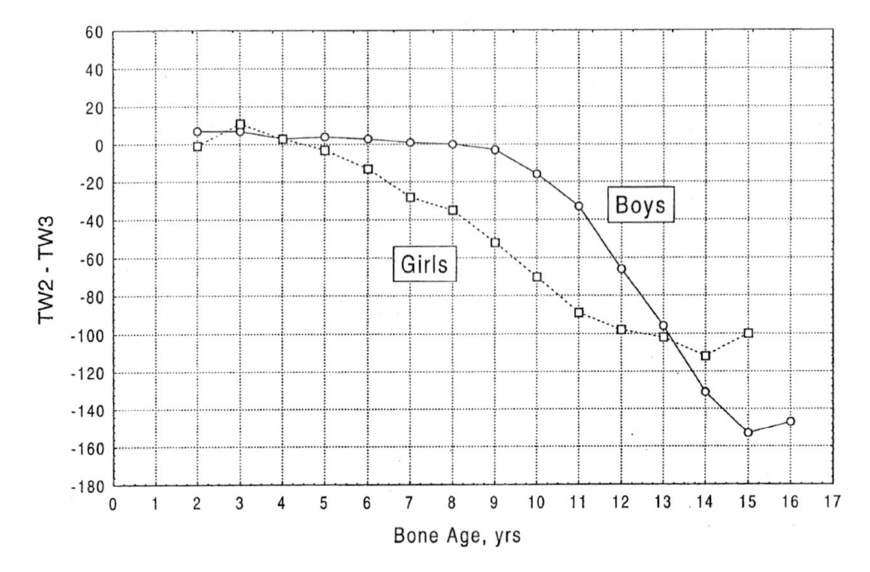

Figure 5.2 Differences between TW2 and TW3 bone maturity scores for whole year chronological age from 2 to 16 years. The TW3 score has been subtracted from the TW2 score thus negative values indicate advancement of TW3 over TW2. For example, an average 12-year-old boy would have bone maturity score of 361 on TW2 and 427 on TW3. The difference is therefore −66 (361 − 427 = −66). Thus in the TW3 sample this level of maturity has been reached at an earlier age than in the TW2 sample and TW3 is advanced in relation to TW2.

of 11.7 years. This difference of 1 year to 18 months is pretty consistent during adolescence reflecting the relative advancement of the EA90 sample compared to the TW2 sample.

The application of the technique is relatively straightforward. Each bone is assessed in a prescribed order; radius, ulna, metacarpals 1,3,5, proximal phalanges 1,3,5, medial phalanges 3,5, distal phalanges 1,3,5. If the carpus is also to be assessed then once again a prescribed order is used; capitate, hamate, triquetral, lunate, scaphoid, trapezium, trapezoid. The assessment involves the judgement of whether a particular bone conforms to descriptive criteria contained within the manual. There is a maximum of three criteria or 'stages' for each appearance of each bone. If only one criterion is described then this must be satisfied for the bone to have reached that stage of maturity. If two criteria are given then only one need be met and if three criteria are given two must be met. However, to proceed from one stage to the next the first criterion of the preceding stage must always be satisfied (e.g. for a rating of stage G to be given, stage F(i) must be met). Each stage has a corresponding score that is gender and method dependent (e.g. boys RUS and boys TW20) and these scores are noted for each bone and summed to produce a bone maturity score (BMS). This BMS is compared to a set of reference BMS for each sex within each maturity system and a bone 'age' is calculated.

The Roche–Wainer–Thissen technique

In 1975 Roche, Wainer and Thissen published a technique to estimate the skeletal maturity of the knee (Roche *et al.*, 1975). Roche in particular was critical of the hand–wrist techniques because the bones of the hand and wrist exhibit few maturational changes over the age ranges of 11 to 15 in boys and 9 to 13.5 in girls (Roche, 1970). In addition, the usefulness of the hand–wrist techniques was limited at early ages when few centres were visible and in later ages when some areas (e.g. the carpus) reach their adult maturity levels prior to others. He chose the knee as an area for assessment because he believed that the area investigated should be closely related to the reason for assessment; maturity of the knee relates closely to growth in height. Thus when one is dealing with growth disorders or height prediction the knee ought to give a more appropriate estimation of skeletal maturity but this may not actually be the case. The method has not been used extensively and Roche was to change his approach in the next decade and with his colleagues Cameron Chumlea and David Thissen he produced a hand–wrist scoring technique in 1988 known as the Fels hand–wrist method (Roche *et al.*, 1988; Chumlea *et al.*, 1989).

The Fels hand–wrist technique

The theoretical basis for the Fels hand–wrist method is little different from that of the earlier Tanner–Whitehouse methods. Roche and his colleagues went through the laborious process of identifying suitable maturity indicators from 13 823 serial radiographs of children from the Fels Longitudinal Growth Study. The radiographs were taken between 1932 and 1972 and thus may appear to be rather dated and susceptible to the problems of secular change. From a possible 130 maturity indicators taken from the literature, 98 were finally selected that conformed to the criteria of universality, discriminative ability, reliability, validity and completeness. In addition to graded indicators Roche and his colleagues also used metric ratios of lengths of radius, ulna, metacarpals and phalanges. Roche *et al.* (1988) maintain that the Fels method differs from previous methods in terms of the observations made, the chronological ages when assessments are possible, the maturity indicators, the statistical methods, and scale of maturity.

In practice the method is applied by identifying predefined maturity indicators appropriate to the sex and chronological age of the child and submitting these grades of each bone (stages) to a computer program (FELShw). The data entry forms reflect the fact that the method can use ratings from the radius, ulna, all carpal bones and, like the TW systems, the phalanges of rays 2, 3 and 5. The output is an 'estimated skeletal age' and an 'estimated standard error' to provide an idea of the confidence of the estimated age. The program does not require that every indicator be graded but obviously the standard error will increase with a reduction in the number of indicators.

Comparability of the atlas and bone-specific methods

The Greulich–Pyle atlas technique, the Fels method and the TW2 scoring systems are the result of attempts to quantify skeletal maturation from different theoretical standpoints. The younger TW2 and Fels systems gained much from studying the disadvantages of the Greulich–Pyle system and attempted to overcome them. There are advantages to each system depending on the context within which they are used, but in order to decide on the most appropriate system the clinician or research worker must have at least a nodding acquaintance with their theoretical bases and indeed their practical use and the reliability to be expected from each system.

In the most recent comprehensive comparison of the atlas and bone-specific techniques Aicardi *et al.* (2000) compared the Greulich–Pyle, Tanner–Whitehouse and Fels methods using a sample of 589 children (250 girls and 339 boys) aged from 2 to 15 years of chronological age. These Italian children from Genoa were admitted to a clinical paediatrics department for investigations of growth, obesity and acute diseases. Whilst the Italian boys were generally delayed in relation to all methods, bone age was closer to chronological age using the Roche *et al.* knee method rather than the hand–wrist methods with an average deviation of −0.11 years. This was followed by the Fels hand–wrist (−0.32 years), Tanner–Whitehouse RUS (−0.35 years) and finally Greulich-Pyle (−0.61 years). Conversely girls were generally advanced with equivalent values between chronological and skeletal age of 0.06 (Roche *et al.*), 0.18 (FELS), 0.23 (Tanner–Whitehouse RUS), and −0.04 (Greulich–Pyle). Naturally such studies are rare because of the generally uncommon situation in which children have both hand–wrist and knee radiographs taken in the course of hospital entry. Aicardi *et al.* (2000) conclude that further comparisons are required to decide on whether hand–wrist or knee methods are most useful in clinical settings and particularly when there is a concern for growth potential.

Reliability

Tanner *et al.* (2001) provide a detailed description of the comparative reliability of the different systems. Experienced raters tend to obtain 95% confidence limits of ±0.5 to ±0.6 'years' based on two standard errors of measurement. Thissen (1989) provides examples that suggest a somewhat greater error when using the Fels hand–wrist technique but, like the reliability of TW methods, that may be because of the chronological age of the specific example chosen; reliability tends to be greater or smaller depending on chronological age because at certain ages small changes in maturity indicators can result in large changes in skeletal age. Thus a miss-rating of one of these indicators will result in an inflated standard error of measurement from a test–retest reliability study.

Population differences

Tanner and his colleagues include a discussion on population differences in skeletal maturity in the most recent TW3 method (Tanner *et al.*, 2001) and good historical data is provided by Eveleth and Tanner's review of growth globally (Eveleth and Tanner, 1976, 1990). The major point is that differences between countries and indeed within countries are to be expected because skeletal maturity, like all aspects of maturation, reflects the interaction of both genetic and environmental forces. Whilst ideally reference values should be developed for each relevant population, in the absence of such developments, it seems reasonable to suggest that the method of choice will depend on the proximity of the child under investigation to the source sample of the particular method, the availability of appropriate radiographs of the hand–wrist and/or knee, and in the case of the Fels hand–wrist method, the availability of the appropriate software. Given the presence of secular changes in the appearance of maturity indicators and therefore in the degree of advancement or delay of the child/sample it would seem sensible to use the most recent methods and those that expose the child to the least radiation dosage. Thus the Tanner–Whitehouse and Fels hand–wrist methods would be preferred over the Greulich–Pyle and the Roche *et al.* knee techniques. The Tanner–Whitehouse technique is most often used within European settings and the Fels hand–wrist technique within North American settings but that situation probably reflects marketing and familiarity rather than scientific considerations.

Secondary sexual development

Secondary sexual development is assessed using maturity indicators that provide discrete stages of development within the continuous process of maturation. The most widely accepted assessment scale is described as the Tanner Scale or the Tanner Staging Technique. It was developed by Tanner (1962) and was based on the work of Reynolds and Wines (1948) and Nicholson and Hanley (1952). Tanner (1962) divided the processes of breast development in girls, genitalia development in boys and pubic hair development in both sexes into five stages and axillary hair development in both sexes into three stages. The usual terminology is to describe breast development in stages B1–B5 (Figure 5.3), genitalia development in stages G1–G5 (Figure 5.4), pubic hair development in stages PH1–PH5 (Figure 5.5) and axillary hair development in stages Al–A3.

Breast development

Stage 1 Preadolescent: elevation of papilla only.
Stage 2 Breast bud stage: elevation of breast and papilla as small mound. Enlargement of areolar diameter.
Stage 3 Further enlargement and elevation of breast and areola, with no separation of their contours.
Stage 4 Projection of areola and papilla to form a secondary mound above the level of the breast.
Stage 5 Mature stage: projection of papilla only, due to recession of the areola to the general contour of the breast.

Genitalia development

Stage 1 Preadolescent: testes, scrotum and penis are of about the same size and proportion as in early childhood.
Stage 2 Enlargement of scrotum and testes: the skin of the scrotum reddens and changes in texture. There is little or no enlargement of penis at this stage.
Stage 3 Enlargement of penis: this occurs first mainly in length. Further growth of testes and scrotum.
Stage 4 Increased size of the penis with growth in breadth and development of glans. Further enlargement of testes and scrotum; increased darkening of scrotal skin.
Stage 5 Genitalia adult in size and shape.

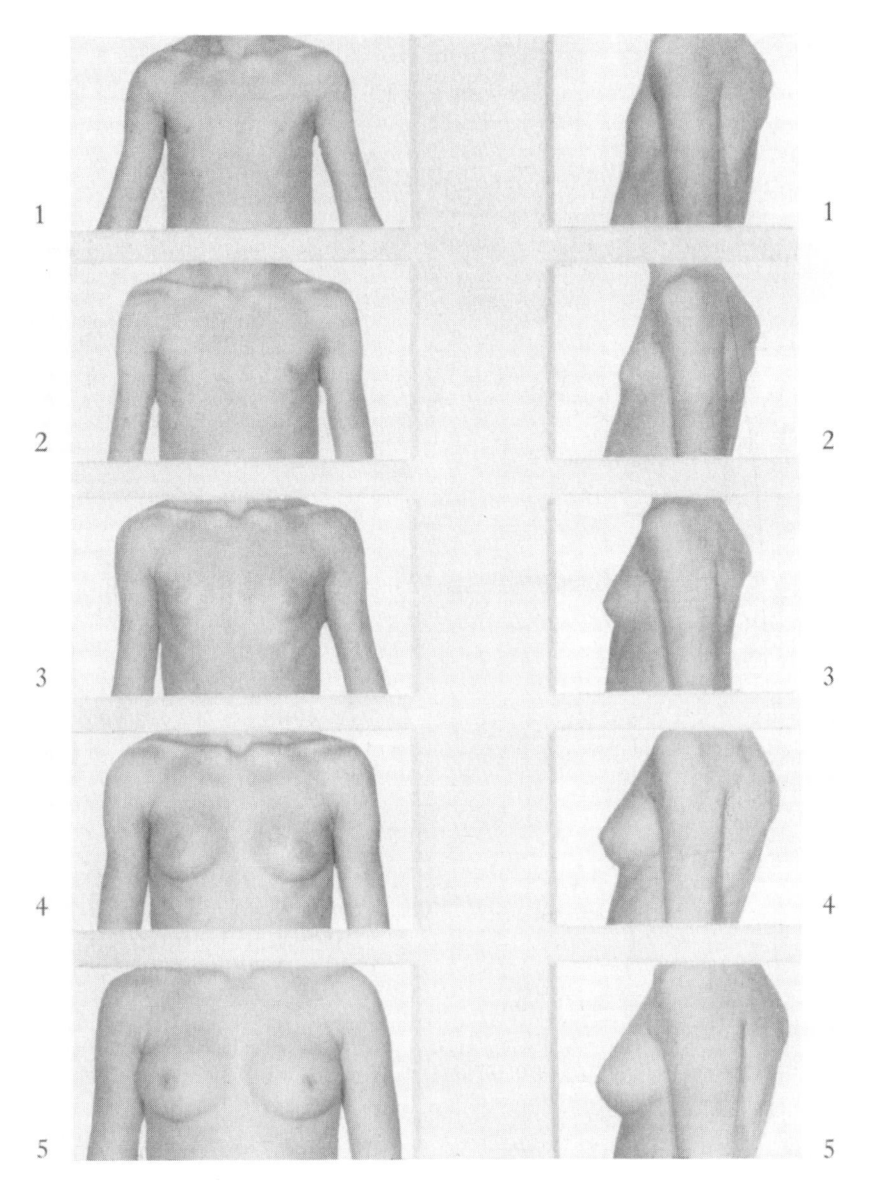

Figure 5.3 Breast development in girls according to the Tanner staging technique. (From Tanner, J. M. (1962). *Growth at Adolescence*, 2nd edn. Oxford, UK: Blackwell Scientific Publications.)

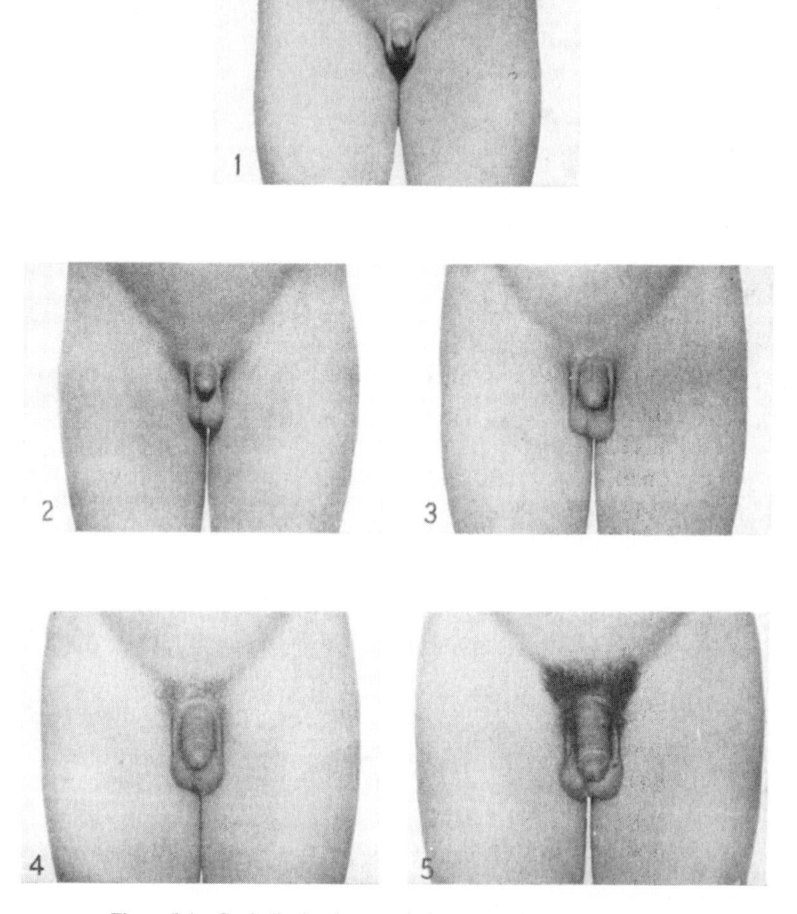

Figure 5.4 Genitalia development in boys according to the Tanner staging technique. (From Tanner, J. M. (1962). *Growth at Adolescence*, 2nd edn. Oxford, UK: Blackwell Scientific Publications.)

Pubic hair development

Stage 1 Preadolescent: the vellus over the pubes is not further developed than that over the abdominal wall, i.e. no pubic hair.

Stage 2 Sparse growth of long, slightly pigmented downy hair, straight or only slightly curled, appearing chiefly at the base of the penis or along the labia.

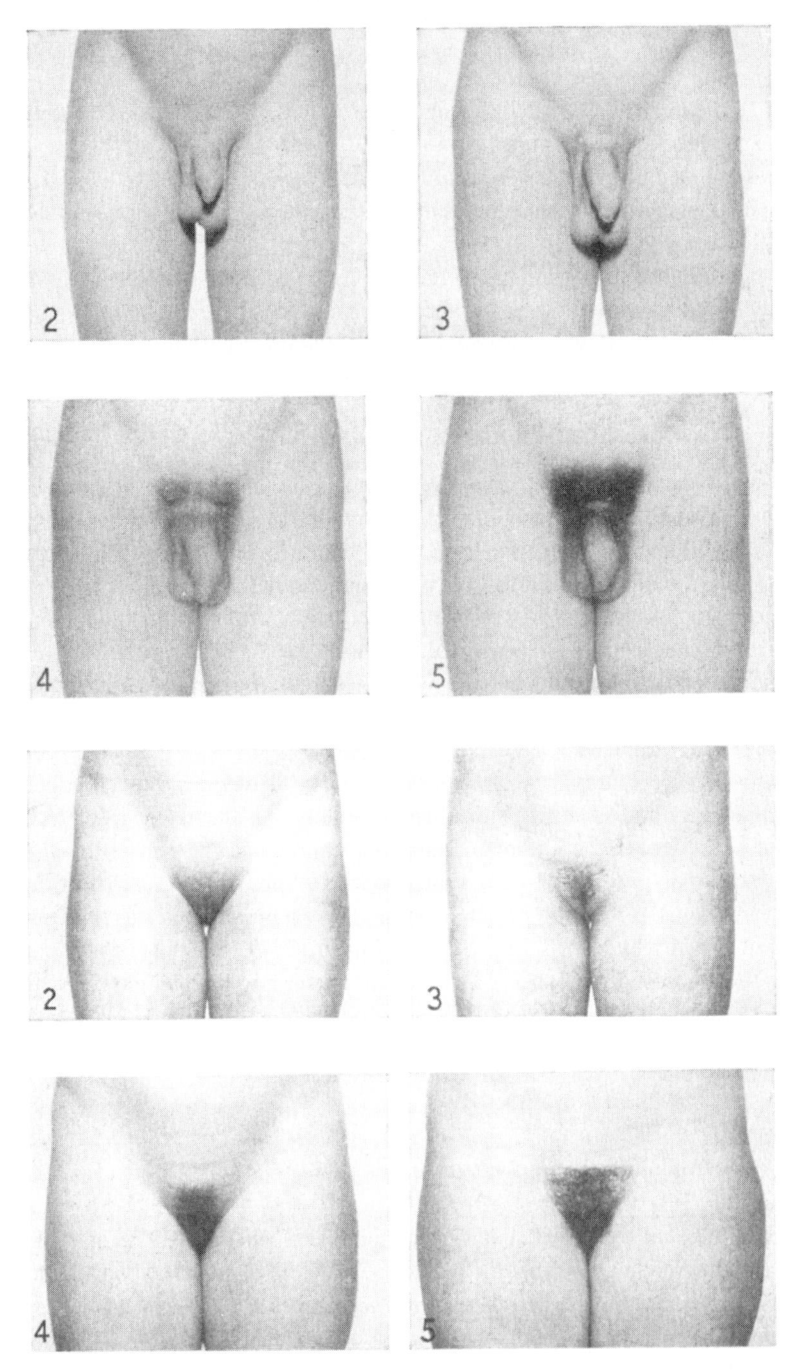

Figure 5.5 Pubic hair development in boys and girls according to the Tanner staging technique. (From Tanner, J. M. (1962). *Growth at Adolescence*, 2nd edn. Oxford, UK: Blackwell Scientific Publications.)

Stage 3 Considerably darker, coarser and more curled. The hair spreads sparsely over the junction of the pubes.
Stage 4 Hair now resembles adult in type, but the area covered by it is still considerably smaller than in the adult. No spread to the medial surface of the thighs.
Stage 5 Adult in quantity and type with distribution of the horizontal or classically feminine pattern. Spread to the medial surface of the thighs but not up the linea alba or elsewhere above the base of the inverse triangle.

Clinical evaluations

The assessment of secondary sexual development is a standard clinical procedure and at such times the full Tanner scale is used. There are some practical problems with the Tanner stages, however, in that the unequivocal observation of each stage is often dependent on having longitudinal observations. In most situations, outside the clinical setting, the observations are cross-sectional. This practical difficulty has led to the amalgamation of some of the stages to create pubertal stages. These pubertal stages are either on a three- or four-point scale and combine breast or genitalia development with pubic hair development (Kulin *et al.*, 1982; Chaning-Pearce and Solomon, 1986). Assessing breast or genitalia development with pubic hair development is obviously much easier than assessing these maturity indicators separately, but inevitably leads to a lack of sensitivity in the interpretation of the timing and duration of the different stages of pubertal development. Indeed the intra-subject variation in the synchronous appearance of pubic hair and breast or genitalia stages, illustrated in British children by Marshall and Tanner (1969, 1970), suggests that it may be misleading to expect stage synchronization in as many as 50% of normal children. Thus the combined assessment is not recommended unless an extremely general measure of maturity is required within the research design.

Self-assessment of pubertal status

The assessment of secondary sexual characteristics is, to some extent, an invasive procedure in that it invades the privacy of the child or adolescent involved. Thus such assessments on normal children who participate in growth studies, as opposed to those being clinically assessed, are problematical from both ethical and subject compliance viewpoints. In order to overcome this problem the procedure of self-assessment has been developed and validated in a number of studies.

The self-assessment procedure requires the child to enter a well-lit cubicle or other area of privacy in which are provided pictorial representations of the Tanner scales and suitably positioned mirrors on the wall(s). The pictures may be either in photographic or line-drawing styles as long as the contents are clear. To each picture of each stage is appended an explanation, in the language of the participant, of what the stage represents. The participant is instructed to remove whatever clothing is necessary in order for them to be able to properly observe their pubic hair and genitalia or pubic hair and breast development in the mirrors. The participant then marks on a separate sheet their stage of development and seals that sheet within an envelope on which is marked the study identity number of the participant. The envelope is either left in the cubicle or handed to the observer on leaving the cubicle.

The results of validation studies vary greatly depending upon the age of the participants (e.g. early or late adolescence) (Varona-Lopez *et al.*, 1988), gender (Varona-Lopez *et al.*, 1988; Sarni *et al.*, 1993), the setting in which assessments are performed (e.g. school or clinic) (Schlossberger *et al.*, 1992; F. C. Wu *et al.*, 1993; W. H. Wu *et al.*, 1993), ethnicity (Hergenroeder *et al.*, 1999), and whether they are a distinct diagnostic group such as those with cystic fibrosis (Boas *et al.*, 1995) or anorexia nervosa (Hick and Kutzman, 1999) or the socially disadvantaged (Hardoff and Tamir, 1993). Younger less-developed children tend to overestimate their development and older more developed children tend to underestimate. Boys have been found to overestimate their development whilst girls have been more consistent with experts (Sarni *et al.*, 1993). The amount of attention given to explaining the required procedure appears to be of major importance. Thus excellent rating agreement between physicians and adolescents has been found in clinical settings, with κ coefficients between 0.66 and 0.91 (Duke *et al.*, 1980; Brooks-Gunn *et al.*, 1987; Schlossberger *et al.*, 1992) but rather less agreement in school settings ($\kappa = 0.35$–0.42; correlations = 0.25–0.52) (Schlossberger *et al.*, 1992; W. H. Wu *et al.*, 1993). Improved agreement in clinical settings probably reflects the more controlled environment of a physician's surgery as opposed to a school. The main reason for low correlations and thus poor validity in any setting with any group of participants is likely to be centred around the amount of explanation that is provided to the child. When the participant has been the subject of a clinical trial, and the scientist or clinician has spent considerable time and effort ensuring that the child is completely appraised of what he/she has to do, then validity is high. Less effort in explaining procedures leads to lower validity.

The procedure that should be adopted is that the observer should explain the procedure thoroughly to the participant using appropriate (non-scientific) language and invite questions to ensure that the participant fully understands the procedure. Only when the observer is sure that understanding is total should the child be allowed to follow the procedure. Randomized reliability assessments by the observer would, of course, be ideal but would also be ethically difficult to substantiate.

A simplified pubertal questionnaire

The difficulties inherent in accurately grading pubertal development according to Tanner's criteria make the search for a reliable but simplified method (i.e. with obvious maturity indicators) desirable. Because the process of secondary sexual development proceeds in a sequential pattern punctuated by specific and obvious events it should be possible to identify children within general pubertal categories (prepuberty, early puberty, etc.) based on the timing and sequence of the events. A simple questionnaire ought to elicit sufficient information to divide adolescents into four divisions: prepubertal, early puberty, late puberty and adult as opposed to Tanner's five stages.

The questions for girls are:

1. Have you started puberty, i.e. do you have any pubic hair or have your breasts enlarged since you were a child?
2. Do you have regular menstrual cycles, i.e. periods?
3. If so, have you been having periods for *more than* two years?
4. How old were you when you started to have periods?

The questions for boys are:

1. Have you started puberty, i.e. do you have any pubic hair or have your genitalia enlarged since you were a child?
2. Has your voice broken, i.e. do you speak in a deeper voice then when you were a child?
3. If so, have you been speaking in a deeper voice for *more than* two years?
4. How old were you when your voice broke?
5. Do you shave?
6. How often do you shave, e.g. per week/month?
7. When did you start shaving?

Question 1 will identify the prepubertal children in both sexes. An affirmative response to question 2 will identify the girls and boys who are at least in the latter part of pubertal development. In the absence of a sensible response from boys it is possible for the observer to note whether he/she believes the voice to have broken. Unpublished observations within my own research programme suggest that this is possible on almost all occasions particularly if the observer and recorder concur. Question 3 will allow the identification of those who are to all intents and purposes adult, i.e. mature as opposed to being in late puberty. If they are more than two years postmenarche or post-voice-breaking then according to the normal sequence and timing of pubertal development they should be fully mature. Question 4 is an optional question but if answered accurately, will allow the calculation of actual age at menarche or voice-breaking as a useful individual or sample statistic. The addition of questions on the appearance of facial hair and the initiation of shaving for boys attempts to use this potentially useful indicator of maturity. It is appreciated that the initiation of regular shaving can be an event determined by culture and religion but where this is the case observation of the unshaven face and identification of the stages described above from Billewicz *et al.* (1981) could be usefully employed.

The language of the questions needs to be adjusted to allow for local terminology that will be better understood by the children. The advantages of this procedure are that it does not require the child to inspect itself, it mostly requires simple yes/no responses, and it requires a minimum of thought by the subject, i.e. memory is not really an issue. Of course, for clinical purposes in which the rate of progression through puberty and the duration of each stage are important the Tanner scale is vital. But for a simple indication of whether puberty has started and how far it has progressed the self-assessment questions detailed above should suffice. In addition they are far less invasive than either an inspection by a third party or a full self-assessment in a private cubicle.

Age at menarche

Age at menarche is usually obtained in one of three ways; status quo, retrospectively, or prospectively. Status quo techniques require the girls to respond to the question, ‘Do you have menstrual cycles (“periods”)?’ The resulting data on a sample of girls will produce a classical dose response sigmoid curve that may be used to graphically define an average age at menarche. More commonly the data are analysed using logit or probit analysis to determine the mean or median age at menarche and the parameters of the distribution such as the standard error of the mean or the standard deviation. Retrospective techniques require the participants to respond to the question, ‘When did you have your first period?’ Most adolescents can remember to within a month, and some to the day, when this event occurred. Others may be prompted to remember by reference to whether the event occurred during summer or winter, whether the girl was at school or on holiday, and so on. One interesting result of such retrospective analyses is that there appears to be a negative association between the age of the women being asked and the age at which they report menarche – the older the women the younger they believe they were. Such results have been found in both developed and developing countries and cast a seed of doubt about the reliability of retrospective methods beyond the teenage and early adult years. Prospective methods are normally only used in longitudinal monitoring situations such as repeated clinic visits or longitudinal research studies. This method requires the teenager to be seen at regular intervals (usually every three months) and to be asked on each occasion whether or not she has started her periods. As soon as the response is positive an actual date on which menarche occurred can be easily obtained.

There is little doubt that the prospective method is the most accurate in estimating menarcheal age but it has the disadvantage of requiring repeated contact with the subjects. That is seldom possible except in clinical situations and it is thus more likely that status quo and retrospective methods are the technique of choice. Status quo techniques that rely on logit or probit analysis require large sample sizes because the analysis requires the data to be grouped according to age classes. With few subjects broader age ranges are required such as whole or half years with a consequent loss of precision in the mean or median value. Retrospective methods result in parametric descriptive statistics but have the problem of the accuracy of recalled ages at this particular event.

Secondary sexual events in boys

While status quo, prospective and retrospective methods may easily obtain age at menarche, assessments of secondary sexual development in boys are complicated by the lack of a similar clearly discernible maturational event. Attempts to obtain information on the age at which the voice breaks, or on spermarche, are complicated by the time taken for the voice to be consistently in a lower register, and the logistical complications involved in the assessment of spermarche (see below). Testicular volume, using the Prader orchidometer (Prader, 1966), is commonly the only measure of male secondary sexual development outside the rating scales previously mentioned, though other measurement techniques have been described to estimate testicular volume (Daniel *et al.*, 1982).

The detection of spermatozoa in the urine has been proposed as a quick, non-invasive method to assess the functional state of the maturing gonad and may be useful as a screening technique in population studies (Baldwin, 1928; Richardson and Short, 1978; Hirsch *et al.*, 1979; Nielson *et al.*, 1986; Kulin *et al.*, 1989; Schaefer *et al.*, 1990). Its use, however, may be limited because longitudinal (Hirsch *et al.*, 1985; Nielson *et al.*, 1986) and cross-sectional (Hirsch *et al.*, 1979) studies have shown that spermaturia is a discontinuous phenomenon.

There have been some attempts to record the timing of first ejaculation, or 'spermarche', as a maturational event. The problems in doing so may be illustrated by the example provided by Buga *et al.* (1996) in which they asked a sample of African boys from the Transkei, South Africa the question, 'How old were you when you had your first wet dream?' Clearly the variety of answers to such a question, which would cover the spectrum of 'What is a wet dream?' to 'Last night!', testify to its dubious usefulness. Indeed the authors maintained that their estimate of mean age at spermarche 'was crude, and relied on recall of a fairly nebulous isolated event that in most cases was difficult to recall precisely'. (G. A. B. Buga, pers. comm.). The one prerequisite of a maturity indicator that was not considered at the start of this chapter is that it must be identifiable, i.e. one must have some unequivocal measure that it is present or has occurred and *when* it has occurred. Clearly this is not the case with a potential indicator such as first ejaculation.

Dental development

Dental development is best assessed by taking panoramic radiographs of the mandible and maxilla and scoring the stages of formation and calcification of each tooth using the method developed by Demirjian *et al.* (1973) and later modified by Demirjian and Goldstein (1976). Scores are assigned to the stages of development of the seven mandibular teeth on the left side (there are no significant between-side differences) and these lead to a dental maturity score comparable to the skeletal maturity scores resulting from the Tanner–Whitehouse (TW) skeletal maturity technique described below. This score can be translated into the dental age. A similar system is available for sets of four teeth, seen on apical radiographs, notably M1, M2, PM1, PM2 or alternatively I1, M2, PM1, PM2. The advantage of using this system, like using the TW bone-specific scoring system, is that if the scores can be obtained on a sufficiently large local or national sample then local or national reference values can be developed that reflect more closely the rate of maturation within countries (see for example Willems *et al.*, 2001). Liversidge (2003) cites numerous studies from France, Italy, Germany, China and Iran that have demonstrated advancement in dental maturation compared to the original Canadian reference values of Demirjian and numerous methods of adapting the method to adjust the total score conversion tables (e.g. Nyström *et al.*, 1986, 1988; Kataja *et al.*, 1989) or recalculate by formulae (Frucht *et al.*, 2000; Teivens and Mornstad, 2001).

Concern over the exposure of normal children to radiation has resulted in tooth emergence as the most commonly used method to obtain estimates of dental maturity. The emergence of the teeth above the level of the gum is recorded either by oral inspection or in a dental impression. Most observers have considered that a tooth has emerged if any part has pierced the gum, but some have used the criterion of the tooth being halfway between gum and final position (Eveleth and Tanner, 1990). Three types of standards have been developed that give either the number of teeth emerged at specific ages, or the average age when one, two, three, etc. teeth have emerged, or the median age in a population for the emergence of a specific tooth or pair of teeth. The latter technique is considered the best for permanent teeth because of the individual variation in the order of emergence of each tooth pair. Barrett *et al.* (1981) provide an acceptable review of the tooth emergence procedures but warn that:

> Dental 'age' can properly be estimated only by reference to standards established on the population to which the subject belongs, in similar environmental conditions, and a short time previously. If no such standard exists, the dental 'age' assignment by reference to foreign standards (for example American or British) can hardly serve to give more than a rough indication of chronological age when this is unknown.

In other words it is important to appreciate that age of emergence of teeth is strongly environmentally determined within any particular population and one must be sensitive to local factors. Whilst there are considerable differences between populations as to the exact ages of eruption of the permanent dentition, Figure 5.6 provides an example of the mean ages of eruption of the secondary teeth for British children (Sinclair, 1989).

Liversidge (2003) highlights the fact that several factors need to be borne in mind when making comparisons of tooth formation data between studies, populations and tooth types. First, the timing of clinical eruption of teeth varies more than the timing of crown and root growth stages, and second, both eruption and formation of teeth take place on a time-related gradient. The age variability for the eruption of deciduous and early permanent teeth is considerably smaller than later emerging teeth and deciduous tooth formation varies less than permanent tooth formation (Moorrees *et al.*, 1963a,b). Liversidge (2003) adds:

> It stands to reason that population differences in tooth formation, and eruption are small in early childhood, become more evident during the mixed dentition, and are probably greatest during root formation of the later forming teeth. Meaningful comparisons are not possible if the age range and sample, or the methodology and analyses are inappropriate or unsound.

Whilst differences exist in both tooth eruption and formation between populations these are usually small (<1 SD) and there appears to be no pattern between populations from geographical areas to the differences in average timing of eruption of teeth. The implication of these findings is clearly that environmental and

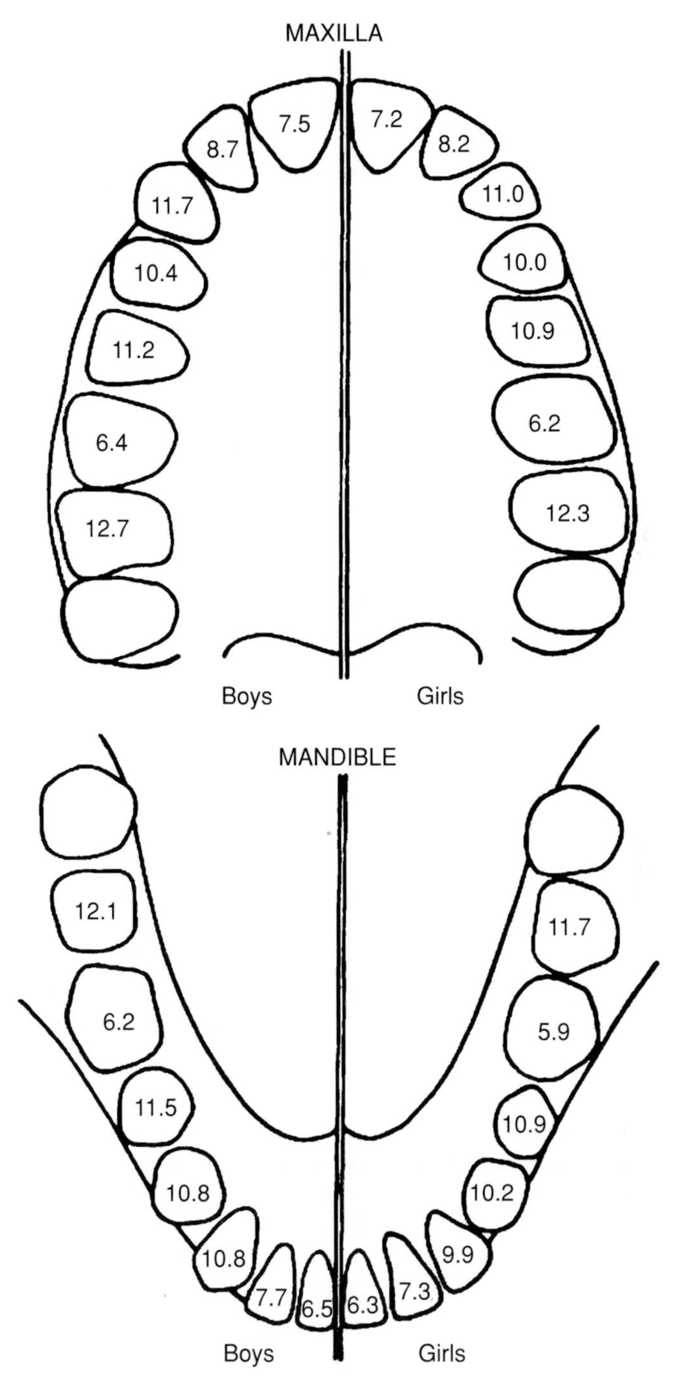

Figure 5.6 Mean times of the eruption of the permanent teeth. (The times of the appearance of the third molar ('wisdom') teeth are so variable that they have been omitted. (From Sinclair, D. (1989). *Human Growth after Birth.* Oxford, UK: Oxford Medical Publications.)

nutritional factors have less of an impact on dental development than on, say, skeletal maturity or secondary sexual development and it is possible, with care, to make valid comparisons between populations for dental maturity.

Landmarks on the growth curve

Many of the chapters in this book deal with the way in which mathematical functions can be used to model the pattern of human growth and to arrive at a predicted growth curve that accurately models an individual child's growth. It is possible to use landmarks on such curves, such as age at take-off, peak height velocity, and at the cessation of growth, and the magnitude of height or weight velocity at these ages, as maturational events. Whilst the reader is encouraged to consult these specific chapters in detail it is worth making a few comments on such procedures.

Initial curve-fitting techniques used only part of the growth curve (e.g. from birth to the start of adolescence) or involved the addition of different functions. The major problem with these early techniques, apart from their mathematical complexity and biological interpretation, was their relative inability to cover the transition between developmental periods such as preadolescence to adolescence. This was solved to a certain extent by the development of single curves that described growth from birth through to adulthood (Bock *et al.*, 1973; Preece and Baines, 1978). However, long-term parametric models have the disadvantage that the researchers preselect the shape of the resulting growth curve. The choice of the model necessitates the acceptance of its form as being representative of the pattern of growth (see also Chapter 8). Individuals or samples departing from the standard pattern of growth in height or weight would not be fitted well by any of these parametric functions. Estimates of landmarks on the growth curve are difficult to determine. Tanner and Davies (1985), for instance, when developing the clinical longitudinal standards for American children, relied on empirically derived values for the magnitude of peak velocity because 'parametric curves are insufficiently flexible to accommodate the full rise of the observed curves' during adolescence (Tanner and Davies, 1985: p. 328). The widely used Preece–Baines curve (Preece and Baines, 1978), for example, is known to underestimate the peak velocity.

Non-parametric models, such as the smoothing spline function (Largo *et al.*, 1978) and kernel estimation (Gasser *et al.*, 1984, 1985), have been proposed to overcome the problems inherent in preselection of a pattern of growth. Non-parametric techniques are usually short-term functions that smooth adjacent data points rather than fit a function to data from birth to adulthood. They have been useful in demonstrating the sensitivity of growth analysis using acceleration (Gasser *et al.*, 1984, 1985) and can result in mathematically derived values for adolescent landmarks such as the age and magnitude of peak velocity. Such landmarks, if taken from curves that have been smoothed using non-parametric techniques, may be more accurately determined than if derived from a parametric function. However, that accuracy is dependent on the frequency of data points during the period of adolescence. The inability of preselected parametric functions to fit abnormal growth and the retrospective nature of growth assessment of non-parametric methods make these techniques useful as research tools but not for diagnosis and monitoring the value of treatment. They are not able to predict, say, the magnitude and timing of catch-up growth following growth hormone treatment because they do not allow for such a dramatic a shift in the pattern of growth.

REFERENCES

Acheson, R. M. (1954). A method of assessing skeletal maturity from radiographs. *Journal of Anatomy (London)*, **88**, 498–508.

(1957). The Oxford method of assessing skeletal maturity. *Clinical Orthopedics*, **10**, 19–39.

Aicardi, G., Vignolo, M., Milani, S., *et al.* (2000). Assessment of skeletal maturity of the hand–wrist and knee: a comparison of methods. *American Journal of Human Biology*, **12**, 610–15.

Baldwin, B. (1928). The determination of sex maturation in boys by a laboratory method. *Journal of Comparative Psychology*, **8**, 39–43.

Barrett, M. J., Brabant, H., Brothwell, D. R., Dahlberg, A. L. and Garn, S. M. (1981). Dental and oral examination. In *Practical Human Biology*, eds. J. S. Weiner and J. A. Lourie, pp. 159–71. London: Academic Press.

Billewicz, W. Z., Fellowes, H. M. and Thomson, A. M. (1981). Pubertal changes in boys and girls in Newcastle-upon-Tyne. *Annals of Human Biology*, **8**, 211–19.

Boas, S. R., Falsetti, D., Murphy, T. D. and Orenstein, D. M. (1995). Validity of self-assessment of sexual maturation in adolescent male patients with cystic fibrosis. *Journal of Adolescent Health,* **17**, 42–5.

Bock, R. D., Wainer, H., Peterson, A., Thissen, L. M. and Roche, A. F. (1973). A parameterization for individual human growth curves. *Human Biology*, **45**, 63–80.

Bogin, B. (1999). *Patterns of Human Growth*, 2nd edn. Cambridge, UK: Cambridge University Press.

Brooks-Gunn, J., Warren, M. P., Russo, J. and Gargiulo, J. (1987). Validity of self-report measures of girls' pubertal status. *Child Development*, **58**, 829–41.

Buga, G. A. B., Amoko, D. H. A. and Ncayiyana, D. J. (1996). Sexual behaviour, contraceptive practice and reproductive health among school adolescents in rural Transkei. *South African Medical Journal,* **86**, 523–7.

Chaning-Pearce, S. M. and Solomon, L. (1986). A longitudinal study of height and weight in black and white Johannesburg children. *South African Medical Journal*, **70**, 743–6.

Chumlea, W. C., Roche, A. F. and Thissen, A. F. (1989). The Fels method for assessing the skeletal maturity of the hand–wrist. *American Journal of Human Biology*, **1**, 175–83.

Daniel, W. A., Feinstein, R. A., Howard-Pebbles, P. and Baxley, W. D. (1982). Testicular volume of adolescents. *Journal of Pediatrics*, **101**, 1010–12.

Demirjian, A. and Goldstein, H. (1976). New systems for dental maturity based on seven and four teeth. *Annals of Human Biology*, **3**, 411–21.

Demirjian, A., Goldstein, H. and Tanner, J. M. (1973). A new system of dental age assessment. *Human Biology*, **45**, 211–27.

Duke, P. M., Litt, I. F. and Gross, R. T. (1980). Adolescents' self-assessment of sexual maturation. *Pediatrics*, **66**, 918–20.

Eveleth, P. B. and Tanner, J. M. (1976). *Worldwide Variations in Human Growth.* Cambridge, UK: Cambridge University Press.

(1990). *Worldwide Variations in Human Growth*, 2nd edn. Cambridge, UK: Cambridge University Press.

Frucht, S., Schnegelsberg, C., Schulte-Monting, J., Rose, E. and Jonas, I. (2000). Dental age in southwest Germany: a radiographic study. *Journal of Orofacial Orthopedics*, **61**, 318–29.

Gasser, T., Kohler, W., Müller, H. G., *et al.* (1984). Velocity and acceleration of height growth using kernel estimation. *Annals of Human Biology,* **11**, 397–411.

Gasser, T., Kohler, W., Müller, H. G., *et al.* (1985). Human height growth: correlational and multivariate structure of velocity and acceleration. *Annals of Human Biology*, **12**, 501–15.

Greulich, W. W. and Pyle, S. I. (1950). *Radiographic Atlas of the Skeletal Development of the Hand and Wrist.* Palo Alto, CA: Stanford University Press.

(1959). *Radiographic Atlas of the Skeletal Development of the Hand and Wrist*, 2nd edn. Palo Alto, CA: Stanford University Press.

Hardoff, D. and Tamir, A. (1993). Self-assessment of pubertal maturation in socially disadvantaged learning-disabled adolescents. *Journal of Adolescent Health*, **14**, 398–400.

Healy, M. J. R. and Goldstein, H. (1976). An approach to the scaling of categorized attributes. *Biometrika*, **63**, 219–29.

Hergenroeder, A. C., Hill, R. B., Wong, W. W., Sangi-Haghpeykar, H. and Taylor, W. (1999). Validity of self-assessment of pubertal maturation in African American and European American adolescents. *Journal of Adolescent Health*, **24**, 201–5.

Hick, K. M. and Kutzman, D. K. (1999). Self-assessment of sexual maturation in adolescent females with anorexia nervosa. *Journal of Adolescent Health*, **24**, 206–11.

Hirsch, M., Shemesh, J. and Modan, M. (1979). Emission of spermatozoa: age of onset. *International Journal of Andrology*, **2**, 289–98.

Hirsch, M., Lunenfeld, B., Modan, M., Oradia, J. and Semesh, J. (1985). Spermarche: the age of onset of sperm emission. *Journal of Adolescent Health Care*, **6**, 35–9.

Kataja, M., Nyström, M. and Aine, L. (1989). Dental maturity standards in southern Finland. *Proceedings of the Finnish Dental Society*, **85**, 187–97.

Kulin, B. E., Bwibo, N., Mutie, D. and Santner, S. J. (1982). The effect of chronic childhood malnutrition on pubertal growth and development. *American Journal of Clinical Nutrition*, **36**, 527–36.

Kulin, H. E., Frontera, M. E., Demers, L. D., Bartholomew, M. J. and Lloyd, T. A. (1989). The onset of sperm production in pubertal boys. *American Journal of Diseases in Children*, **143**, 190–3.

Largo, R. H., Gasser, T. H., Prader, A., Stuetzle, W. and Huber, P. J. (1978). Analysis of the human growth spurt using smoothing spline functions. *Annals of Human Biology*, **5**, 421–34.

Liversidge, H. (2003). Worldwide variation in human dental development. In *Growth and Development in the Genus* Homo, eds. J. L. Thompson, A. Nelson and G. Krovitz, pp. 53–88. Cambridge, UK: Cambridge University Press. (In press.)

Marshall, W. A. and Tanner, J. M. (1969). Variations in the pattern of pubertal changes in girls. *Archives of Diseases in Childhood*, **44**, 291–303.

(1970). Variations in the pattern of pubertal changes in boys. *Archives of Diseases in Childhood,* **45**, 13–23.

Moorrees, C. F. A., Fanning, E. A. and Hunt, E. E. (1963a). Formation and resorption of three deciduous teeth in children. *American Journal of Physical Anthropology*, **21**, 99–108.

(1963b). Age variation of formation stages for ten permanent teeth. *Journal of Dental Research*, **42**, 1490–502.

Nicholson, A. B. and Hanley, C. (1952). Indices of physiological maturity. *Child Development*, **24**, 3–38.

Nielson, C. T., Skakkebaek, N. S. and Richardson, D. W. (1986). Onset of the release of spermatozoa (spermarche) in boys in relation to age, testicular growth, pubic hair, and height. *Journal of Clinical Endocrinology and Metabolism*, **62**, 532–5.

Nyström, M., Haataja, J., Kataja, M., *et al.* (1986). Dental maturity in Finnish children, estimated from the development of seven permanent mandibular teeth. *Acta Odontologica Scandinavica*, **44**, 193–8.

Nyström, M., Ranta, R., Kataja, M. and Silvola, H. (1988). Comparisons of dental maturity between the rural community of Kuhmo in northeastern Finland and the city of Helsinki. *Community Dentistry and Oral Epidemiology*, **16**, 215–17.

Prader, A. (1966). Testicular size: assessment and clinical importance. *Triangle,* **7**, 240–3.

Preece, M. A. and Baines, M. I. (1978). A new family of mathematical models describing the human growth curve. *Annals of Human Biology*, **5**, 1–24.

Pyle, S. I. and Hoerr, N. L. (1955). *A Radiographic Standard of Reference for the Growing Knee*, 2nd edn. Springfield, IL.: C. C. Thomas.

Reynolds, E. L. and Wines, J. V. (1948). Physical changes associated with adolescence in boys. *American Journal of Diseases in Children*, **75**, 329–50.

Richardson, D. and Short, R. (1978). Time of onset of sperm production in boys. *Journal of Biosocial Science*, **5**, 15–25.

Roche, A. F. (1970). Associations between rates of maturation of the bones of the hand–wrist. *American Journal of Physical Anthropology*, **33**, 341–8.

Roche, A. F., Wainer, H. and Thissen, D. (1975). *Skeletal Maturity: The Knee Joint as a Biological Indicator*. New York: Plenum Press.

Roche, A. F., Chumlea, W. C. and Thissen, D. (1988). *Assessing the Skeletal Maturity of the Hand–Wrist: Fels Method*. Springfield, IL: C. C. Thomas.

Sarni P., de Toni, T. and Gastaldi, R. (1993). Validity of self-assessment of pubertal maturation in early adolescents. *Minerva Paediatric*, **45**, 397–400.

Schaefer, F., Marr, J., Seidel, C., Tilden, W. and Scherer, K. (1990). Assessment of gonadal maturation by evaluation of spermaturia. *Archives of Diseases in Childhood*, **65**, 1205–7.

Schlossberger, N. M., Turner, R. A. and Irwin C. E., Jr (1992). Validity of self-report of pubertal maturation in early adolescents. *Journal of Adolescent Health*, **13**, 109–13.

Sinclair, D. (1989). *Human Growth after Birth*, 5th edn. Oxford, UK: Oxford Medical Publications.

Tanner, J. M. (1962). *Growth at Adolescence*, 2nd edn. Oxford, UK: Blackwell Scientific Publications.

Tanner, J. M. and Davies, P. S. W. (1985). Clinical longitudinal standards for height and height velocity for North American children. *Journal of Pediatrics*, **107**, 317–29.

Tanner, J. M., Whitehouse, R. H. and Healy, M. J. R. (1962). *A New System for Estimating the Maturity of the Hand and Wrist, with Standards Derived from 2600 Healthy British Children*, Part II, *The Scoring System*. Paris: International Children's Centre.

Tanner, J. M., Whitehouse, R. H., Marshall, W. A., Healy, M. J. R. and Goldstein, H. (1975). *Assessment of Skeletal Maturity and Prediction of Adult Height*. London: Academic Press.

Tanner, J. M., Whitehouse, R. H., Cameron, N., *et al.* (1983). *Assessment of Skeletal Maturity and Prediction of Adult Height (TW2 Method)*, 2nd edn. London: Academic Press.

Tanner, J. M., Healy, M. J. R., Goldstein, H. and Cameron, N. (2001). *Assessment of Skeletal Maturity and Prediction of Adult Height (TW3 Method)*, 3rd edn. London: Academic Press.

Teivens, A. and Mornstad, H. (2001). A modification of the Demirjian method for age estimation in children. *Journal of Forensic Odontostomatology*, **19**, 26–30.

Thissen, D. (1989). Statistical estimation of skeletal maturity. *American Journal of Human Biology*, **1**, 185–92.

Todd, T. W. (1937). *Atlas of Skeletal Maturation*, Part I, *The Hand*. St Louis, MO: C. V. Mosby.

Varona-Lopez, W., Guillemot, M., Spyckerelle, Y. and Deschamps, J. P. (1988). Self-assessment of the stages of sex maturation in male adolescents. *Pédiatrie*, **43**, 245–9.

Willems, G., Van Omen, A., Spouses, B. and Carrels, C. (2001). Dental age estimation in Belgian children: Demirjian technique revisited. *Journal of Forensic Science*, **46**, 893–5.

Wu, F. C., Brown, D. C., Butler, G. E., Stirling, H. F. and Kelnar, C. J. (1993). Early morning plasma testosterone is an accurate predictor of imminent pubertal development in pre-pubertal boys. *Journal of Clinical Endocrinology and Metabolism*, **76**, 26–31.

Wu, W. H., Lee, C. H. and Wu, C. L. (1993). Self-assessment and physician's assessment of sexual maturation in adolescents in Taipei. *Chung Hua Min Kuo Hsiao Erh Ko I Hsueh Hui Tsa Chih*, **34**, 125–31.

6 *Measuring body composition*

Babette Zemel
University of Pennsylvania School of Medicine

and

Elizabeth Barden
Massachusetts Department of Public Health

Introduction

The field of body composition research is a multidisciplinary effort that serves multiple goals. It is very technology oriented, and a great deal of effort is expended on discussion of the techniques, and their validation and cross-calibration. For studies involving infants, children and adolescents, this is of particular concern, because techniques that are of proven accuracy in adults may be fraught with error in children due to changes in the size and composition of body compartments during development. Consequently, standard approaches used in adults may not be applicable or safe for use with children. This review is targeted at techniques for assessing body composition, and includes a disproportionate discussion of those techniques most feasible and practical for use with children. Considerations of the methodological assumptions and of the analysis and presentation of body composition results for children also are reviewed.

Basic concepts

Body composition assessment involves quantification of the amount and relative proportions of fat, muscle and bone, and their chemical components. Significant changes in body composition occur during growth and development, especially during infancy and puberty. Thus, body composition assessment in children is far more challenging than in adults, and serves a variety of purposes. First, it provides a better understanding of growth processes by describing changes

Methods in Human Growth Research, eds. R. C. Hauspie, N. Cameron and L. Molinari.
Published by Cambridge University Press.

in the size of body compartments and the chemical composition of the body. Infants and children are not miniature adults, and knowledge of their unique differences in body composition is key to understanding the biology of the human species. Second, determination of nutrient requirements is informed by body composition assessments because of the chemical substrate necessary for appropriate growth and maintenance of body compartments (Fomon and Nelson, 2002). The need for revision of the recommended energy requirements is one such example that has become evident in the recent decades which have witnessed the rapidly growing epidemic of childhood obesity (Food and Nutrition Board, 2002). Concerns about osteoporosis later in life also have raised interest in the adequacy of calcium and vitamin D intake to achieve optimal total body bone mineral during growth and development (Kalkwarf *et al.*, 2003). Third, in the clinical setting, health and nutritional status screening, monitoring and treatment is sometimes dependent on measurements of body composition (Kehayias and Valtuena, 1999). For example, interventions to treat nutritional wasting or obesity aim to achieve appropriate-for-age proportions of fat mass, muscle and bone. Dosage regimens for medical therapies are often based on estimated lean body mass, and age- or disease-related deviations from assumptions inherent in standard formulas to estimate lean body mass can affect the delivery of therapy. Although body composition assessment has many applications, techniques for assessment in children are not widely used except in the research setting. Here we will describe the conceptual approaches to body composition assessment and provide detailed descriptions of the methods used as well as their advantages and limitations.

Techniques used to assess body composition range from relatively simple, prediction-based, descriptive methods to more mechanistic methods based on stable component associations (Pietrobelli *et al.*, 1998), such as the ratio of total body potassium to total body cell mass. With the exception of cadaveric studies, body composition methods are indirect and involve assumptions that may bias the results. Estimates of the relative and absolute size of body compartments vary across methods. For example, as shown in Table 6.1, body composition of children with sickle-cell disease were assessed by multiple methods (Barden *et al.*, 2002). The mean values differ by method, but the pattern of body composition differences between groups remains the same across methods. Body composition methods also differ in terms of accuracy, availability, expense, inconvenience and risk. In addition, for infants and children, some methods are impractical or simply not feasible. All methods involve varying degrees of inherent deviations in measurements that are due to natural biological or clinical variance, or measurement errors (Guo *et al.*, 2000). Usually there is a trade-off between precision and practicality, with indirect, prediction-based methods that are easily applied on an epidemiologic scale being less precise

Table 6.1 *Body composition differences in children with sickle-cell disease using four assessment methods*

	Pubertal males		Pubertal females	
	Sickle-cell disease group	Control group	Sickle-cell disease group	Control group
Total body electrical conductivity	$n = 5$	$n = 10$	$n = 11$	$n = 9$
Fat-free mass (kg)	31.4 ± 3.3	36.2 ± 11.1	34.5 ± 6.7	36.7 ± 4.6
Fat mass (kg)	6.7 ± 3.0	9.5 ± 5.3	11.9 ± 5.9	14.0 ± 7.5
Percent body fat	17.0 ± 6.4	20.0 ± 6.2	24.7 ± 7.0	26.3 ± 8.5
Two skinfolds[a]	$n = 6$	$n = 10$	$n = 11$	$n = 10$
Fat-free mass (kg)	37.4 ± 7.5	39.5 ± 11.3	36.0 ± 6.6	39.5 ± 6.1
Fat mass (kg)	3.6 ± 1.2	6.3 ± 4.8	10.4 ± 6.0	12.5 ± 6.9
Percent body fat	8.7 ± 2.2	12.6 ± 5.2	21.4 ± 6.6	22.9 ± 8.0
Four skinfolds[b]	$n = 6$	$n = 10$	$n = 11$	$n = 10$
Fat-free mass (kg)	35.6 ± 7.5	37.6 ± 10.9	35.1 ± 7.2	38.7 ± 6.5
Fat mass (kg)	5.4 ± 1.6	8.1 ± 4.6	11.3 ± 4.8	13.3 ± 6.0
Percent body fat	13.2 ± 3.1	16.9 ± 3.8	23.6 ± 4.4	24.7 ± 6.3
Total body water	$n = 6$	$n = 8$	$n = 9$	$n = 8$
Fat-free mass (kg)	33.4 ± 6.6	31.6 ± 7.8	32.1 ± 5.9	36.2 ± 5.6
Fat mass (kg)	7.6 ± 2.7	10.6 ± 6.4	12.5 ± 4.0	15.3 ± 8.4
Percent body fat	18.2 ± 4.8	23.9 ± 7.0	27.6 ± 5.3	28.2 ± 9.1

[a] Anthropometric prediction of body composition using two skinfolds (Slaughter *et al.*, 1988).
[b] Anthropometric prediction of body composition using four skinfolds (Durnin and Ruhaman, 1967; Durnin and Womersley, 1974).
Source: Barden *et al.* (2002).

than more direct laboratory-based methods. Methods should be chosen following consideration of the requirement for accuracy and precision dictated by the research question or clinical concern, and should be evaluated in the context of the level of cost, inconvenience, risk and developmental stage of the subjects involved.

Body composition models

Levels of organization

Conceptually, the composition of the body can be defined at many levels of biological organization, as described by Wang *et al.* (1992). The five levels,

Table 6.2 *Conceptual models of body composition*

Level	Kind of information obtained	Methods of investigating body composition
Atomic	Elemental: O, C, H, Ca, P, S, K, Cl, Na	Neutron activation, total body potassium counting with ^{40}K
Molecular	Water, protein, lipid, bone mineral, soft tissue mineral, glycogen	Total body water, neutron activation, magnetic resonance spectroscopy, in combination with other methods and assumptions
Cellular	Fat, body cell mass, extracellular fluid, extracellular solids	Total body water, isotope dilution for sodium bromide, total body potassium counting with ^{40}K, in combination with other methods and assumptions
Tissue	Adipose, skeletal, muscle, organs	Densitometry, dual-energy X-ray absorptiometry, bioeletrical methods, computed tomography, magnetic resonance imaging
Whole body	Height, weight	Anthropometry

atomic (e.g. carbon, oxygen and hydrogen), molecular, cellular, tissue-system (e.g., fat, muscle and bone) and whole body, are described in Table 6.2. For example, determining the amount of water in the body represents the molecular level, whereas determining the amount of calcium in the body represents the atomic level of organization.

This conceptual model of body composition is useful in deciphering the assumptions involved in the practical approaches to body composition. Examples of the methodologies used at different levels of this conceptual model are shown in Table 6.2. In general, the difficulty, expense, risk and accuracy associated with body composition assessment increases along with the level of detail in the hierarchy of biological organization. Methods from different levels in this hierarchy can be used in combination to generate the multicompartment models of body composition described in the next section. Translational equations can be used for conversion from the elemental to the molecular and cellular levels as shown in Table 6.2.

Multicomponent models of body composition

In practice, most body composition approaches are organized according to the number of compartments described. Regardless of the model used, body weight represents the sum of the individual compartments. Compartments are distinct within a given level, although relationships exist between components at the same or different levels. For instance, fatness can be characterized by total body carbon at the most complex, atomic level, and by adipose tissue at the least complex, tissue-system level (Heymsfield and Matthews, 1994).

Most commonly, two-compartment models of body composition are used in which the body is divided into the compartments of fat mass (FM) and fat-free mass (FFM), such that total body mass is equal to the sum of FM and FFM. FM is the ether-extractable lipid component of the body, and excludes the supporting tissues of adipose cells (Forbes, 1987). FFM represents the remainder of the body and is a complex tissue compartment composed of skeletal muscle, organs, bone and supporting tissues. Chemically, it is composed of water, protein and osseous and non-osseous minerals. These two compartments are estimated using methods such as hydrodensitometry, anthropometry, bioelectrical methods or isotope dilution for total body water (TBW). Two-compartment models are problematic in studies of children because of the changing composition of the FFM with growth and maturation. In particular, changes in the hydration of FFM and the contribution of osseous minerals to FFM represent the greatest sources of uncertainty in estimating FFM (Lohman *et al.*, 1984; Lohman, 1986; Wells *et al.*, 1999; Ellis *et al.*, 2000).

Three- and four-component models are based on methods or combinations of methods that can measure FM along with two or more components of FFM. Examples of three-compartment models are shown in Table 6.3. For instance, body composition assessed by dual-energy X-ray absorptiometry (DXA), measures three components:

$$\text{Total body mass} = \text{fat mass} + \text{lean body mass} \\ + \text{total body bone mineral mass}$$

In addition, measurement of total body water in combination with another method, such as hydrodensitometry, provides estimates of the aqueous and non-aqueous fractions of FFM, as well as FM:

$$\text{Total body mass} = \text{fat mass} + \text{water} + \text{non-aqueous solids}$$

However, this approach assumes a constant density for non-aqueous solids. For children and other groups, this assumption can be problematic due to differences in bone mineralization. Four-compartment models require measurement of at least three body compartments. Several four-compartment models are described in Table 6.3.

The accuracy of body composition assessment improves with the number of components measured (Hewitt *et al.*, 1993; Nielsen *et al.*, 1993; Wells *et al.*, 1999; Fields and Goran, 2000). For instance, body composition can be derived using conversion formulas from measures of body density, determined as body mass divided by volume. Total body density (D_b) is assumed to be the sum

Table 6.3 *Multicompartment models of body composition*

	Components measured	Techniques used	Assumptions
Two-compartment models	Fat mass + fat-free mass	Anthropometry, underwater weighing, isotope dilution for total body water, bioelectrical impedance analysis, total body electrical conductivity	Fixed densities of fat mass and fat-free mass
Three-compartment models	Fat mass + total body water + non-aqueous solids	Isotope dilution for total body water, underwater weighing/Bod Pod	Fixed ratio of protein to minerals
	Fat mass + lean body mass + bone	Dual-energy X-ray absorptiometry	Constant density of fat mass and lean body mass
Four-compartment models	Fat mass + body cell mass + extracellular water + extracellular solids	Isotope dilution for total body water and for sodium bromide, total body potassium counting with ^{40}K, dual-energy X-ray absorptiometry	Fixed ratio of K, $^{2}H_2O$ or $^{2}H^{18}O$ and NaBr in cellular compartments
	Fat mass + lean body mass + intracellular water + extracellular water	Dual-energy X-ray absorptiometry, isotope dilution for total body water and for sodium bromide, bioelectrical impedance analysis	
	Fat mass + isotope dilution for total body water + lean body mass + protein	Dual-energy X-ray absorptiometry, underwater weighing, isotope dilution for total body water	
Five-compartment models	Glycogen + total body nitrogen + total body protein + fat mass + lean body mass	Neutron activation	Glycogen estimated from another component such as total body protein or by magnetic resonance spectroscopy
	Total body protein + total bone mineral + fat + extracellular water + intracellular water	Neutron activation	

of the densities of individual compartments (such as FM and FFM). In a two-compartment model, this conceptual relationship can be described by the formula:

$$1/D_b = \text{FM}/d_{fm} + \text{FFM}/d_{ffm}$$

where FM/d_{fm} and FFM/d_{ffm} represent the relative proportions of the fat and fat-free compartments divided by their respective densities (Going, 1996), which are assumed to be constant. Since the composition of FFM varies through growth and development, greater accuracy can be achieved through utilization of three- and four-compartment models which expand the components of FFM in the formula, and are thereby less dependent on these assumptions (Wells *et al.*, 1999). The formula reflecting a four-compartment model would include the respective densities of fat, water, mineral and protein (0.9007 g/ml, 0.9937 g/ml, 3.038 g/ml and 1.34 g/ml, respectively (Brozek *et al.*, 1963)).

Assessment methodologies

Anthropometry

Anthropometry provides the most technologically simple approach to body composition. It is quite feasible for all age ranges and has the distinct advantage of being the only body composition method with widely available reference ranges. It is therefore extremely useful for epidemiological and clinical studies, and for screening purposes. Height and weight are the most fundamental measures. Additional measures, such as circumferences, skinfold thicknesses and lengths and breadths of bony tissue, provide further detail for estimation of body composition. Anthropometric equipment is inexpensive and portable, but proper training of anthropometrists is essential. Several comprehensive publications describing anthropometric techniques are available (Weiner and Lourie, 1981; Cameron, 1986; Lohman *et al.*, 1988; Gibson, 1990). Stylistic differences can develop over time, even among well-trained anthropometrists, so periodic checks on intra- and inter-observer reliability are necessary. Intra- and inter-observer reliability for anthropometric techniques have been described elsewhere (Ulijaszek and Lourie, 1994).

Weight-for-height ratios and assessment of obesity

Ratios have been developed to assess weight while taking into account a child's length or height as indicators of nutritional status. Weight-for-height and body mass index (BMI) for-age reference charts are excellent screening tools to identify children with excess adiposity (Cole *et al.*, 2000; Kuczmarski *et al.*,

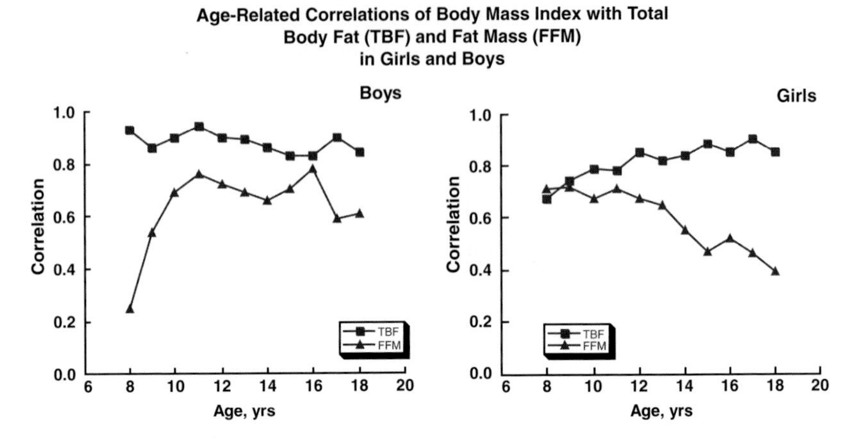

Figure 6.1 BMI correlations with body compartments during growth. The correlations between body mass index (BMI) and the body compartments total body fat (TBF) and fat-free mass (FFM) vary by age and gender. (Data from Mei *et al.*, 2002.)

2000). In children greater than 2 years of age, BMI is highly correlated with percentage of body fat (Norgan, 1994; Maynard *et al.*, 2001), and performs better than other weight-for-height measures for this purpose (Mei *et al.*, 2002). However, BMI does have some unique limitations in children. In particular, the association of BMI with body compartments changes with age and differs by gender (Figure 6.1). In adults, BMI is a good indicator of weight independent of height. In children, BMI is correlated with height at some ages, most notably in 10- to 14-year-old boys (Maynard *et al.*, 2001). Consequently, in this age range, BMI is partly an indicator of overall size as well as body composition. The use of a triceps skinfold measurement in combination with BMI is likely to give a more accurate categorization of adiposity and has been proposed for determination of overweight in children 10 years of age and older (Himes and Dietz, 1994). The utility of BMI to identify children at risk for undernutrition has not been assessed. In developing nations where chronic undernutrition is prevalent, weight-for-height measures can appear normal to high in the presence of stunting and low body fat reserves (Trowbridge *et al.*, 1987).

Body composition based on anthropometric prediction equations

Total body fat mass, FFM and percent body fat can be estimated using anthropometric prediction equations developed from samples of healthy children (Durnin and Rahaman, 1967; Brook, 1971; Durnin and Womersley, 1974; Slaughter *et al.*, 1988). Several prediction models are shown in Table 6.4. Some prediction

Table 6.4 *Equations for predicting body composition from anthropometry*

Two-skinfold[a] method for prediction of percent body fat	
Prepubescent white males	Percent body fat = 1.21 (T + S) – 0.008 $(T + S)^2$ – 1.7
Prepubescent black males	Percent body fat = 1.21 (T + S) – 0.008 $(T + S)^2$ – 3.2
Pubescent white males	Percent body fat = 1.21 (T + S) – 0.008 $(T + S)^2$ – 3.4
Pubescent black males	Percent body fat = 1.21 (T + S) – 0.008 $(T + S)^2$ – 5.2
Postpubescent white males	Percent body fat = 1.21 (T + S) – 0.008 $(T + S)^2$ – 5.5
Postpubescent black males	Percent body fat = 1.21 (T + S) – 0.008 $(T + S)^2$ – 6.8
All females	Percent body fat = 1.33 (T + S) – 0.013 $(T + S)^2$ – 2.5
When sum of triceps and subscapular folds is greater than 35 mm, use	
All males	0.783 (T + S) + 1.6
All females	0.546 (T + S) + 9.7
Four-skinfold[b] method for prediction of percent body fat	
Percent body fat = [(4.95/body density) – 4.5] × 100, where body density is defined as	
Prepubertal children (1 to 11 years)	
Males	Density = 1.1690 – 0.0788 log sum of four skinfolds
Females	Density = 1.2063 – 0.0999 log sum of four skinfolds
Adolescent children (12 to 16 years)	
Males	Density = 1.1533 – 0.0643 log sum of four skinfolds
Females	Density = 1.1369 – 0.0598 log sum of four skinfolds
Ages 17 to 19 years	
Males	Density = 1.1620 – 0.0630 log sum of four skinfolds
Females	Density = 1.1549 – 0.0678 log sum of four skinfolds
Ages 20 to 29 years	
Males	Density = 1.1631 – 0.0632 log sum of four skinfolds
Females	Density = 1.1599 – 0.0717 log sum of four skinfolds

[a] T, triceps; S, subscapular.
[b] Sum of four skinfolds equals triceps plus biceps plus subscapular plus suprailiac.
Source: From Brook (1971), Durnin and Rahaman (1967), Durnin and Womersley (1974) and Slaughter *et al.* (1988); table adapted from Zemel and Barden (2001).

equations include race, obesity and pubertal status. Selection of the appropriate prediction equation is based on the child's age and gender. Estimates based on these prediction equations yield results that are highly correlated with results from other methods for body composition determination. Figure 6.2 illustrates the correspondence between body composition estimated by a skinfold prediction equation and by DXA for healthy children. Anthropometric prediction equations are best used to evaluate groups of children, rather than individuals. For clinical purposes, use of FFM and FM information is limited because of the absence of reference data in children and the known gender- and age-related changes in these measures.

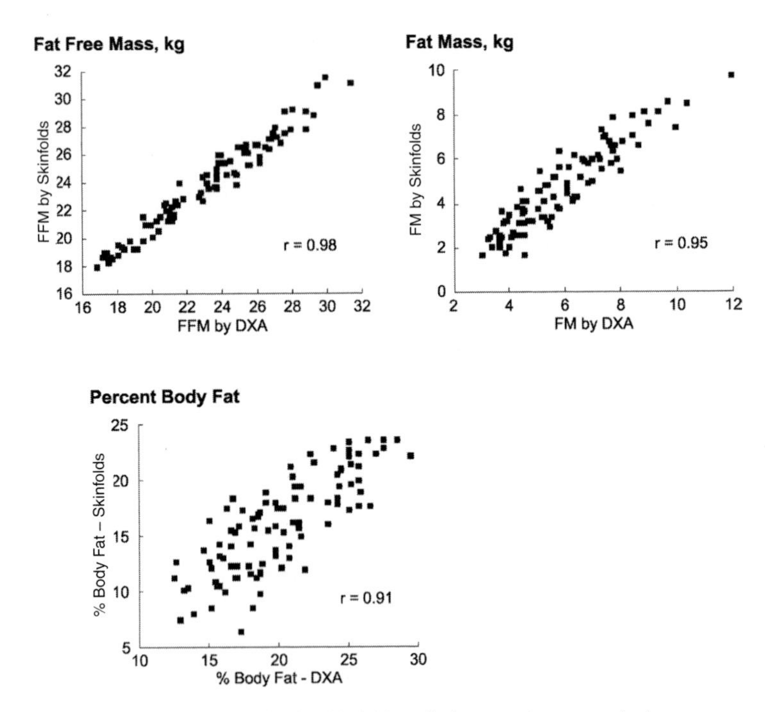

Figure 6.2 Body composition by skinfold prediction equation versus dual-energy X-ray absorptiometry (DXA). When anthropometric data is collected by trained anthropometrists, there is excellent correspondence between body composition predicted by skinfold prediction equations and body composition measured by dual-energy X-ray absorptiometry. Data presented are from 139 healthy children, ages 7 to 10 years, using equations by Brook (1971) (B. Zemel, unpublished data).

The greatest shortcoming of anthropometric measures of body composition is that these prediction equations were derived from samples of healthy children, and thereby carry several important assumptions, such as normal patterns of sexual and skeletal maturation, fat distribution, and mineral and water composition of the FFM. When applied to children with chronic diseases or from other ethnic backgrounds, validation of anthropometric prediction equations often is necessary to determine the effect of deviations from these assumptions on the estimation of body composition. For example, for children with cystic fibrosis, comparisons of FFM predicted by skinfolds with FFM assessed by other methods such as isotope dilution for total body water, bioelectrical impedance analysis or dual-photon absorptiometry have yielded inconsistent results (Lands *et al.*, 1993; Spicher *et al.*, 1993; de Meer *et al.*, 1999). Consequently, body composition results based on skinfold prediction equations should be interpreted with caution in children with cystic fibrosis.

Other applications of anthropometric data in body composition assessment

Skinfold thicknesses can be used as indicators of body fatness without predicting the size of body compartments. The triceps skinfold thickness is particularly useful because of its sensitivity to fluctuations in nutritional status, its high correlation with total body fat stores, and the availability of excellent reference data for individuals from 1 year of age through adulthood (Frisancho, 1981; Shepard, 1991). The subscapular skinfold measure is informative because it is a good measure of adiposity on the trunk, and it is highly correlated with health risk (Rolland-Cachera, 1993).

The sum of triceps and subscapular skinfolds is a good indicator of overall fatness, and reference data are available (Frisancho, 1981, 1990). This measure also can be used to estimate fat distribution using the centripetal fat ratio (subscapular skinfold/[triceps + subscapular skinfold]) (Bogin and Sullivan, 1986; Spender *et al.*, 1988). When mid-upper arm circumference is combined with the triceps skinfold measurement, upper arm fat area, muscle area, and muscle circumference can be estimated. These measures correlate well with whole body measures of FM and FFM even though they are measured at a single site. The availability of reference data for children enhances the utility of these measures (Frisancho, 1981, 1990). These and other simple anthropometric measures may perform better than BMI in estimating body composition at the population level (Himes and Dietz, 1994; Boye *et al.*, 2002).

It is important to note that soft tissues generally are more difficult to measure reliably and reproducibly than measurements of linear growth and weight. In addition, the accuracy and reproducibility of skinfold thickness measures decline with increasing adiposity. For obese children, accurate measurement of skinfold thickness is not always possible. Indices of fatness and fat distribution, such as skinfold ratios, have even lower precision than direct measurements due to compounding of measurement errors (Mueller and Kaplowitz, 1994). However, in using skinfold measures to predict body composition, Durnin and colleagues (Durnin *et al.*, 1971, 1997) have shown that inter-observer differences in measurement technique have only a small effect on the prediction of body fat.

Densitometric methods

Densitometric methods, such as hydrodensitometry and air displacement plethysmography, utilize the principle that body composition can be derived using conversion formulas from measures of body density (Table 6.5). The classic two-compartment models of Siri (1961) and Brozek *et al.* (1963) were based

Table 6.5 *Prediction of body fat from body density in children; to calculate percent body fat* = $(C_1/D_b - C_2) \times 100$, *where* D_b *is measured body density*

	Males		Females	
Age, years	C_1	C_2	C_1	C_2
1	5.72	5.36	5.69	5.33
1–2	5.64	5.26	5.65	5.26
3–4	5.53	5.14	5.58	5.20
5–6	5.43	5.03	5.53	5.14
7–8	5.38	4.97	5.43	5.03
9–10	5.30	4.89	5.35	4.95
11–12	5.23	4.81	5.25	4.84
13–14	5.07	5.64	5.12	4.69
15–16	5.03	4.59	5.07	4.64
Young adult	4.95	4.50	5.05	4.62

Source: Lohman (1989).

on the assumption that the densities of fat (0.900 kg/l) and FFM (1.100 kg/l) do not vary among individuals or populations (Lohman, 1992). However, it has been recognized that the density of FFM is not constant for all individuals, particularly among certain groups such as the elderly, children and blacks (Lohman, 1992; Wagner and Heyward, 2001). Differences in the density of FFM can be attributed largely to differences in the proportion of bone mineral. Presently, recommended formulas for estimating percent body fat among adults include those published by Siri (1961) and Brozek *et al.* (1963) for whites, and Schutte *et al.* (1984) and Wagner and Heyward (2001) for blacks. For children, changes in the density of FFM were estimated by Lohman (1986) using combined data from Fomon *et al.* (1982), Haschke (1983), Boileau *et al.* (1984, 1985) and Lohman *et al.* (1984). The revised Siri formula, using age- and sex-specific constants for children to account for the developmental changes in chemical maturity is shown in Table 6.5. As noted above, studies that use a four-compartment approach by estimating percent body fat from total body density corrected for measured total body water and total body bone mineral greatly improve the accuracy of body composition estimates based on densitometric methods, especially among children.

Hydrodensitometry

Several techniques have been developed to measure body volume for determination of body density. Hydrodensitometry, also known as underwater weighing

Table 6.6 *Prediction of body fat in adults using body density measurements*

Siri (1961)[a]	Percent body fat = $(4.95/D_b - 4.50) \times 100$
Brozek *et al.* (1963)[a]	Percent body fat = $(4.570/D_b - 4.142) \times 100$
Lohman (1986)[b]	Percent body fat = $(6.386/D_b + 3.961\ m - 6.090) \times 100$

[a] D_b, the measured density of the body.
[b] m, the measured bone mineral.

or hydrostatic weighing, was at one time the routine reference method for assessment of body composition. In this technique, body volume is determined according to Archimedes' principle, utilizing measurement of body mass in air and while immersed in water according to the following formula:

$$\text{body volume (l)} = (\text{mass in air (kg)} - \text{mass in water (kg)})/\text{density of water (kg/l)}$$

Corrections are needed for the volume of air in the lungs and intestines and for the density of air and water at given temperatures. The volume of air in the gastrointestinal tract is usually assumed to be a constant 100 ml. Residual lung volume has been measured by several techniques. Most commonly, a closed-circuit oxygen dilution method has been used, and open-circuit nitrogen dilution and closed-circuit helium dilution are additional techniques (see Going (1996) for a review). Errors in the measurement of residual lung volume and variation in the density of FFM are the greatest sources of error in hydrodensitometry.

Once body volume is determined, D_b is calculated as body mass divided by body volume, and percent body fat is calculated using assumed densities of FM and FFM as described above (p. 000) and shown in Table 6.5 for children and Table 6.6 for adults. Body density measured by hydrodensitometry is also used in the multicompartment models described in Table 6.3. Hydrodensitometry techniques are acceptable for healthy adults and adolescents, and can be used in healthy children 8 years of age or older. For subjects with physical or cognitive disabilities, this method usually is not feasible.

Air-displacement plethysmography

Acoustic plethysmography, helium displacement, photogrammetry and air-displacement plethysmography are among the other techniques used to estimate body volume. The commercial availability of a whole-body air-displacement plethysmograph has promoted the use of this technique for body composition assessment. The Bod Pod Body Composition System (Life Measurement

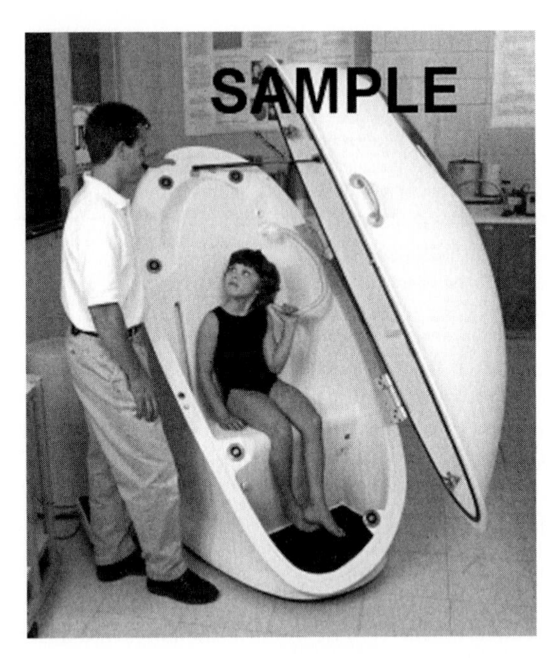

Figure 6.3 The Bod Pod air displacement plethysmograph manufactured by Life Measurement Instruments (Concord, CA).

Instruments, Concord, CA) shown in Figure 6.3 contains a two-compartment fibreglass chamber of known size. One compartment serves as a reference chamber and the other is a test chamber in which the subjects sits during the test. A pulsating diaphragm separates the two compartments and is used to create small changes in pressure. The ratio of the changes in pressure (P) in the two chambers is proportional to the volume (V) of the two chambers according to Poisson's law (Dempster and Aitkens, 1995; Fields *et al.*, 2002):

$$P_1/P_2 = (V_2/V_1)^{\gamma}$$

where γ is a gas specific constant equal to 1.4 for air. The volume of air in the reference chamber remains constant, but the volume in the test chamber changes by displacement when the subject sits in the chamber. Thus, body volume is measured indirectly using pressure changes to determine the volume of air within the chamber with the subject sitting in the test chamber compared to the empty chamber. However, the raw body volume measurement ($V_{b_{raw}}$) must be obtained under specific conditions and with several adjustments in order to achieve accurate results. First, the air close to the surface of skin, clothing and hair responds differently to pressure changes than the rest of the air in the chamber. To minimize the effects of hair and clothing, it is recommended that subjects wear tight-fitting swimsuits and swim caps. To adjust for the artefact that occurs from the temperature at the skin surface, the Bod Pod software automatically estimates a surface area artefact (SAA), which is proportional to body surface area (estimated from height and weight). And finally, the thoracic gas volume (V_{TG}) is measured with a breathing apparatus within the Bod Pod, or estimated by proprietary prediction equations in the Bod Pod software (Fields *et al.*, 2001). Ultimately, body density is calculated as:

$$D_b = \text{mass} / \left(V_{b_{raw}} + 0.40 V_{TG} - \text{SAA} \right)$$

The Bod Pod offers an appealing alternative to underwater weighing for several reasons. First, it requires less time and training on the part of the subject to complete the test. Although a bathing suit and cap are required for accurate measurement, it does not involve immersion in water, so it is acceptable to a broader range of subjects in terms of age, cognitive status and physical disability. In addition, the reliability is similar to that of underwater weighing (Fields *et al.*, 2002), with coefficients of variation (CVs) ranging from 1.7% to 4.5% for percent body fat in adults. Two studies have reported the CV for percent body fat in children, and the technical error of measurement is small (less than 2%) (Wells and Fuller, 2001; Demerath *et al.*, 2002). Nicholson and colleagues found that Bod Pod measurements of body fat corresponded well with DXA measures of body fat, independently of age, gender, ethnicity (African-American vs. Caucasian) and puberty stage, and were better than other 'field' methods of body composition assessment such as bioelectrical impedance analysis or anthropometry (Nicholson *et al.*, 2001). Further studies are needed to determine the reliability of this method in children, especially those less than 8 years of age.

Isotope dilution methods

Water is the largest component of the body, comprising 40% to 60% of total body mass. As the majority of body water is associated with fat-free tissue, its assessment is informative as a proxy for quantifying FFM. Stable isotopes are used to estimate the amount of total body water using the dilution principle. The dilution principle asserts that the concentration of a compound (e.g. a stable isotope) in a solvent (in this case, body water) is dependent upon the amount of the compound added to the solvent, and the solvent's volume. When the amount and concentration of the compound are known, the volume of the solvent can be computed:

$$N = d \times f \times E_{dose} / E_{bw}$$

where d is moles of water given in the dose, f is the fractionation factor for physiological sample relative to body water, and E_{dose} and E_{bw} are the enrichments of the dose and body water, respectively (Schoeller, 1996). Although other compounds have been used in human studies, the stable isotopic tracers deuterium oxide (^2H_2O), and oxygen-18 hydride, are preferred as they present low risk in small concentrations, are evenly distributed in body water, and are not extensively metabolized (Schoeller *et al.* 1985). Tritiated water is not recommended for use with children, as it is mildly radioactive.

Because they occur naturally, these isotopes are already present in the body and a baseline body fluid sample (such as urine, serum or saliva) must be obtained. Following this sample collection, a concentrated dose of isotopic tracer is administered which elevates the concentration of the isotopes in the body above that observed from drinking water. After an equilibration period during which the isotope mixes with the total body water pool (approximately 4 hours), further sample collections are obtained. The most sensitive method for measuring the isotopic enrichment of the fluid sampled is isotope-ratio mass spectrometry.

FFM is derived from the TBW measurement using hydration factors that estimate the fraction of the TBW in fat-free mass (Schoeller *et al.*, 1980; Fomon *et al.*, 1982; Boileau *et al.*, 1984). Once fat-free mass is estimated, fat mass and percent body fat can then be derived (fat mass = body weight − FFM; percent body fat = fat mass/body weight). TBW is also used in multicompartment models as described in Table 6.3.

The TBW method is very accurate and precise (1–2%), provided proper sampling, dosing and storage procedures are followed. Depending on the protocol, corrections may be necessary for the amount of isotopic tracer lost from the body (if the tracer was given at night prior to collection of the first morning void) in urine, perspiration and other gases such as water vapour in breath or CO_2 production in the case of ^{18}O, or for the amount of water entering the body (if the tracer was given in the morning prior to a 4-hour equilibration period), such as from water ingested at sample administration, ingested food or beverages, metabolic water formation, or exchange with moisture in the air. Tritium and ^2H_2O overestimate TBW by about 4% in adults and children, and 2% in premature infants, and ^{18}O overestimates TBW by approximately 1% in adults and 0.5% in premature infants, due to mixing of the isotopes with non-aqueous fractions of the body. Measurement error is mainly related to the laboratory analysis of the isotopic concentrations (Schoeller, 1996).

A limitation of the isotope dilution method is that estimations assume a constant relationship between total body water and FFM, which often fails to account for the chemical maturity of the subject or other sources of biological variability. Other limitations include high isotope and analysis cost, and time necessary for preparation and administration of the sample, equilibration of the tracer in the body, and analysis.

A related method can be used to measure the volume of extracellular water (ECW), which can be used in combination with TBW to derive intracellular water (ICW):

$$ICW = TBW - ECW.$$

Intracellular water is strongly associated with the metabolic properties of the body, and therefore is a useful measure of body composition at the cellular level. The extracellular space most commonly is measured using the dilution of bromide (given as sodium bromide in water), although other compounds have been used. It should be noted, however, that dilution by tracer of the various compartments of extracellular water (such as plasma, bone, interstitial, dense connective tissue, and transcellular water) is less uniform than in TBW, so the method is more imperfect. Assumptions about tracer distribution, equilibration and metabolism are the same as for TBW. The technique also is similar to that for TBW, in which a baseline sample is acquired, and a postequilibration sample is obtained 3–4 hours later in healthy subjects, and 5–6 hours later in subjects with expanded extracellular spaces. However, the sampling must be restricted to blood plasma due to differences in bromide concentrations in different fluids. The bromide dilution space is calculated from the dose of bromide and the change in concentration of plasma bromide according to the dilution principle. The concentration must be corrected for overestimation due to the Gibbs–Donnan effect on bromide's concentration in various extracellular fluids and the penetration of bromide in the intracellular space (Schoeller, 1996).

Isotope dilution methods are generally safe, reliable, accurate and quite feasible in infants and children. They can be administered in a clinical or field setting. The expense and expertise associated with the determination of isotopic enrichment of biological samples is a consideration in choosing this method. Measurement of TBW has been used as a criterion method in many of the classic studies of body composition in children (Haschke *et al.*, 1981; Fomon *et al.*, 1982; Forbes, 1987; Fomon and Nelson, 2002).

Bioelectrical methods

Bioelectrical impedance analysis

Bioelectrical impedance analysis (BIA) is a technique based on the principle that the impedance of a small electrical current as it passes through the body is proportional to the water and electrolytes in the body. Physiologically, an electrical current flows through the body by the movement of ions through body fluids. The resistance to flow is high for adipose tissue and bone compared to the resistance of muscle tissue. Cell membranes and other interfaces become charged, acting like capacitors, impeding the flow of the electrical current. The impedance (Z) of an electrical current is the combined effect of resistance (R) (the inverse of conductance which is proportional to the ions in aqueous solution) and reactance (X_c) (the reciprocal of capacitance) (Foster and Lukaski, 1996; Lukaski, 1996). The relationship among these is described as:

$$Z^2 = R^2 + X_c^2$$

In practice, impedance is dependent on overall size and the distance the current needs to travel from the source to the detector. This is described by the equation

$$Z = \rho L^2 / V$$

where L is the conductor length, V is volume and ρ is resistivity. L is assumed to be equivalent to stature and V is total body water.

Single-frequency BIA systems usually operate with a frequency of 50 kHz. The body is assumed to be a cylinder with a uniform cross-section. When measuring the total body, two source and two detector electrodes are used on the wrist and ipsilateral ankle to estimate the TBW of the entire body. Other BIA models measure leg impedance using a stand-on scale with electrodes positioned under the heel and foot. Positioning the electrodes over smaller segments of the body is used for regional estimates of body composition.

One of the major limitations of single frequency BIA systems is that 28% of the whole body resistance occurs in the arm, and 33% in the lower leg, even though they represent only 1.5% to 3% of body weight compared to the trunk (Foster and Lukaski, 1996). Further, sources of measurement error can arise from the skin surface where the electrodes are placed, and leakage along the surface between terminals (Oldham, 1996). For small children with small hands and feet, this is of particular concern. An additional concern is the estimation of TBW, FFM, FM and percent body fat derived from BIA measurements. Prediction equations are required to estimate FFM and FM from BIA measurements (Chumlea and Guo, 1994). Numerous equations have been developed for use with children (Davies *et al.*, 1988; Deurenberg *et al.*, 1989; Houtkooper *et al.*, 1989; Kushner *et al.*, 1992; Danford *et al.*, 1992). Differences in the prediction equations may be attributable to the characteristics of the reference sample in terms of age and body proportions, or the reference method used to generate the prediction equation (Wells *et al.*, 1999). Compared to a four-compartment model, BIA measures of FFM and fatness do not perform much better than using height and weight to estimate body composition (Wells *et al.*, 1999).

Bioelectrical impedance spectroscopy (BIS), or multifrequency BIA, is a more detailed approach that utilizes the intrinsic bioelectrical properties of cell membranes to distinguish intracellular and extracellular spaces. At low frequencies, a current must flow through the extracellular space because cells do not conduct electricity well. At higher frequencies, the current can easily pass through cell membranes as well as the intracellular space (Foster and Lukaski, 1996). BIA uses a range of frequencies, from 1kHz to 1.35 MHz (Ellis *et al.*, 1999). Several studies have been conducted in children using BIS. In general, these studies show that this technique can be used to describe body composition of groups of children. However, due to wide limits of agreement with criterion methods, it may not be suitable for body composition assessment of individuals (Ellis *et al.*, 1999; Treuth *et al.*, 2001; Fors *et al.*, 2002).

Total body electrical conductivity

Total body electrical conductivity (TOBEC) measures the displacement caused by the electrical properties of water and electrolytes in the body when passed through a low-energy electromagnetic field. The measurement chamber is a large (approximately 2.1 m long), cylindrical electromagnetic coil driven by a 2.5 MHz source to create an electromagnetic field of specific frequency in the enclosed space. The subject passes through the coil, causing alterations in conductance in the coil (Baumgartner, 1996; Puig, 1997). The change in the electrical signal due to the absorption of a small amount of energy in the body (E) is converted to estimates of TBW or FFM by computerized prediction equations developed for this methodology (Van Loan and Mayclin, 1987; Van Loan, 1990; Van Loan *et al.*, 1990; Fiorotto *et al.*, 1995). The technique assumes that the body is essentially a cylinder of known length (L), and the E is proportional to the volume (V) of the conductive component of the body as:

$$V \propto (E \times L)^{0.5}$$

The conductive component is theoretically equivalent to TBW, and the TOBEC method was calibrated using TBW measurements in infants, children, adolescents, and adults. Paediatric and adult TOBEC devices (EM-Scan, Springfield, IL) provide accurate, rapid, safe, non-invasive estimates of body composition in healthy subjects from infancy through to old age (Boileau, 1988; Van Loan, 1990; Baumgartner, 1996). Due to the size and cost of these devices, it never has been a widely used method in human body composition studies.

Dual-energy X-ray absorptiometry methods

Absorptiometric methods are based on the premise that the thickness, densities and chemical composition of the fat, lean and bone tissue compartments differentially affect the attenuation of an energy beam as it passes through the body. For a given energy level, tissues have unique attenuation properties, such that the attenuation is a function of a constant specific to that tissue and the tissue mass. The relationship between attenuation of an energy beam and the mass of two tissues through which it passes is explained by the natural log equation:

$$\ln(I/I_0) = -\mu_a m_a - \mu_b m_b$$

where I_0 is the initial intensity of the beam, I is the final intensity, μ_a is the mass attenuation coefficient and m_a the mass of tissue a, and μ_b is the mass attenuation coefficient and m_b the mass of tissue b. In single-energy absorptiometry methods, the mass of soft tissue surrounding the bone is assumed to be constant so that the equation is used to solve for bone mass. In dual-energy absorptiometry methods, the use of two energy beams differing in intensity generates two sets of equations and allows for solution of two tissue compartments. As the beam passes over regions of soft tissue, the respective masses of lean and fat tissues are determined. As the beam passes over regions of bone, the composition of the soft tissue surrounding the bone is assumed to be equivalent to that of the adjacent soft tissue. The mass of lean, fat and skeletal tissue are thereby calculated based on the attenuation data collected over the entire body (Prentice, 1995; Ellis, 2000).

Dual photon absorptiometry (DPA) was the first dual-energy system developed. DPA used a radionuclide source and digital detector to determine body composition. This method has become obsolete with the development of dual-energy X-ray absorptiometry (DXA), which is faster and more accurate than the older DPA systems. DXA uses a very low-energy X-ray source; the X-ray exposure is less than a day of background radiation exposure. Earlier DXA models used a ‘pencil beam’ which scanned across the body one line at a time. Whole-body scan times were 30 to 45 minutes. More recent models use a ‘fan beam’ with multiple detectors, whereby a larger region is measured simultaneously, and scan times are 3 to 5 minutes. Whole-body and regional estimates of body composition can be obtained for infants, children and adolescents (Figure 6.4). It should be noted that regional estimates of body composition derived from whole body scans are influenced by the positioning of the subject in the scan field and the placement of subregion lines by the DXA operator. So, subregion estimates of body composition are somewhat operator dependent. All DXA machines use site-specific scans to measure the bone mass and bone density of subregions of interest, namely the lumbar spine, hip and forearm.

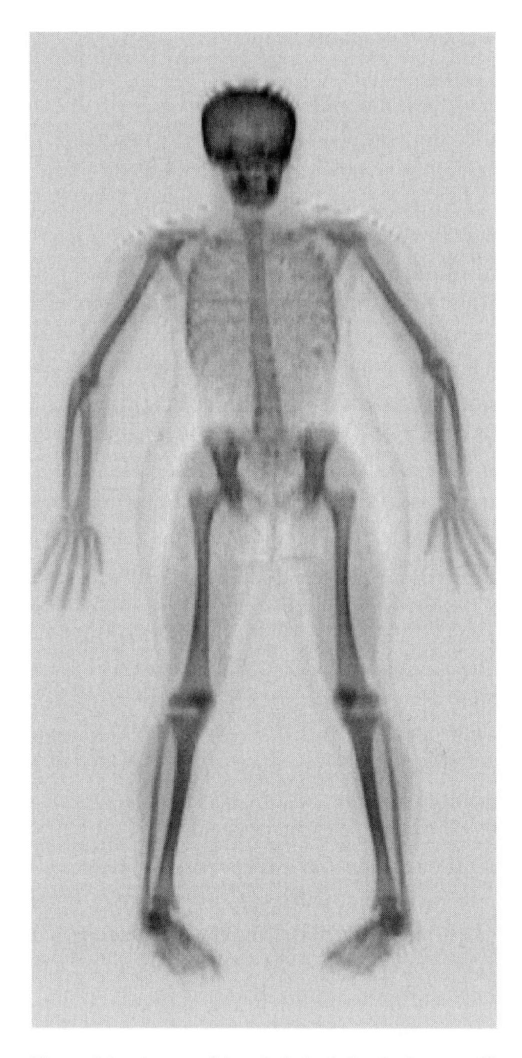

Figure 6.4 A scan of the whole body by dual energy X-ray absorptiometry (DXA) illustrating the fat, lean and bone tissue detected in subregions and the entire body.

DXA is becoming a widely used technique in body composition studies. It is rapid, safe, accurate and reproducible. Using spine phantoms, the coefficient of variation for bone mineral density measurements is usually less than 1%. For human subjects, reproducibility is somewhat larger and varies by anatomic site of measurement, with reported values ranging from 1% to 5% (Prentice, 1995; Laskey, 1996). Only one study has reported the reproducibility of body composition estimates by DXA for children (Figueroa-Colon *et al.*, 1998). They reported CVs of 2.08 to 6.55 over a 6-week period for FFM, FM and percent body fat. Other limitations of DXA should also be recognized. The equipment is expensive, rarely mobile, and requires a fair amount of space and technician training. The technique assumes that bone is a homogeneous tissue, and it is unable to distinguish variability in the amounts of cortical and trabecular bone, or the size or composition of the marrow space within the bone. The bone mineral density derived from DXA is an areal density (gm/cm^2) that comes from a two-dimensional image of the body, rather than a volumetric density such as that obtained by computed tomography (gm/cm^3), and has thus led to some erroneous interpretations (Leonard and Zemel, 2002). The calibration of devices varies by manufacturer and, within the same manufacturer, by model and software version. Body composition analysis for infants is available for research purposes on the Hologic, Inc. devices (Brunton *et al.*, 1997; Picaud *et al.*, 1999). The validity of body composition results for older infants and toddlers, as they transition from the infant mode to the adult mode, is uncertain. Further, body composition analysis by DXA for young children may require further refinement. For example, Hologic, Inc. developed a paediatric analysis software program for their DXA devices for use in children under age 10 to 12 years old. However, the guidelines for use of this analysis option, such as the appropriate age or body size range, were not clearly delineated, and the results were sufficiently different that data derived from children analysed in paediatric mode and adolescents analysed in adult mode could not be combined (Zemel *et al.*, 2000). Further refinement of the paediatric analysis software is under way (Kelly, 2002).

Other limitations of DXA include the exclusion of subjects with metal implants, those who are unable to complete a measurement in a standard position or those who can not co-operate without movement (such as young children, or subjects with seizures or cerebral palsy (Henderson *et al.*, 2001)). The scan time continues to decrease as the technology improves, so that the issue of movement artefact may become negligible in the future. In addition, the increasing availability of DXA for body composition assessment has expanded the potential for three- and four-compartment models in body composition research. The inclusion of bone mineral density in these models eliminates one of the primary sources of error in the estimation of the density of FFM in classic two-compartment models. Unlike some other methods, such as BIA and anthropometric methods of body composition assessment, DXA estimates of body composition are based on assumptions about the attenuation properties of tissue compartments and are not dependent on sample-specific prediction equations.

Total body potassium

Measurement of whole-body potassium is used to estimate the body cell mass. The body cell mass is the fat-free intracellular space, consisting of the intracellular fluids and a smaller proportion of intracellular solids of the organs and muscles. It excludes extracellular fluids, solids (such as bone mineral and collagen) and fat. Whole-body potassium is an index of the body cell mass because it is present only in the intracellular fluid space; a constant ratio of intracellular fluid to body cell mass is assumed (body cell mass = total body K (mmol) $\times$ 0.0083). Potassium (^{40}K) is a naturally occurring stable isotope that occurs in the human body as a small percentage of total body potassium (0.0118%), and it emits a strong gamma ray. Using a gamma ray detector within a lead-shielded room (^{40}K counter), the whole-body content of ^{40}K can be determined. The total body potassium is calculated as ^{40}K/0.0118%. Since potassium is within the intracellular space, ^{40}K also can be used in combination with other methods, such as TBW, in multilevel models of body composition (see Table 6.3).

In vivo ***neutron activation analysis***

About 95% of the body is comprised of O, C, H and N (Wang *et al.*, 1992). With the addition of Na, K, P, Cl, Ca, Mg and S, over 99.5% of the elemental composition of the body can be described (Heymsfield *et al.*, 1997). *In vivo* neutron activation analysis is a highly specialized method for measuring many of the atomic level (see Table 6.2) components of the body (Ca, C, Cl, H, N, Na, O and P, as well as trace elements of Al, Cd, Cu, Fe and Si). Translational equations (see Table 6.7) can be used to convert these measures to molecular or tissue level compartments.

This technique is based on the principle that controlled neutron irradiation generates a known amount of radioactivity in a given substance of known mass. An element within that substance can be quantified by the characteristic energy it emits as well as the decay rate. While resting in a shielded chamber, the subject is bombarded with a dose of fast neutrons derived from either a neutron source, such as ^{238}PuBe, or from a neutron accelerator. The neutrons interact with body tissues creating unstable isotopes, which emit gamma radiation. A whole-body gamma radiation counter determines the total quantity of the element in the body (Ryde, 1995).

In vivo neutron activation is a methodology that is available only at a few research centres, so it is not widely used. In addition, this technique has not

Table 6.7 *Translational equations for measurements of elements and other components in the body*

Total body measure	Translation equation
Calcium	0.340 bone mineral content
Nitrogen	0.161 total body protein
Carbon	0.759 total body lipid + 0.532 total body protein + 0.018 bone mineral content
Potassium	120 body cell mass
Chloride	111 extracellular water
Lipid	1.318 total body carbon − 4.353 total body nitrogen + 0.070 total body calcium
Protein	6.21 total body nitrogen
Osseous Mineral	2.941 total body calcium
Soft Tissue Mineral	2.75 total body potassium + total body sodium + 1.43 + total body chloride − 0.038 total body calcium

Source: Adapted from Ellis (2000).

been used for the assessment of body composition in infants and children due to the radiation exposure and the great expense involved (Puig, 1997). More detailed descriptions of the technique are available (Ryde, 1995; Ellis, 1996, 2000; Heymsfield *et al.*, 1997).

Computed tomography

Computed tomography (CT) is an X-ray based imaging modality that is used to generate three-dimensional images of body compartments. A single CT slice can be used to estimate cross-sectional areas of body compartments, such as visceral adipose tissue (Borkan *et al.*, 1982, 1983; Tershakovec *et al.*, 2003) or cortical versus trabecular bone (Gilsanz, 1998); multiple slices can be used to construct organ volumes (Heymsfield *et al.*, 1979) and regional or total body estimates of body composition (Sjöstrom, 1993; Mitsiopoulos *et al.*, 1998). It has been used in body composition studies of adults and children; however, the expense and radiation exposure limits the broad use of this technique.

The technique involves an X-ray source and detectors located at opposite ends of a ring. As the ring rotates around the subject, the detectors measure the attenuation of the X-ray beam as it passes through tissues of varying density. The information from each detector at each rotation is digitized, and sent to a computer in order to construct a two-dimensional image of the tissue area. CT systems vary in speed and slice thickness, and newer systems are able to

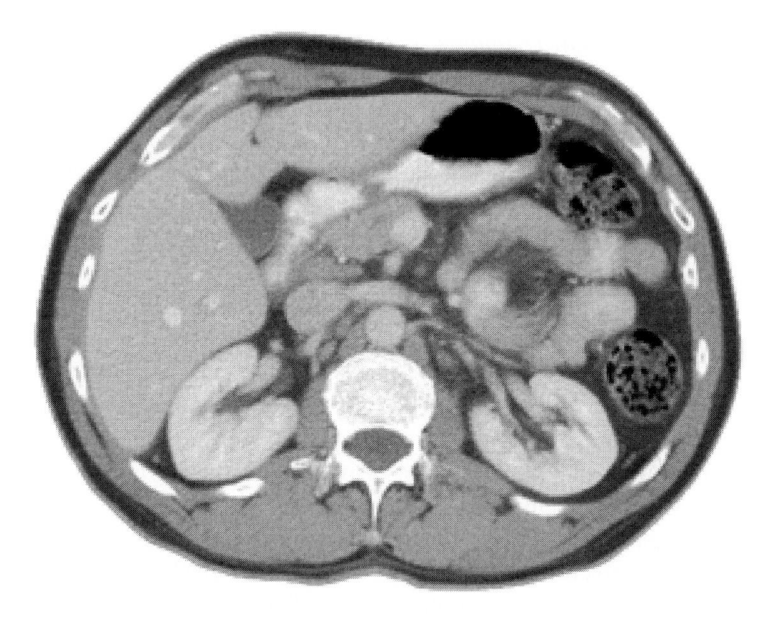

Figure 6.5 Computed tomography (CT) scan: a cross-sectional image of the abdomen. Cross-sectional CT images are used to estimate the volume of tissues such as subcutaneous and intra-abdominal fat, and organ sizes.

obtain multiple slices simultaneously. Helical scan modes are also available (Kobayashi *et al.*, 2002). Peripheral CT devices are used for imaging the bone (especially cortical versus trabecular bone), muscle and fat at appendicular sites such as the distal radius or tibia (Gilsanz, 1998; Leonard and Zemel, 2002).

CT imaging has been used most widely to estimate abdominal fat distribution. An example of an abdominal scan is shown in Figure 6.5. In adults, the reproducibility of results is excellent. Thaete *et al.* (1995) reported a high correlation between duplicate measurements ($r = 0.99$, $P < 0.01$), and small precision errors for adipose tissue areas (1.2% for total cross-sectional area, 1.9% for subcutaneous area, and 3.9% for visceral area). It is important to note that these measurements are highly sensitive to the slice position; Greenfield *et al.* (2002) measured four adjacent areas between the second and fourth lumbar vertebrae and observed an intra-subject variability of 28% (8–61%) for intra-abdominal fat area, and 26% (14–38%) for subcutaneous fat area. The reproducibility in children is uncertain, but studies in rats have documented the accuracy of CT in determining adipose tissue volumes in small animals (Ross *et al.*, 1991).

Magnetic resonance imaging

Magnetic resonance imaging (MRI) is a recent advancement in body composition technology that generates three-dimensional images and quantifies the volume and distribution of tissues or organs (Despres *et al.*, 1996; Pietrobelli *et al.*, 1998). In general, the information generated by MRI scans is preferred for body composition research since it does not involve the ionizing radiation associated with CT scans. Limiting factors are the expense and availability of MRI devices. A consideration in some applications of body composition research is that MRI assesses only tissue size and distribution, so under certain circumstances, ascertainment of tissue density by CT is preferred. The need to reduce motion artefacts, even those due to respiration and circulation, are additional considerations for its application in young children.

The MRI technique is based on the principle that the hydrogen protons in the body will align themselves within a high-powered magnetic field. A radiofrequency pulse specific to hydrogen atoms is then applied through a coil, causing the protons to absorb energy. When the pulse is turned off, the protons return to their natural state within the magnetic field, releasing their stored energy. Measurement of this energy release is computerized and used to generate the MRI images. Altering the local magnetic field during the measurement process permits imaging of different tissue compartments, since the proton density and relaxation times are tissue dependent.

Magnetic resonance imaging of tissue compartments has been validated in animal and cadaver studies (Fowler *et al.*, 1992; Ross, 1996; Mitsiopoulos *et al.*, 1998). It has been used successfully in studies of children, mainly in the assessment of intra-abdominal fat (Brambilla *et al.*, 1994; Yanovski *et al.*, 1996; Fox *et al.*, 2000).

Presentation and interpretation of body composition data in children

In addition to the challenges of obtaining body composition information in growing infants and children, there remains the problem of interpretation of body composition data. It would be ideal if body composition reference data were available that captured the normal (and healthy) range of variability in body compartments relative to gender, age and maturation. It would be relatively simple, for example, to determine if a child or group of children has too much body fat, reduced lean body mass, or expanded extracellular water. However, at present, such reference data do not exist and there are many obstacles to their development. Most measures of body composition are method dependent.

For example, FM measured by underwater weighing and DXA, or by DXA using equipment from different manufacturers, will give different results. The results may be highly correlated but different in scale, and the difference, or bias, is often size dependent. Therefore, the results are not interchangeable (Treuth *et al.*, 2001; Fors *et al.*, 2002) and reference data would need to be method specific. Even less is known about inter-laboratory differences in body composition techniques because of the time and expense involved in investigating such differences. Certainly, small deviations in mass spectroscopy techniques or equipment, or methods for measuring gas volumes in underwater weighing laboratories can result in significant inter-laboratory differences.

Given the many obstacles to the development of body composition reference data for children, it is important that studies of healthy children carefully delineate recruitment techniques so that sampling bias can be evaluated, and be of adequate sample size to approximate normal variability. Body compartments vary in relative and absolute size, and chemical composition in children relative to age, gender and maturation. These important biological sources of variability should be kept in mind when attempting to characterize body composition in children.

The size of body compartments is dependent, in large part, on overall body size. In growing children, particularly among those with growth abnormalities or health-related problems that impact growth, this is of particular concern. Typically, percent body fat is used as a single indicator of body composition, with the assumption that it is adequately adjusted for body size. However, this approach assumes that the relationship of fat mass to body weight is equivalent to a linear relationship with an intercept of zero and a constant slope. Certainly, for children this is not the case, as shown in Figure 6.6. The use of ratios such as percent body fat can result in an overestimation or underestimation of normal ranges of body fat relative to body mass depending on the size of the child. Thus, regression techniques are a more appropriate statistical approach for determining differences in body composition. In addition, it should be recognized that there is no single indicator of body composition in children. Inadequate size of body compartments relative to age and gender may represent a deficit in body composition and failure to achieve normal growth.

An alternative to the regression approach described above has been described which involves indexing body compartments to height, as is done with the body mass index. The fat mass index (FMI, calculated as FM/HT^2) and fat-free mass index (FFMI, FFM/HT^2) have been suggested but are not widely used. As shown by VanItallie *et al.* (1990) in a reanalysis of the Minnesota starvation experiments, FFMI and FMI were more sensitive indicators of nutritional status during chronic starvation. The utility of this index in children has been suggested (Wells, 2001), but requires further evaluation. The ideal index would have a zero

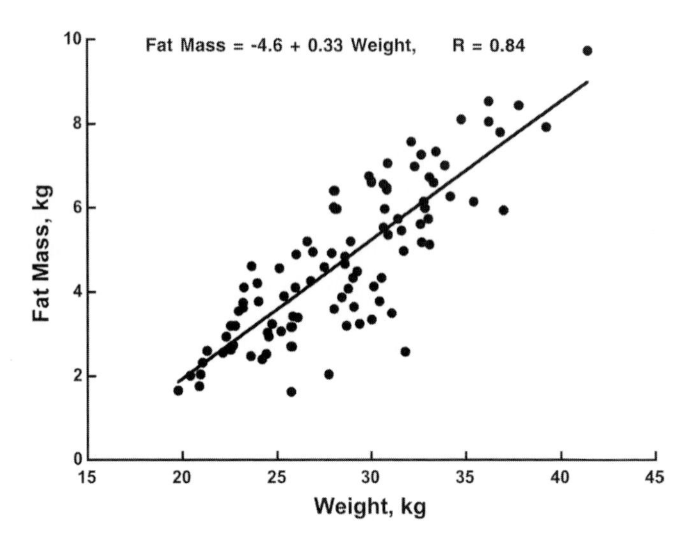

Figure 6.6 Fat mass versus body weight. The relationship between fat mass and weight can be described by a regression equation with a non-zero intercept. A simple ratio like percent body fat (fat mass/body weight × 100) misrepresents this important relationship. Data presented are from 139 healthy children, ages 7 to 10 years. (B. Zemel, unpublished data).

correlation with height, so as to minimize the effects of frame size on a measure of FM or FFM. The use of FMI and FFMI in children requires further evaluation due to the changes in body proportions and body composition with age.

Lastly, the changes in body composition associated with puberty represent an enormous challenge in the interpretation of body composition results in children. Typically, girls gain in fatness and muscle mass as puberty progresses. Boys undergo a prepubertal fat spurt, followed by distinctive increases in muscle mass with no change or loss in fat mass (see Figure 6.7). For both boys and girls, fat distribution changes with puberty. Thus, the timing of puberty must always be taken into account in interpreting the size and relative proportions of body compartments at both the individual and population level. While there are no clear guidelines on the analysis, presentation or interpretation of these complex data, it is important to keep these developmental patterns in mind in evaluating body composition in children.

Summary

The assessment of body composition in children and adolescents is complicated by physiological variability in tissue composition associated with growth and

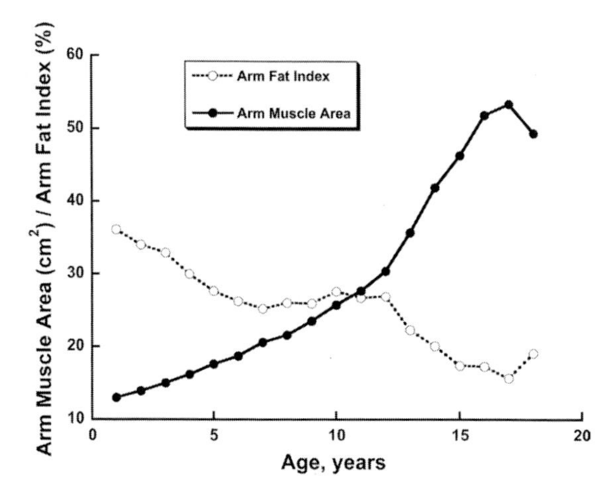

Figure 6.7 Changes in body composition during puberty in boys. Rapid changes in body composition occur during puberty. In boys, muscle mass increases while fat mass remains the same or decreases. Shown here are the 50th centile values for boys for anthropometric measures of body composition from the upper arm (Frisancho, 1990). The arm fat index (upper arm fat area/upper arm area) illustrates the prepubertal fat spurt and subsequent decline in relative fatness in boys. Because of these dynamic changes, multiple measures are needed to interpret body composition information.

maturation. Several techniques are available for use with children, although careful consideration must be given to the level of complexity required, underlying assumptions inherent in a technique's computations, and the availability of appropriate reference data for comparison. Techniques based on prediction equations derived from healthy subjects may not be suitable for use with children who have disease-associated alterations in the composition or proportions of fat, lean or skeletal tissues.

REFERENCE

Barden, E. M., Kawchak, D. A., Ohene-Frempong, K., Stallings, V. A. and Zemel, B. S. (2002). Body composition in children with sickle cell disease. *American Journal of Clinical Nutrition*, **76**, 218–25.

Baumgartner, R. (1996). Electrical impedance and total body electrical conductivity. In *Human Body Composition*, eds. A. F. Roche, S. B. Heymsfield and T. G. Lohman, pp. 79–107. Champaign, IL: Human Kinetics.

Bogin, B. and Sullivan, T. (1986). Socioeconomic status, sex, age, and ethnicity as determinants of body fat distribution for Guatemalan children. *American Journal of Physical Anthropology*, **69**, 527–35.

Boileau, R. (1988). Utilization of total body electrical conductivity in determining body composition. In *Designing Foods: Animal Product Options in the Marketplace*,

National Research Council, pp. 251–7. Washington, DC: National Academy Press.

Boileau, R., Lohman, T. G., Slaughter, M. H., *et al.* (1984). Hydration of the fat-free body in children during maturation. *Human Biology*, **56**, 651–66.

Boileau, R., Lohman, T. G. and Slaughter, M. H. (1985). Exercise and body composition in children and youth. *Scandinavian Journal of Sports Science*, **7**, 17–27.

Borkan, G. A., Gerzof, G. S., Robbins, A. H., Hults, D. E. and Silbert, C. K. (1982). Assessment of abdominal fat content by computerized tomography. *American Journal of Clinical Nutrition*, **36**, 172–7.

Borkan, G. A., Hults, D. E., Gerzof, S. G., Burrows, B. A. and Robbins, A. H. (1983). Relationships between computed tomography tissue areas, thicknesses and total body composition. *Annals of Human Biology*, **10**, 537–45.

Boye, K. R., Dimitriou, T., Manz, F., *et al.* (2002). Anthropometric assessment of muscularity during growth: estimating fat-free mass with two skinfold-thickness measurements is superior to measuring mid upper arm muscle area in healthy prepubertal children. *American Journal of Clinical Nutrition*, **76**, 628–32.

Brambilla, P., Manzoni, P., Sironi, S., *et al.* (1994). Peripheral and abdominal adiposity in childhood obesity. *International Journal of Obesity and Related Metabolic Disorders*, **18**, 795–800.

Brook, C. (1971). Determination of body composition of children from skinfold measurements. *Archives of Disease in Childhood*, **46**, 182–4.

Brozek, J., Grande, F., Anderson J. T. and Keys, A. (1963). Densitometric analysis of body composition: revision of some quantitative assumptions. *Annals of the New York Academy of Science*, **110**, 113–40.

Brunton, J. A., Weiler, H. A. and Atkinson, S. A. (1997). Improvement in the accuracy of dual energy X-ray absorptiometry for whole body and regional analysis of body composition: validation using piglets and methodologic considerations in infants. *Pediatric Research*, **41**, 590–6.

Cameron, N. (1986). The methods of auxological anthropometry. In *Human Growth: A Comprehensive Treatise*, eds. F. Falkner and J. M. Tanner, vol. 2, pp. 3–46. New York: Plenum Press.

Chumlea, W. C. and Guo, S. S. (1994). Bioelectrical impedance and body composition: present status and future directions. *Nutrition Reviews*, **52**, 123–31.

Cole, T., Bellizzi, M. C., Flegal, K. M. and Dietz, W. H. (2000). Establishing a standard definition for child overweight and obesity worldwide: international survey. *British Medical Journal*, **320**, 1240–3.

Danford, L. C., Schoeller, D. A. and Kushner, R. F. (1992). Comparison of two bioelectrical impedance analysis models for total body water measurement in children. *Annals of Human Biology*, **19**, 603–7.

Davies, P. S., Preece, M. A., Hicks, C. J. and Halliday, D. (1988). The prediction of total body water using bioelectrical impedance in children and adolescents. *Annals of Human Biology*, **15**, 237–40.

de Meer, K., Gulmans, V. A., Westerterp, K. R., Houwen, R. H. and Berger, R. (1999). Skinfold measurements in children with cystic fibrosis: monitoring fat-free mass and exercise effects. *European Journal of Pediatrics*, **158**, 800–6.

Demerath, E. W., Guo, S. S., Chumlea, W. C., *et al.* (2002). Comparison of percent body fat estimates using air displacement plethysmography and hydrodensitometry in adults and children. *International Journal of Obesity and Related Metabolic Disorders*, **26**, 389–97.

Dempster, P. and Aitkens, S. (1995). A new air displacement method for the determination of human body composition. *Medicine and Science in Sports and Exercise*, **27**, 1692–7.

Despres, J.-P., Ross, R. and Lemieux, S. (1996). Imaging techniques applied to the measurement of human body composition. In *Human Body Composition*, eds. A. F. Roche, S. B. Heymsfield and T. G. Lohman, pp. 149–66. Champaign, IL: Human Kinetics.

Deurenberg, P., van der Kooy, K., Paling, A. and Withagen, P. (1989). Assessment of body composition in 8–11-year-old children by bioelectrical impedance. *European Journal of Clinical Nutrition*, **43**, 623–9.

Durnin, J. and Rahaman, M. (1967). The assessment of the amount of fat in the human body from measurements of skinfold thickness. *British Journal of Nutrition*, **21**, 681–9.

Durnin, J. and Womersley, J. (1974). Body fat assessment from total body density and its estimation from skinfold thickness measurements on 481 men and women aged 16 to 72 years. *British Journal of Nutrition*, **32**, 77–97.

Durnin, J. V., Armstrong, W. H. and Womersley, J. (1971). An experimental study on the variability of measurements of skinfold thicknesses by three observers on twenty-three young women and twenty-seven young men. *Proceedings of the Nutrition Society*, **30**, 9A–10A.

Durnin, J. V., de Bruin, H. and Feunekes, G. I. (1997). Skinfold thicknesses: is there a need to be very precise in their location? *British Journal of Nutrition*, **77**, 3–7.

Ellis, K. J. (1996). Whole-body counting and neutron activation analysis. In *Human Body Composition*, eds. A. F. Roche, S. B. Heymsfield and T. G. Lohman, pp. 45–61. Champaign, IL: Human Kinetics.

(2000). Human body composition: *in vivo* methods. *Physiological Reviews*, **80**, 649–80.

Ellis, K. J., Shypailo, R. J., Abrams, S. A. and Wong, W. W. (2000). The reference child and adolescent models of body composition: a contemporary comparison. *Annals of the New York Academy of Sciences*, **904**, 374–82.

Ellis, K. J., Shypailo, R. J. and Wong, W. W. (1999). Measurement of body water by multifrequency bioelectrical impedance spectroscopy in a multi-ethnic pediatric population. *American Journal of Clinical Nutrition*, **70**, 847–53.

Fields, D. A. and Goran, M. I. (2000). Body composition techniques and the four-compartment model in children. *Journal of Applied Physiology*, **89**, 613–20.

Fields, D. A., Wilson, G. D., Gladden, L. B., *et al.* (2001). Comparison of the Bod Pod with the four-compartment model in adult females. *Medicine and Science in Sports and Exercise*, **33**, 1605–10.

Fields, D. A., Goran, M. I. and McCrory, M. A. (2002). Body-composition assessment via air-displacement plethysmography in adults and children: a review. *American Journal of Clinical Nutrition*, **75**, 453–67.

Figueroa-Colon, R., Mayo, M. S., Treuth, M. S., Aldridge, R. A. and Weinsier, R. L. (1998). Reproducibility of dual-energy X-ray absorptiometry measurements in prepubertal girls. *Obesity Research*, **6**, 262–7.

Fiorotto, M., de Bruin, N. C., Brans, Y. W., Degenhart, H. J. and Visser, H. K. A. (1995). Total body electrical conductivity measurements: an evaluation of current instrumentation for infants. *Pediatric Research*, **37**, 94–100.

Fomon, S. J. and Nelson, S. E. (2002). Body composition of the male and female reference infants. *Annual Review of Nutrition*, **22**, 1–17.

Fomon, S., Haschke, F., Ziegler, E. E. and Nelson, S. E. (1982). Body composition of reference children from birth to age 10 years. *American Journal of Clinical Nutrition* **35**, 1169–75.

Food and Nutrition Board (2002). *Dietary Reference Intakes for Energy, Carbohydrate, Fiber, Fat, Fatty Acids, Cholesterol, Protein, and Amino Acids (Macronutrients)*. Washington, DC: National Academic Press.

Forbes, G. (1987). *Human Body Composition: Growth, Aging, Nutrition and Activity*. New York: Springer-Verlag.

Fors, H., Gelander, L., Bjarnason, R., Albertsson-Wikland, K. and Bosaeus, I. (2002). Body composition, as assessed by bioelectrical impedance spectroscopy and dual-energy X-ray absorptiometry, in a healthy paediatric population. *Acta Paediatrica*, **91**, 755–60.

Foster, K. R. and Lukaski, H. C. (1996). Whole-body impedance: what does it measure? *American Journal of Clinical Nutrition*, **64**, 388S–396S.

Fowler, P. A., Fuller, M. F., Glasbey, C. A., Cameron, G. G. and Foster, M. A. (1992). Validation of the *in vivo* measurement of adipose tissue by magnetic resonance imaging of lean and obese pigs. *American Journal of Clinical Nutrition*, **56**, 7–13.

Fox, K. R., Peters, D. M., Sharpe, P. and Bell, M. (2000). Assessment of abdominal fat development in young adolescents using magnetic resonance imaging. *International Journal of Obesity and Related Metabolic Disorders*, **24**, 1653–9.

Frisancho, A. (1981). New norms of upper limb fat and muscle areas for assessment of nutritional status. *American Journal of Clinical Nutrition*, **34**, 2450–5.

Frisancho, A. (1990). *Anthropometric Standards for the Assessment of Growth and Nutritional Status*. Ann Arbor, MI: University of Michigan Press.

Gibson, R. (1990). *Principles of Nutritional Assessment*. Oxford, UK: Oxford University Press.

Gilsanz, V. (1998). Bone density in children: a review of the available techniques and indications. *European Journal of Radiology*, **26**, 177–82.

Going, S. (1996). Densitometry. In *Human Body Composition*, eds. A. F. Roche, S. B. Heymsfield and T. G. Lohman, pp. 3–23. Champaign, IL: Human Kinetics.

Greenfield, J. R., Samaras, K., Chisholm, D. J. and Campbell, L. V. (2002). Regional intra-subject variability in abdominal adiposity limits usefulness of computed tomography. *Obesity Research*, **10**, 260–5.

Guo, S. S., Siervogel, R. M. and Chumlea, W. C. (2000). Epidemiological applications of body composition: the effects and adjustment of measurement errors. *Annals of the New York Academy of Sciences*, **904**, 312–16.

Haschke, F. (1983). Body composition of adolescent males. I. Total body water in normal adolescent males. II. Body composition of the male reference adolescent. *Acta Paediatrica Scandinavica (Supplement)*, **307**, 1–23.

Haschke, F., Fomon, S. J. and Ziegler, E. E. (1981). Body composition of a nine-year-old reference boy. *Pediatric Research*, **15**, 847–9.

Henderson, R. C., Lark, R. K., Renner, J. B., *et al.* (2001). Dual X-ray absorptiometry assessment of body composition in children with altered body posture. *Journal of Clinical Densitometry*, **4**, 325–35.

Hewitt, M. J., Going, S. B., Williams, D. P. and Lohman, T. G. (1993). Hydration of the fat-free body mass in children and adults: implications for body composition assessment. *American Journal of Physiology*, **265**, E88–E95.

Heymsfield, S. B. and Matthews, D. (1994). Body composition: research and clinical advances. *Journal of Parenteral and Enteral Nutrition*, **18**, 91–103.

Heymsfield, S. B., Fulenwider, T., Nordlinger, B., *et al.* (1979). Accurate measurement of liver, kidney, and spleen volume and mass by computerized axial tomography. *Annals of Internal Medicine*, **90**, 185–7.

Heymsfield, S. B., Wang, Z., Baumgartner, R. N. and Ross, R. (1997). Human body composition: advances in models and methods. *Annual Review of Nutrition*, **17**, 527–58.

Himes, J. and Dietz, W. H. (1994). Guidelines for overweight in adolescent preventive services: recommendations from an expert committee. *American Journal of Clinical Nutrition*, **59**, 307–16.

Houtkooper, L. B., Lohman, T. G., Going, S. B. and Hall, M. C. (1989). Validity of bioelectric impedance for body composition assessment in children. *Journal of Applied Physiology*, **66**, 814–21.

Kalkwarf, H. J., Khoury, J. C. and Lanphear, B. P. (2003). Milk intake during childhood and adolescence, adult bone density, and osteoporotic fractures in US women. *American Journal of Clinical Nutrition*, **77**, 257–65.

Kehayias, J. J. and Valtuena, S. (1999). Neutron activation analysis determination of body composition. *Current Opinion in Clinical Nutrition and Metabolic Care*, **2**, 453–63.

Kelly, T. L. (2002). Pediatric whole body measurements. *American Journal of Bone Mineral Research*, **17** (Suppl. 1), S297.

Kobayashi, J., Tadokoro, N., Watanabe, M. and Shinomiya, M. (2002). A novel method of measuring intra-abdominal fat volume using helical computed tomography. *International Journal of Obesity and Related Metabolic Disorders*, **26**, 398–402.

Kuczmarski, R., Ogden, C. L. and Grummer-Strawn, L. M. (2000). *CDC Growth Charts: United States*. Hyattsville, MD: National Center for Health Statistics.

Kushner, R. F., Schoeller, D. A., Fjeld, C. R. and Danford, L. (1992). Is the impedance index (ht2/R) significant in predicting total body water? *American Journal of Clinical Nutrition*, **56**, 835–9.

Lands, L. C., Gordon, C., Bar-Or, O., *et al.* (1993). Comparison of three techniques for body composition analysis in cystic fibrosis. *Journal of Applied Physiology*, **75**, 162–6.

Laskey, M. A. (1996). Dual-energy X-ray absorptiometry and body composition. *Nutrition*, **12**, 45–51.

Leonard, M. B. and Zemel, B. S. (2002). Current concepts in pediatric bone disease. *Pediatric Clinics of North America*, **49**, 143–73.

Lohman, T. G. (1986). Applicability of body composition techniques and constants for children and youths. *Exercise Sport Science Reviews*, **14**, 325–57.

(1989). Assessment of body composition in children. *Pediatric and Exercise Science*, **1**, 19–30.

(1992). *Advances in Body Composition Assessment*. Champaign, IL: Human Kinetics.

Lohman, T. G., Slaughter, M. H., Boileau, R. A., Bunt, J. and Lussier, L. (1984). Bone mineral measurements and their relation to body density in children, youth and adults. *Human Biology*, **56**, 667–79.

Lohman, T. G., Roche, A. F. and Martorell, R. (1988). *Anthropometric Standardization Reference Manual*. Champaign, IL: Human Kinetics.

Lukaski, H. C. (1996). Biological indexes considered in the derivation of the bioelectrical impedance analysis. *American Journal of Clinical Nutrition*, **64**, 397S–404S.

Maynard, L. M., Wisemandle, W., Roche, A. F., *et al.* (2001). Childhood body composition in relation to body mass index. *Pediatrics*, **107**, 344–50.

Mei, Z., Grummer-Strawn, L. M., Pietrobelli, A., *et al.* (2002). Validity of body mass index compared with other body-composition screening indexes for the assessment of body fatness in children and adolescents. *American Journal of Clinical Nutrition*, **75**, 978–85.

Mitsiopoulos, N., Baumgartner, R. N., Heymsfield, S. B., *et al.* (1998). Cadaver validation of skeletal muscle measurement by magnetic resonance imaging and computerized tomography. *Journal of Applied Physiology*, **85**, 115–22.

Mueller, W. H. and Kaplowitz, H. J. (1994). The precision of anthropometric assessment of body fat distribution in children. *Annals of Human Biology*, **21**, 267–74.

Nicholson, J. C., McDuffie, J. R., Bonat, S. H., *et al.* (2001). Estimation of body fatness by air displacement plethysmography in African American and white children. *Pediatric Research*, **50**, 467–73.

Nielsen, D., Cassady, S. L., Janz, K. F., *et al.* (1993). Criterion methods of body composition analysis for children and adolescents. *American Journal of Human Biology*, **4**, 211–23.

Norgan, N. (1994). Relative sitting height and the interpretation of the body mass index. *Annals of Human Biology*, **21**, 79–82.

Oldham, N. M. (1996). Overview of bioelectrical impedance analyzers. *American Journal of Clinical Nutrition*, **64**, 405S–412S.

Picaud, J. C., Nyamugabo, K., Braillon, P., *et al.* (1999). Dual-energy X-ray absorptiometry in small subjects: influence of dual-energy X-ray equipment on assessment of mineralization and body composition in newborn piglets. *Pediatric Research*, **46**, 772–7.

Pietrobelli, A., Wang, Z. and Heymsfield, S. B. (1998). Techniques used in measuring human body composition. *Current Opinion in Clinical Nutrition and Metabolic Care*, **1**, 439–48.

Prentice, A. (1995). Application of dual energy X-ray absorptiometry and related techniques to the assessment of bone and body composition. In *Body Composition Techniques in Health and Disease*, eds. P. Davies and T. J. Cole, pp. 1–13. Cambridge, UK: Cambridge University Press.

Puig, M. (1997). Body composition and growth. In *Nutrition in Pediatrics: Basic Science and Clinical Applications*, eds. W. A. Walker and J. B. Watkins, pp. 44–62. London: B. C. Decker.

Rolland-Cachera, M. (1993). Body composition during adolescence: methods, limitations and determinants. *Hormone Research*, **39**, 25–40.

Ross, R. (1996). Magnetic resonance imaging provides new insights into the characterization of adipose and lean tissue distribution. *Canadian Journal of Physiology and Pharmacology*, **74**, 778–85.

Ross, R., Leger, L., Guardo, R., De Guise, J. and Pike, B. G. (1991). Adipose tissue volume measured by magnetic resonance imaging and computerized tomography in rats. *Journal of Applied Physiology*, **70**, 2164–72.

Ryde, S. (1995). *In vivo* neutron activation analysis: past, present and future. In *Body Composition Techniques in Health and Disease*, eds. P. Davies and T. J. Cole, pp. 14–37. Cambridge, UK: Cambridge University Press.

Schoeller, D. (1996). Hydrometry. In *Human Body Composition*, eds. A. F. Roche, S. B. Heymsfield and T. G. Lohman, pp. 25–43. Champaign, IL: Human Kinetics.

Schoeller, D., van Santen, E., Peterson, D. W., *et al.* (1980). Total body water measurements in humans using ^{18}O and ^{2}H labeled water. *American Journal of Clinical Nutrition*, **33**, 2686–93.

Schoeller, D., Kushner, R., Taylor, P., Dietz, W. and Bandini, L. (1985). Measurement of total body water: isotope dilution techniques. In *Body-Composition Assessments in Youth and Adults: Report of the Sixth Ross Conference on Medical Research*, pp. 124–9. Columbus, OH: Ross Laboratories.

Schutte, J., Townsend, E. J., Hugg, J., *et al.* (1984). Density of lean body mass is greater in blacks than in whites. *Journal of Applied Physiology*, **56**, 1647–9.

Shepard, R. (1991). *Body Composition in Biological Anthropology*. Cambridge, UK: Cambridge University Press.

Siri, W. (1961). Body composition from fluid spaces and density: analysis of methods. In *Techniques for Measuring Body Composition*, eds. J. Brozek and A. Henschel, pp. 223–4. Washington, DC: National Academy of Sciences, National Research Council.

Sjöstrom, L. (1993). Body composition studies with CT and with CT-calibrated anthropometric techniques. In *Recent Developments in Body Composition Analysis: Methods and Applications*, eds. J. Kral and T. B.VanItallie, pp. 17–34. London: Smith-Gordon Nishimura.

Slaughter, M., Lohman, T. G., Boileau, R. A., *et al.* (1988). Skinfold equations for estimation of body fatness in children and youth. *Human Biology*, **60**, 709–23.

Spender, Q. W., Cronk, C. E., Stallings, V. A. and Hediger, M. L. (1988). Fat distribution in children with cerebral palsy. *Annals of Human Biology*, **15**, 191–6.

Spicher, V., Roulet, M., Schaffner, C. and Schutz, Y. (1993). Bioelectrical impedance analysis for estimation of fat-free mass and muscle mass in cystic fibrosis patients. *European Journal of Pediatrics*, **152**, 222–5.

Tershakovec, A. M., Kuppler, K. M., Zemel, B. S., *et al.* (2003). Body composition and metabolic factors in obese children and adolescents. *International Journal of Obesity and Related Metabolic Disorders*, **27**, 19–24.

Thaete, F. L., Colberg, S. R., Burke, T. and Kelley, D. E. (1995). Reproducibility of computed tomography measurement of visceral adipose tissue area. *International Journal of Obesity and Related Metabolic Disorders*, **19**, 464–7.

Treuth, M. S., Butte, N. F., Wong, W. W. and Ellis, K. J. (2001). Body composition in prepubertal girls: comparison of six methods. *International Journal of Obesity and Related Metabolic Disorders*, **25**, 1352–9.

Trowbridge, F. L., Marks, J. S., Lopez de Romana, G., *et al.* (1987). Body composition of Peruvian children with short stature and high weight-for-height. II. Implications for the interpretation for weight-for-height as an indicator of nutritional status. *American Journal of Clinical Nutrition*, **46**, 411–18.

Ulijaszek, S. J. and Lourie, J. (1994). Intra-and inter-observer error in anthropometric measurement. In *Anthropometry: The Individual and the Population*, eds. S. J. Ulijaszek and C. G. N. Mascie-Taylor, pp. 30–54. Cambridge, UK: Cambridge University Press.

Van Loan, M. (1990). Assessment of fat-free mass in teen-agers: use of TOBEC methodology. *American Journal of Clinical Nutrition*, **52**, 586–90.

Van Loan, M. and Mayclin, P. (1987). A new TOBEC instrument and procedure for the assessment of body composition: use of Fourier coefficients to predict lean body mass and total body water. *American Journal of Clinical Nutrition*, **45**, 131–7.

Van Loan, M., Keim, N. L. and Belko, A. Z. (1990). Body composition assessment of a general population using total body electrical conductivity ToBEC. In *Sports, Medicine and Health*, ed. G. Hermans, pp. 665–70. Geneva, Switzerland: Elsevier.

VanItallie, T. B., Yang, M. U., Heymsfield, S. B., Funk, R. C. and Boileau, R. A. (1990). Height-normalized indices of the body's fat-free mass and fat mass: potentially useful indicators of nutritional status. *American Journal of Clinical Nutrition*, **52**, 953–9.

Wagner, D. R. and Heyward, V. H. (2001). Validity of two-component models for estimating body fat of black men. *Journal of Applied Physiology*, **90**, 649–56.

Wang, Z., Pierson, R. N. and Heymsfield, S. B. (1992). The five level model: a new approach to organizing body composition research. *American Journal of Clinical Nutrition*, **56**, 19–28.

Weiner, J. and Lourie, J. A. (1981). *Practical Human Biology*. New York: Academic Press.

Wells, J. C. (2001). A critique of the expression of paediatric body composition data. *Archives of Disease in Childhood*, **85**, 67–72.

Wells, J. C. and Fuller, N. J. (2001). Precision of measurement and body size in whole-body air-displacement plethysmography. *International Journal of Obesity and Related Metabolic Disorders*, **25**, 1161–7.

Wells, J. C., Fuller, N. J., Dewit, O., *et al.* (1999). Four-component model of body composition in children: density and hydration of fat-free mass and comparison with simpler models. *American Journal of Clinical Nutrition*, **69**, 904–12.

Yanovski, J. A., Yanovski, S. Z., Filmer, K. M., *et al.* (1996). Differences in body composition of black and white girls. *American Journal of Clinical Nutrition*, **64**, 833–9.

Zemel, B. S. and Barden, E. M. (2001). Assessment of obesity. In *Obesity, Growth and Development*, eds. F. E. Johnston and G. D. Foster, pp. 143–67. London: Smith-Gordon.

Zemel, B. S., Leonard, M. B. and Stallings, V. A. (2000). Evaluation of the Hologic experimental pediatric whole body analysis software in healthy children and children with chronic diseases. *American Journal of Bone Mineral Research*, **15** (Suppl 1), 400.

Part II

Non-parametric and parametric approaches for individual growth

7 *Kernel estimation, shape-invariant modelling and structural analysis*

Theo Gasser
University of Zürich
Daniel Gervini
University of Zürich
and
Luciano Molinari
Kinderspital, Zürich

Introduction

There are a multitude of reasons for studying human growth, but two main areas of interest can be distinguished. First, one is interested in obtaining age-dependent reference curves for various somatic variables, usually based on cross-sectional data (see Chapters 2, 3 and 10 for details). The clinical value is evident, but reference curves are also needed for clinical research to obtain comparable values at different ages. A second motivation is mainly scientific and relates to an understanding of the growth process over certain phases, or, more ambitiously, from birth to adulthood (compare Chapters 1 and 2). Such studies are usually longitudinal and they are central in this chapter. In addition to the scientific interest, there are also clinical implications, e.g. for a qualitative understanding of growth disorders. The aim is to get a clear quantitative picture of prepubertal and pubertal growth. Further, we want to assess differences in growth in different parts of the body, and between boys and girls. An assessment of inter-individual variation of growth patterns is also needed. Whereas the distance curve is of utmost interest for reference curves, it does not offer detailed information about the dynamic and the intensity of the growth process (see also Chapter 2). Thus, interest focuses on the velocity curve and to a much lesser

Methods in Human Growth Research, eds. R. C. Hauspie, N. Cameron and L. Molinari.
Published by Cambridge University Press.

extent on the acceleration curve (while it is appealing, it is difficult to estimate reliably).

The methods presented in this chapter have proved their value for analysing growth longitudinally. From the viewpoint of statistics, regression methods based on an adequate parametric function are the methods of choice to analyse growth longitudinally. While polynomial models would be computationally convenient, they can only be used for short periods and are inadequate for longer periods or for periods including the pubertal growth spurt or other fast changes. Non-linear models need more expertise from the data analyst; the main problem in their application is, however, the choice of an appropriate model function. While for example the logistic function proved to be adequate for fitting the pubertal spurt of height (Marubini *et al.*, 1972), finding a good model for the growth process from birth to adulthood proved to be a thorny problem, since the model should have at most five or six parameters; even the best models missed some of the many different facets of the entire growth process, leading to biased results. This spurred interest in non-parametric methods of function fitting (also called 'smoothing methods') which work without prescribing a parametric model. The idea is that they should bring out the shape of the growth curve which is hidden in the noisy data. The fact that derivatives rather than distance functions are of interest now needs special attention, since derivatives are from the point of view of mathematics more difficult to estimate. The recent statistical literature offers a variety of methods for non-parametric function fitting (splines, kernel estimators, etc.). Here, we concentrate on kernel estimators, but the concepts are rather similar for other approaches. This method does not only render a curve but allows also to determine the age of occurrence of different features, such as the pubertal spurt, and other features of interest.

A characteristic of growth studies is that we have a sample of individual longitudinal data sets. While different subjects are different in quantitative terms, their growth has a similar shape, and this common shape should give information about the underlying model. It would then be of interest to quantify this common shape, and at the same time the individual parameters which stand for inter-individual differences. This can be achieved with a method called 'shape-invariant modelling' ('non-linear semiparametric modelling' in statistical terminology), to be presented here as the second method. The third method ('structural analysis') is also based on the whole sample of growth curves. Our first goal is to determine a valid average velocity curve. A naive average at chronological ages is certainly not valid, due to the different maturational tempo in different subjects: one boy at age 14 might be at the peak of the pubertal spurt, another might have just passed the spurt, whereas a third boy might still be experiencing prepubertal growth. Consequently, one has to align individual curves to an average tempo before averaging them. Determining this alignment function is based on particular features or milestones of the growth process. All three methods discussed in this chapter are of interest beyond the field of growth studies, since they are applicable to many types of noisy curves. In fact they have become, or are going to become, part of the statistical toolbox.

The First Zürich Longitudinal Study

The examples are mainly drawn from the First Zürich Longitudinal Study, which was part of an extensive multicentre growth study (Falkner, 1960). The first visit of the participants was at 4 weeks, and the following at 0.25, 0.5, 1, 1.5 and 2 years. Then the visits were at yearly intervals up to age 9 for girls and age 10 for boys. Afterwards, measurements took place half-yearly until further growth became small. Yearly measurements were then continued at least until age 18, but mostly until 20 (for further details see Prader *et al.*, 1989). In order to quantify infant growth more accurately, it proved to be beneficial to add a 'measurement' with value 0 at −0.75 years. We have almost complete data for 120 boys and 112 girls at our disposal (roughly 32 measurements per child).

Here, we will analyse primarily variables reflecting the growth of the skeleton: height, leg and trunk height, arm length, and biiliac and bihumeral width. They are mainly genetically determined and their growth shows a similar pattern in normal children. The mixed variables arm and calf circumference and weight – which also reflect growth of muscle mass and fat – can be analysed similarly. The development of skinfolds shows a variable pattern from subject to subject, due to a substantial non-genetic influence, and they may thus need a different strategy for analysis (see below, p. 201).

Modelling longitudinal growth

Assume that measurements of some variable y have been taken n times for some child at ages $t_1, \ldots, t_n$. In the examples below $t_1, \ldots, t_n$ will cover the whole growth process. Determining the underlying true growth curve r of that particular child is a standard regression problem:

$$y_i = r(t_i) + \varepsilon_i, \quad i = 1, \ldots, n \tag{7.1}$$

The so-called residuals ε_i are assumed to have mean zero and residual variance σ_ε^2 which may or may not depend on age. For skeletal growth the residual variance does depend on age: it is high in infancy and decreases then continuously, interrupted by a transient increase in puberty (see p. 189). These residuals stand for measurement error but also for biological random variation due to illness, seasonal and other exogenous factors. The regression function r represents the true underlying growth curve following the genetic potential. Fitting a linear function for r is easy, but totally inadequate for growth. Fitting a polynomial would also be easy, but is still inadequate, except for short periods. Thus, we truly need non-linear regression models, which are more difficult to treat statistically (see Chapter 9).

Over the years, a lot of work has gone into the development of adequate growth models. Here, we mention only a few:

- Double logistic model (Bock *et al.*, 1973):

$$r(t) = \frac{a_1}{1 + \exp[-b_1(t - c_1)]} + \frac{AH - a_1}{1 + \exp[-b_2(t - c_2)]}$$

  where AH is adult height.
- Model by Preece and Baines (1978):

$$\begin{aligned} r(t) = 4[a - H(c)]\langle\{b_1 \exp[b_1(t - c)] \\ + b_2 \exp[b_2(t - c)]\}\{1 + \exp[b_3(t - c)]\} \\ + \{\exp[b_1(t - c)] + \exp[b_2(t - c)]\}b_3 \exp[b_3(t - c)]\rangle \\ \div \langle\{\exp[b_1(t - c)] + \exp[b_2(t - c)]\}\{1 + \exp[b_3(t - c)]\}\rangle^2 \end{aligned}$$

  where $H(c)$ is height at age c (treated as an independent parameter).
- IPC model by Karlberg (1987):

$$\begin{aligned} r_1(t) &= a_1 + b_1[1 - \exp(-c_1 t)] && \text{(infancy, exponential)}, \\ r_2(t) &= a_2 + b_2 t + c_2 t^2 && \text{(childhood, quadratic)}, \\ r_3(t) &= a_3/\{1 + \exp[-b_3(t - c_3)]\} && \text{(puberty, logistic)}. \end{aligned}$$

  These submodels are fitted sequentially.
- Model by Jolicoeur *et al.* (1988):

$$r(t) = a_1\{1 - 1/[(t/b_1)^{c_1} + (t/b_2)^{c_2} + (t/b_3)^{c_3}]\}.$$

Note that the first two models need six parameters, the third nine and the fourth seven. The Preece–Baines model for $r(t)$ is actually the derivative of that proposed in the article cited above, because the authors model the distance curve instead of the velocity. The parameters of these models are usually estimated by non-linear least-squares methods, which involve iterative algorithms.

This list is, of course, not exhaustive but covers some of the popular models. These models do not provide a theoretical justification of measurements as do many models in the physical sciences – they are largely descriptive. Often they may describe a number of children adequately, but have difficulties in catching somewhat unusual individual curves. This problem can be avoided by using

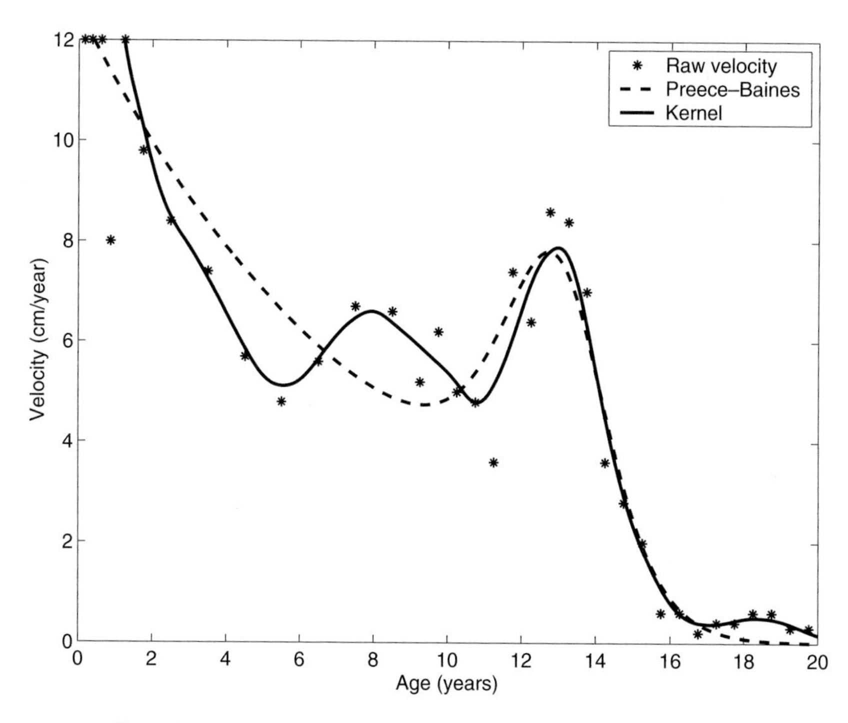

Figure 7.1 Height velocity of a girl in the First Zürich Longitudinal Study, from 0 to 20 years. Stars, raw velocity; solid line, kernel fit; dashed line, Preece–Baines fit.

non-parametric function fitting (see below, p. 184). Choosing models with more parameters allows more flexibility, but at the price of increased variability. For this reason it does not make sense to include more than five or six parameters, given the fact that we have about 30 to 34 measurements.

The main problem with these non-linear models is model bias: they deviate systematically from the underlying biological growth pattern. One example is the lack of the mid-growth spurt in most models which may also affect modelling the pubertal spurt, in particular for girls (as shown in Gasser *et al.* 1984a). This is already a problem for growth in height, where the mid-growth spurt is smallest, but much more so for other variables such as growth in the shoulder or trunk. The parametric models have been evaluated for height growth, which is easier to model than some other somatic variables. Thus, more problems might arise when analysing for example growth of the shoulder or the legs, both due to a relatively higher residual variance and more abrupt changes in the shape of the velocity curve. The general case does not present particular problems for the methods discussed in the sections 'Kernel estimation' (p. 184) and 'Structural analysis' (p. 197). Figure 7.1 illustrates the problem for the Preece–Baines model. This is a particularly regularly growing girl, and from the raw velocities one notes easily a pubertal spurt at about age 13, and a mid-growth spurt peaking around 8 years. The kernel-estimated velocity curve (see p. 185) reflects this pattern nicely, while the Preece–Baines model intrinsically lacks a mid-growth spurt which leads to an inadequate fit in this period. The lack of a mid-growth spurt is not the only bias; another one is a too early take-off for the pubertal spurt, which is a consequence of the first bias (identified as a general phenomenon in Gasser *et al.*, 1984a).

Some models may also show a bias for the pubertal spurt by enforcing a symmetric peak when it is, in fact, asymmetric. Models have typically also some difficulty to formalize the rapid transition from a steep decline in velocity in infancy to an almost constant or steadily decreasing velocity trend in childhood. The estimation is performed for the distance curve where, unfortunately, a poor fit is not easy to spot in comparison to the velocity or acceleration curve (thus, one should always inspect the goodness-of-fit in the velocity curve as well). Using the above models, the velocity curve is obtained by taking the first derivative of the model and plugging in the estimated individual parameters.

What are the peculiarities of analysing growth data by some suitable regression method? In contrast to most applications of regression in various fields, we are mainly interested in the velocity curve (and maybe in the acceleration curve). Derivatives of a function are notoriously more difficult to determine from noisy data than the function itself. Then we have a sample of subjects – and two genders – for which the method should be valid (in many disciplines there is just one curve to model). On one hand this multitude of curves represents a challenge, but on the other hand it is an opportunity, since we might be able to pool information from different subjects to obtain information on the 'true shape' of growth velocity. In what follows we will illustrate how these problems can be dealt with.

Kernel estimation

Motivation and definition

As discussed above, parametric non-linear models often fail to capture the true shape of the regression function underlying the noisy data. This problem is not exclusive to the analysis of growth data. For this reason, methods of curve estimation that work without an a priori fixed parametric model have been developed in statistics. They are therefore called *non-parametric*. The most important non-parametric function estimators are splines, kernel estimators and local polynomials (for a review see for example Gasser *et al.* (1993a), or the book by Fan and Gijbels (1996)). The different methods are closely related: an estimate of the regression function r at age t_0 is obtained by computing a local weighted average of the data y_i falling in an interval $(t_0 - h, t_0 + h)$ around t_0. Such weight functions or kernels are called 'compact' because they are null outside a given interval; non-compact kernels exist but for convenience we will use only compact ones (besides, the optimal kernels – in a sense explained below – are compact). Smoothing splines were applied in growth studies quite early (Largo *et al.*, 1978), but since kernel estimators have some advantages we will concentrate on these. The statistical theory has been presented in Gasser and Müller (1984) and Gasser *et al.* (1985a), while aspects related to growth can be found in Gasser *et al.* (1984a,b).

In what follows we will assume that $0 \leq t_1 \leq t_2 \leq \cdots \leq t_n \leq 1$ for ease of notation. A kernel estimator $\hat{r}(t_0)$ for $r(t_0)$ can be written as follows:

$$\hat{r}(t_0) = \sum_{i=1}^{n} w_i(t_0; t_1, \ldots, t_n; h) y_i \qquad (7.2)$$

where the weights w_i are given by

$$w_i(t_0; t_1, \ldots, t_n; h) = \frac{1}{h} \int_{s_{i-1}}^{s_i} K\left(\frac{t_0 - u}{h}\right) du,$$

$$s_i = \frac{t_i + t_{i+1}}{2}, \quad s_0 = 0, \quad s_n = 1 \qquad (7.3)$$

While the sum goes from 1 to n, only those y_i lying in the interval $(t_0 - h, t_0 + h)$ contribute to $\hat{r}(t_0)$. The above weighting scheme automatically accounts for unequal spacings between subsequent ages t_i, by giving more weight when the interval between two measurements is large. When applying this estimator, the decisive questions are how to choose the kernel (or weight function) K and how to choose the bandwidth (or smoothing parameter) h. This will be the central theme of this section.

The velocity curve $r'(t_0)$ or the acceleration curve $r''(t_0)$ (in general, the νth derivative $r^{(\nu)}(t_0)$) can be estimated using a formula similar to Eqn 7.2 with modified weights. The modifications needed in Eqn 7.3 are: first, the factor h^{-1} has to be replaced by $h^{-\nu-1}$, which implicitly produces a larger bandwidth for derivatives; second, a kernel K_ν with specific properties (see below) has to be used. For numerical reasons the data should be interpolated to an equally spaced dense grid before estimating derivatives. (For our growth data we have used simple linear interpolation between measurements based on 200 interpolated points.) It is important to note that bandwidth selection (see p. 187) has to be performed on the original data, not on the interpolated one.

While parametric methods – in the ideal case where the model is true – have null or negligible bias, kernel estimators, as all smoothing methods, lead to bias. Intuitively it is clear that strong smoothing incurs a strong bias and low variability, while little smoothing leads to little bias and strong variability; in between there should be a compromise balancing the two types of error. To make these intuitive arguments more precise, consider the following approximate (or asymptotic) expressions for bias and variance of the estimator of $r(t_0)$, for equally spaced data:

$$\text{Bias}[\hat{r}(t_0)] = h^2 r''(t_0) \int K(u) u^2 du$$

$$\text{Var}[\hat{r}(t_0)] = \frac{\sigma_\varepsilon^2}{2nh} \int K(u)^2 du$$

The assumption of equal spacing is made for simplicity and can be relaxed, and similar expressions can be obtained for the estimators of the derivatives (Gasser and Müller, 1984). Observe that while the bias increases as a square of the bandwidth h, the variance decreases hyperbolically with h. Moreover, the variance depends on the data only via the residual variance σ_ε^2, and the bias only on the second derivative of the underlying regression function r (or on $r^{(\nu+2)}$ when estimating the νth derivative). Qualitatively, then, bias leads to a flattening of peaks and valleys (where $|r''(t_0)|$ is large), and little distortion in flat parts of the curve (where $|r''(t_0)|$ is small). Thus the bias is 'conservative' insofar as it dampens true structure but does not generate artificial structure.

Ideally from a statistical perspective, the optimal choices of K (or K_ν for derivatives) and h should lead simultaneously to low bias and low variability. The following two subsections explain how to achieve this.

Choice of kernel

To find a compromise between bias and variance one usually minimizes the mean square error MSE: MSE = Bias^2 + Variance. Minimizing the MSE leads to different optimal kernels for $\nu = 0, 1, 2$, which do not depend on the data or on h. Certain conditions for K, K_1 and K_2 are necessary to guarantee that appropriately normalized sums (for K) and appropriately normalized differences (for K_1 and K_2) are formed so that r, r' and r'' are estimated correctly. For $\nu = 0$ one has to postulate that

$$\int K(u) du = 1, \quad K(u) = K(-u) \,(\text{symmetric}), \quad K(u) \geq 0$$

For $\nu = 1$ we require that

$$\int K_1(u) du = 0, \quad \int K_1(u) u\, du = -1, \quad K_1(u) = -K_1(-u)$$

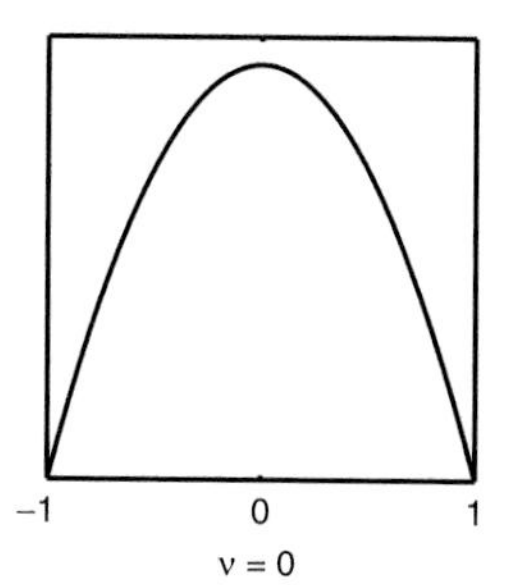

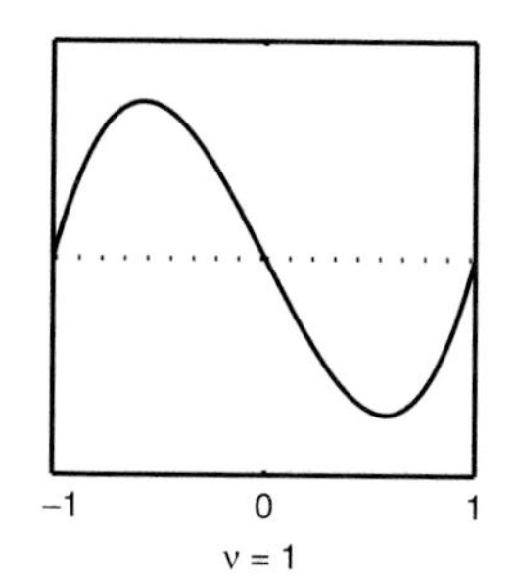

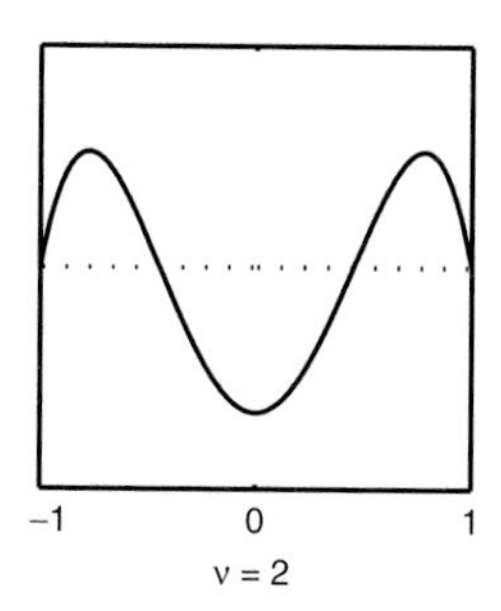

Figure 7.2 Optimal kernels for estimating r (left), for r' (middle) and for r'' (right).

and for $\nu = 2$:

$$\int K_2(u)du = 0, \quad \int K_2(u)udu = 0,$$
$$\int K_2(u)u^2 = 1, \quad K_2(u) = K_2(-u)$$

By non-elementary mathematical arguments, it is possible to explicitly construct optimal kernels K, K_1 and K_2 which minimize the MSE for estimating r, r' and r'' respectively (Gasser *et al.*, 1985b). The shape of these optimal kernels is depicted in Figure 7.2. These kernels are simple polynomials of degree 2, 3 and 4, respectively. As a consequence, the integral in Eqn 7.3 can be computed analytically and does not need to be computed numerically.

A little thinking shows that problems arise when these kernels are naively used at the boundary (e.g. to compute the velocity at birth, when data start at birth – truly a difficult problem). Special 'boundary kernels' have been constructed to solve those problems. In growth studies, one can also alleviate boundary problems by introducing 'data points' at age −0.75 years with value zero (representing height at conception).

Choice of bandwidth

It is a nice feature that the same kernels can be used irrespective of the data set at hand. Unfortunately this is not true for the optimal bandwidth, which depends heavily on the data. Figure 7.3 shows, for simulated data, the effect of mis-specifying the bandwidth: it results in under- or oversmoothing. Too small a bandwidth leads to a rough appearance of the estimated curve, with many random wiggles – a sign of undersmoothing. Choosing too large a bandwidth leads to a very smooth appearance, but a distortion of the curve at peaks. The

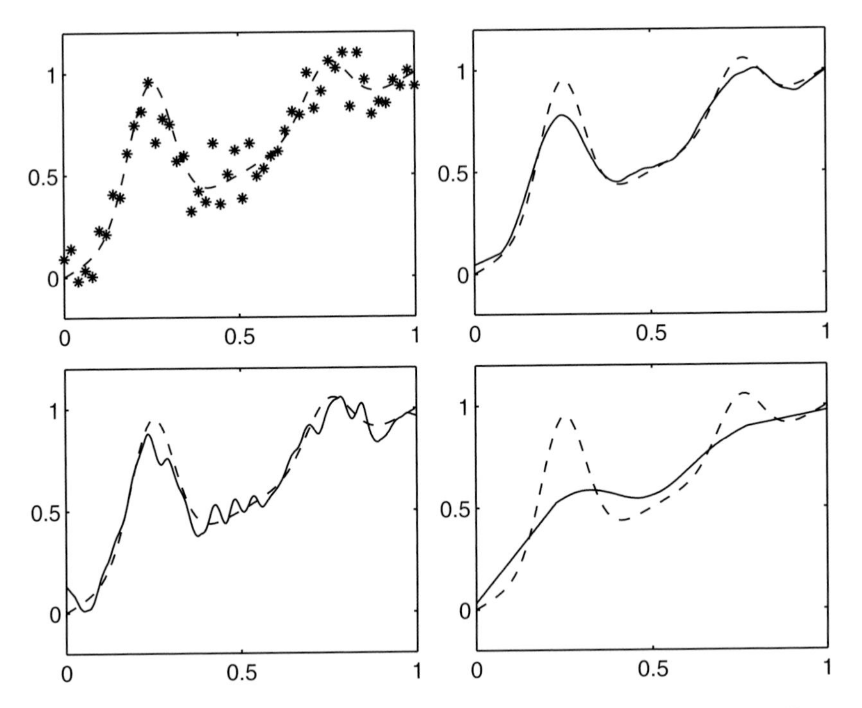

Figure 7.3 Kernel smoothing with various bandwidths for synthetic data ($n = 50$, above left; stars, data, dashed line, true function), undersmoothing (below left), oversmoothing (below right) and optimal smoothing (above right).

optimal bandwidth is a compromise tolerating some bias for an overall smooth estimated curve.

Formulae for an approximately optimal bandwidth can be obtained by mathematical means both locally (at one point t_0) and globally (the same bandwidth for all t). They are given here for the estimator of the regression function r (similar expressions hold for r' and r''):

- Local bandwidth:

$$h_{\text{opt}}(t_0) = \left(\frac{\sigma_\varepsilon^2}{c \cdot n \cdot r''(t_0)^2} \right)^{1/5}$$

- Global bandwidth:

$$h_{\text{opt}} = \left(\frac{\sigma_\varepsilon^2}{c \cdot n \int r''(u)^2 du} \right)^{1/5}$$

Here c is a known constant that depends on the kernel function. These optimal bandwidths result from minimizing the MSE. They therefore achieve a compromise between bias and variability. The optimal local bandwidth adapts to the curvature in various regions of the regression function: it is large in flat parts (with little bias) and small at peaks (thus avoiding a large bias). The optimal global bandwidth is used in all regions and thus makes a compromise between different aspects of the curve. Note also that both optimal bandwidths increase as the noise variance σ_ε^2 increases.

Unfortunately, these formulae cannot be applied to real data as they stand, since neither the residual variance σ_ε^2 nor the second derivative of r are known (for estimating $r^{(\nu)}$ we would need knowledge of $r^{(\nu+2)}$). Since $r''(t)$ is much more difficult to estimate than $r(t)$, it seems unrealistic to use the above result to determine h_{opt} or $h_{\mathrm{opt}}(t_0)$ from the data in an adaptive way. However, there are algorithms that successfully solve these problems. In general, we can say that it is easier to approximate the integral $\int r''(u)^2 du$ than to estimate $r''(t_0)^2$; a solution for the first problem has been proposed by Gasser *et al.* (1991a) and for the second by Brockmann *et al.* (1993). As for the estimation of σ_ε^2, there is a relatively simple non-parametric method proposed by Gasser *et al.* (1986). It consists in taking successive triplets of points, computing a linear function through the two outer points and taking the residual for the middle point; the estimate of σ_ε^2 is obtained by adding the squares of these pseudo-residuals (appropriately scaled). Note that besides its role in the optimal bandwidth, the quantity σ_ε^2 can also be of biological interest, giving information about the amount of random variation.

According to theory, simulations and practical applications, these methods of bandwidth choice (called 'plug-in' methods) work well. They lead to data-driven estimates of the global or the local optimal bandwidth and thus to rational rather than subjective smoothing. This is particularly relevant for derivatives, but also more difficult to achieve. There exist other methods for estimating the optimal bandwidth; a popular one is cross-validation, but it has been shown in the statistical literature that it is inferior to the plug-in method.

The question that comes up when we have a sample of curves as in growth studies, is whether we should use an optimal bandwidth for each curve or the same bandwidth for the whole sample. We favour the latter, and suggest taking as the bandwidth the average of the individually estimated optimal bandwidths. This is advisable essentially for two reasons: first, because treating all subjects statistically in the same way leads to results that are comparable from subject to subject (for the same reason, the same bandwidth should be used for both genders); and second, because by averaging the individual bandwidths one gets a more reliable bandwidth, in particular for the local choice and for derivatives.

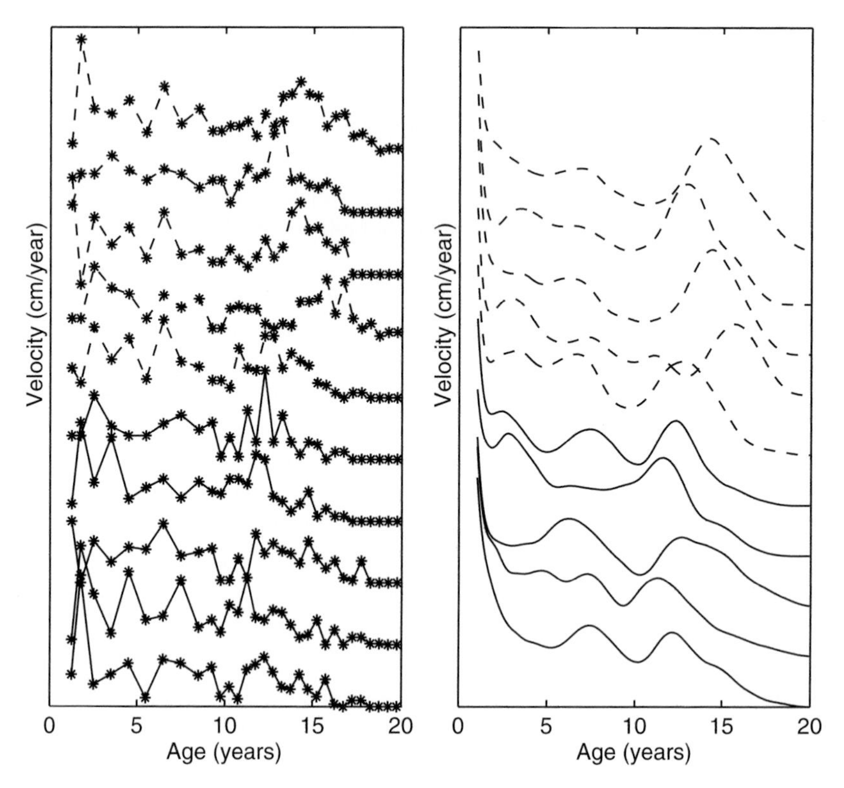

Figure 7.4 Velocities of growth in trunk length for five boys (above) and five girls (below). Left, raw velocities; right, kernel estimated velocities.

Application to growth data

An illustration of kernel estimated velocities of trunk length is given in Figure 7.4. The average optimal local bandwidth was used. While raw velocities are really noisy, estimated velocities based on kernel estimation show a smooth distinct pattern with a pubertal spurt and usually a discernible mid-growth spurt. One also notes visually two types of variability: the intensity of growth – for example the amplitude of the pubertal spurt – varies from child to child, and the timing of the spurt varies too, due to an individual maturational tempo. While the first type of variability is classical for statisticians and has been widely analysed, e.g. by analysis of variance or principal component analysis, the problem of time variability has largely been neglected. But it is time variability which makes many naive approaches – such as cross-sectional averaging of velocity curves – seriously flawed. The section on 'Structural analysis' (below) will deal with this problem.

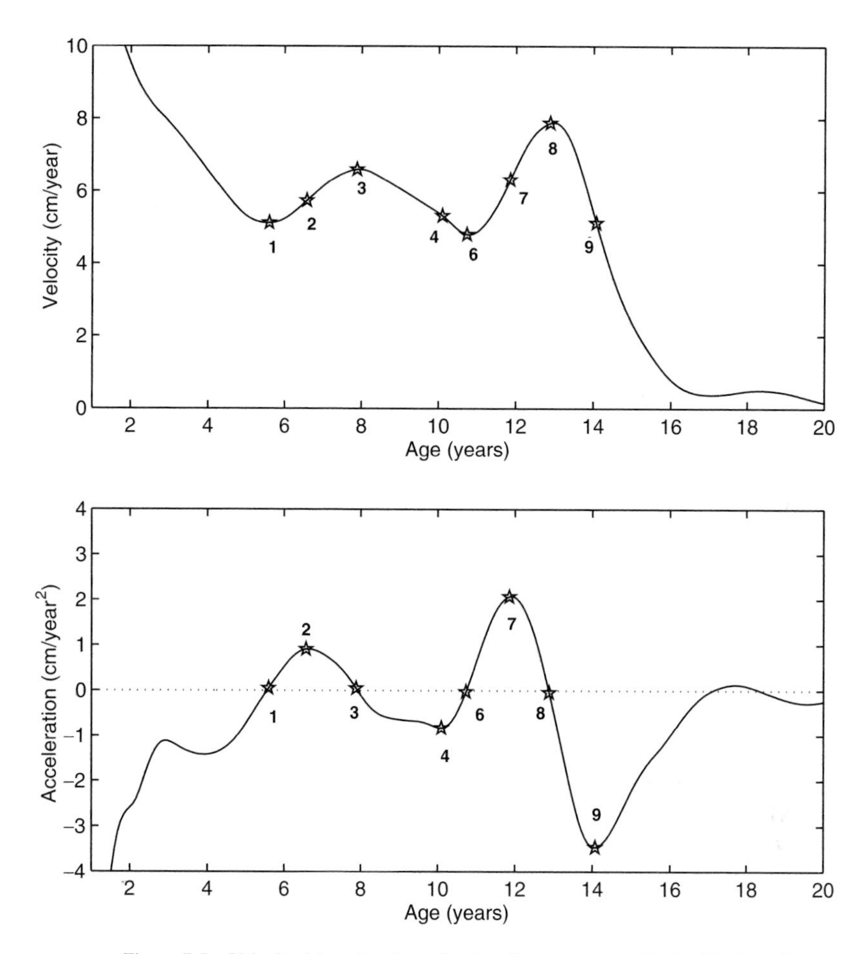

Figure 7.5 Velocity (above) and acceleration (below) curves obtained by kernel estimation, together with features or landmarks.

The estimated velocity and acceleration curves can also be used to define individual features (or parameters) characterizing the individual growth process (see Figure 7.5). The occurrence of these features is not always as clear-cut as it is for the child depicted in this figure; some rules to solve problems of non-uniqueness are discussed below (p. 198). The timing of the mid-growth spurt and of the pubertal spurt, and the tempo of transition through these spurts, can be assessed by the ages of maximal and minimal velocity and acceleration. A statistical evaluation showed that the ages of maximal acceleration (T_7) and in particular of minimal acceleration (T_9) are surprisingly well defined, but age at take-off (T_6) is often more ambiguous (Gasser *et al.*, 2001a). In the same spirit, we can determine maximal velocity and/or maximal acceleration to characterize the intensity of these spurts. The size of the velocity peak above prepubertal level is a more reliable indicator of intensity than maximal acceleration. Once such parameters are estimated for each individual, classical methods of statistics can be used to study various biomedical questions. In Sheehy *et al.* (1999), repeated-measures analysis of variance was used to assess differences between boys and girls, and between different parts of the body, in the whole growth process. These parameters were also useful to investigate which aspects of the growth determine adult size, and in which way these influences act (Sheehy *et al.*, 2000). Depending on the problem at hand, other sets of parameters might be useful: in Gasser *et al.* (2001a), various increments over the growth process have been suggested as an alternative, again using the timings of the pubertal spurt.

Based on these methods, the mid-growth spurt for height could be quantified for the first time (Gasser *et al.*, 1985b); using acceleration was crucial, since the mid-growth spurt is often riding on a falling trend and then not necessarily a true maximum. It showed negligible differences for gender with respect to intensity and duration in all skeletal variables; the different parts of the body showed, however, large differences in intensity and partly also in timing (Sheehy *et al.*, 1999). Height had the least intense mid-growth spurt, bihumeral width the most intense one. (Note that most of classical research was based on height, so that the mid-growth spurt was often overlooked.) Both legs and trunk had a more intense mid-growth spurt than height, but since it was displaced in age, it became rather flat in the height velocity curve (see Figure 7.9). The pubertal spurt shows the differences in timing and intensity that are well known for gender, and partly known for different somatic variables. For boys, the pubertal spurt was not only more intense in absolute terms (except for biiliac width) but also in relative terms when accounting for the larger adult size of males. These are just a few results; further details are given in the papers cited above.

Software for kernel estimation, including choice of an optimal global or local bandwidth, can be downloaded from our homepage (University of Zürich, 2004).

Shape-invariant modelling

While kernel estimation is an attractive method for exploring individual dynamics and intensity of growth, the ultimate goal is to model growth, with the guidance of biomedical knowledge. (In contrast, the known parametric models are purely descriptive and the parameters have little biological relevance – meaningful parameters such as peak velocity and others, must be extracted from the fitted curves.) Kernel estimation yields an individual velocity curve irrespective of the shape of all the other curves, and the individuals can be analysed one by one. In reality, there are general mechanisms that regulate normal growth and result in a common shape for a sample of velocity curves. Fitting an adequate model which reflects this common shape would lead to more consistent results across subjects by disregarding inconsistent, individual-specific features (formalized as random in Eqn 7.1).

In those fields where one can get only one data set to explore a relationship (e.g. in geophysics, where there is only one Earth), one can either guess at a parametric regression model or construct a model based on physical, chemical or biological knowledge. In growth studies, on the other hand, our present knowledge of endocrinological mechanisms and of bone growth does not allow derivation of a functional growth model guided by basic biomedical principles, but we can obtain a multitude of individual realizations of some growth process, e.g. of height, that differ individually in quantitative but not in qualitative terms. Combining the information across subjects in a clever way might thus lead to a valid model.

A one-component shape-invariant model (SIM) is of the following form:

$$y_{ij} = a_i\, s\left(\frac{t_{ij} - b_i}{c_i}\right) + d_i + \varepsilon_{ij},$$
$$i = 1, \ldots, m \quad \text{(subjects)}; \quad j = 1, \ldots, n_i \quad \text{(ages)}.$$

The function s is called 'shape function' and is common to all subjects, whereas a_i, b_i, c_i, d_i are individual parameters. Shape functions and parameters are both assumed unknown and have to be estimated from the data. Note that the logistic and the Gompertz models – often used for analysing the pubertal spurt (Marubini *et al.*, 1972) – can be seen as parametric versions of the SIM where s is fixed:

- Logistic model: $s(x) = 1/[1 + \exp(-x)]$
- Gompertz model: $s(x) = \exp[-\exp(-x)]$

Stützle *et al.* (1980) suggested the following two-component SIMs for raw height velocity data $v_i(t_{ij})$:

- Additive two-component SIM:

$$v_i(t_{ij}) = a_{1i}\, s_1\left(\frac{t_{ij} - b_{1i}}{c_{1i}}\right) + a_{2i}\, s_2\left(\frac{t_{ij} - b_{2j}}{c_{2i}}\right) + \varepsilon_{ij}$$

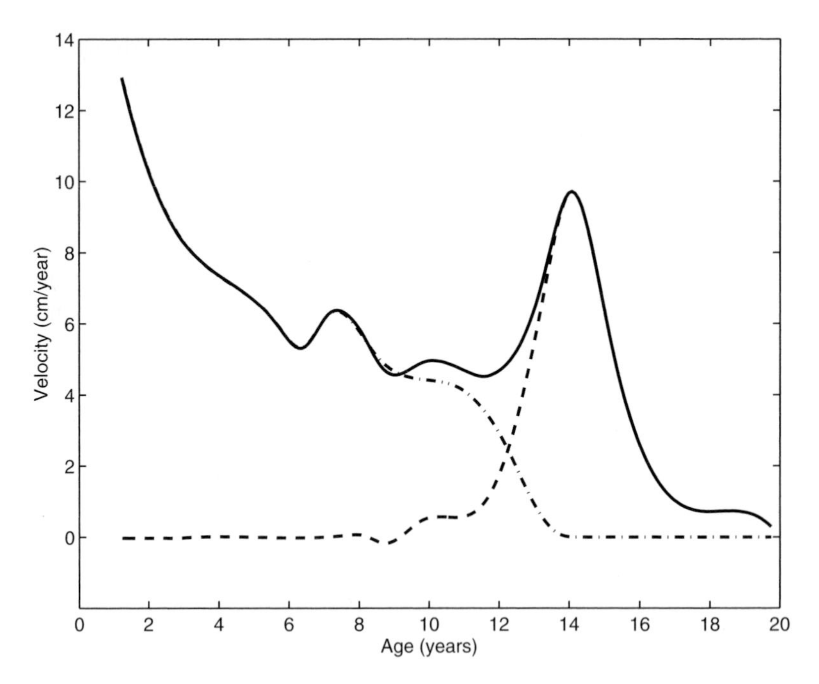

Figure 7.6 Height velocity of the 'median boy' according to the switch-off two-component shape-invariant model (solid curve). Dash–dotline, growth due to first component; dashed line, growth for second component.

- Switch-off two-component SIM:

$$v_i(t_{ij}) = a_{1i}\, s_1\left(\frac{t_{ij} - b_{1i}}{c_{1i}}\right)\varphi\left(\frac{t_{ij} - b_{2i}}{c_{2i}}\right) + a_{2i}\, s_2\left(\frac{t_{ij} - b_{2i}}{c_{2i}}\right) + \varepsilon_{ij}$$

where

$$\varphi(y) = 1 - \frac{\int_{-\infty}^{y} \exp(-u^2)du}{\int_{-\infty}^{+\infty} \exp(-u^2)du}$$

The switch-off function φ is not determined from the data but fixed a priori. It is required that φ descends smoothly from 1 to 0. An evaluation of different switch-off functions showed that the particular choice is not crucial, and that the above choice is plausible. The idea for both models is that the first component represents prepubertal growth and the second component represents pubertal growth. The introduction of a switch-off function, which stops further growth of the prepubertal component when pubertal growth climaxes, is based on biomedical knowledge about normal and pathological growth.

Figure 7.6 shows the resulting height velocity curve for the 'median boy', using the switch-off model ('median boy' means that the median of the estimated

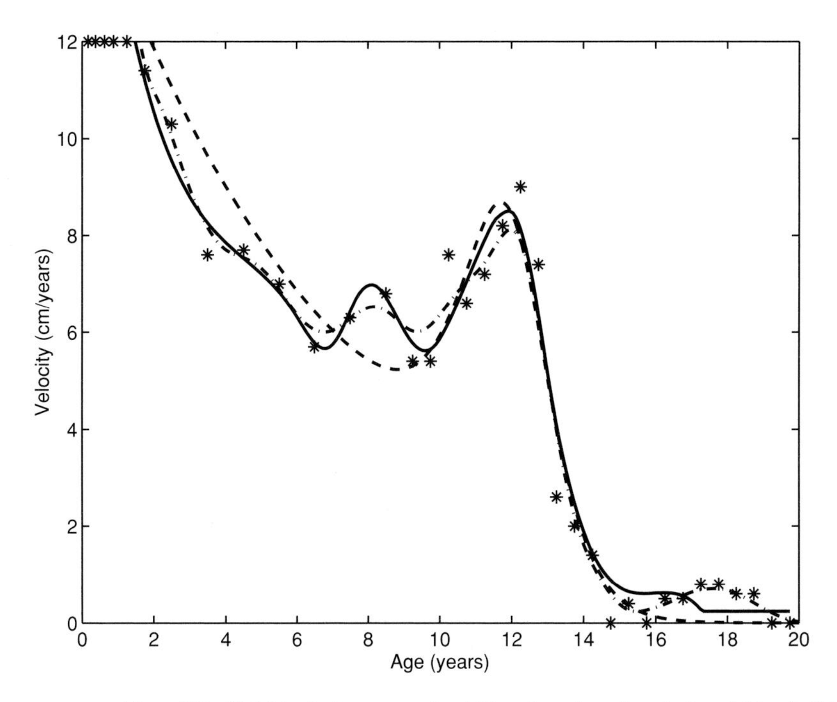

Figure 7.7 Height velocity curves of a girl based on shape-invariant modelling (solid curve), Preece–Baines fitting (dashed) and kernel fitting (dash–dot). Asterisks show raw data.

parameters has been used). The pubertal spurt is clearly asymmetric, with stopping being faster than accelerating growth. A small mid-growth spurt around age 7 is clearly identifiable (compare also Figure 7.9). The switch-off function provides a visually plausible separation into a prepubertal and a pubertal component. Figure 7.7 allows a comparison between the shape-invariant model, the Preece–Baines model and the kernel method for height growth of a girl. The Preece–Baines model leads to a substantially different fit compared to the other two methods.

What highlights about growth have emerged when using SIMs for analysing height growth? These:

- Growth is qualitatively similar in boys and girls, since the same shape functions apply (quantitatively, growth is of course significantly dissimilar, e.g. the pubertal spurt is earlier and smaller in girls).
- The switch-off model is much more plausible than the additive model, as shown for example by goodness-of-fit criteria.
- A distinct mid-growth spurt at around 7 years of age appears in the first component.

That the two sexes grow similarly in qualitative terms cannot easily be deduced by other statistical methods. Based on Kneip and Gasser (1988), the switch-off two-component SIM also holds for other somatic variables, and the above findings generalize. The success of the switch-off model demonstrates that the introduction of biological knowledge in the modelling process is a good strategy. The double-logistic model (Bock *et al.*, 1973), which performs badly on goodness-of-fit criteria, is a special case of an additive model and would as such also profit from the introduction of a switch-off function.

To estimate the shape functions s_1 and s_2, and at the same time the individual parameters a_{1i}, b_{1i}, c_{1i} and a_{2i}, b_{2i}, c_{2i} based on the estimated s_1 and s_2, a relatively sophisticated algorithm is needed. The major steps are as follows:

1. Select approximate models $s_1^{(0)}$ and $s_2^{(0)}$ as starting points. These could be e.g. logistic or exponential functions, or pieces of the structural average (defined below; see p. 198). Put $k = 0$.
2. Determine parameters $a_1^{(k)}, \ldots, c_2^{(k)}$ for $s_1^{(k)}$ and $s_2^{(k)}$ by non-linear least squares.
3. Compute residuals

$$\Delta y_{ij} = y_{ij} - a_{1i}^{(k)} s_1^{(k)} \left(\frac{t_{ij} - b_{1i}^{(k)}}{c_{1i}^{(k)}} \right) \varphi \left(\frac{t_{ij} - b_{2i}^{(k)}}{c_{2i}^{(k)}} \right) - a_{2i}^{(k)} s_2^{(k)} \left(\frac{t_{ij} - b_{2i}^{(k)}}{c_{2i}^{(k)}} \right)$$

4. Improve the model by fitting regression splines to the Δy_{ij} ($i = 1, \ldots, m$; $j = 1, \ldots, n_i$) resulting in functions $\Delta s_1^{(k)}$ and $\Delta s_2^{(k)}$.
5. Compute $s_l^{(k)} = s_l^{(k)} + \Delta s_l^{(k)}$ for $l = 1, 2$ as updated shape functions. Put $k = k + 1$ and go to step 2, or stop, if convergence is reached.

Note that for the model improvement stage (step 4) all measurements for all subjects are used, which leads to powerful results. Under suitable conditions, it can be mathematically shown that the method does what it is expected to do. The regression splines method used in this step is a well-known and convenient statistical tool for flexible function approximation. Regression splines consist of smooth piece-wise polynomials; that is, they are polynomials on each of the several intervals in which the age range is divided, and each polynomial is smoothly connected to the adjacent ones. The demarcations of these intervals are called knots. In our case we used cubic polynomials, which is the standard choice. Due to the many measurements available (more than 6000), the placement and the number of knots is not critical. For further details on the algorithm see Stützle *et al.* (1980). A more general statistical foundation is given in Kneip and Gasser (1988). At present (year 2002) our software implementing the algorithm described above is not offered on the World Wide Web but can be requested from the authors.

As mentioned earlier, an advantage of modelling via SIM is that one can bring in a priori knowledge, e.g. biomedical knowledge. A further advantage is that one gets a parametric model without the need of prescribing from the onset a rather arbitrary regression function (as in the logistic or the Gompertz models, or the parametric models mentioned above). Disadvantages are that the method is quite sophisticated and thus not so easy to implement, and that it still can lead to some misspecification due to the postulated model structure.

Structural analysis

What we discuss in this section is conceptually related to shape-invariant modelling: again, we use the information from the whole sample of growth velocity curves to determine a consistent growth pattern free of individual fluctuations. Methodologically, the approach of structural analysis is based on kernel estimation (or some other non-parametric function estimator), particularly for the extraction of time of occurrence of some events (e.g. the timings of the pubertal spurt). These timings are often called landmarks or milestones and serve to align individual data to an average growth tempo.

In any context, estimation of the mean of an object of interest is a natural first step in data analysis. As already noted by Shuttleworth (1937), a naive cross-sectional mean curve does not provide a valid picture of growth: a cross-sectional average of velocity curves would for example render a 'smeared' appearance of the pubertal spurt due to the variability in the timing of the spurt. To make curves comparable before averaging, some correction has to be made for the differences in timing. We chose a continuous transformation of the age axis to synchronize similar events (Gasser *et al.*, 1990) at an average age (see below for details). Figure 7.8 illustrates this for a small sample of velocity curves of sitting height. Visually and from the average curves obtained, one notes the improvement achieved by the age transformation. The size of the pubertal velocity peak in particular would be severely underestimated without aligning to an average tempo.

Since the need for curve alignment arises in various fields – a technical example being analysis of speech – various data-analytic methods have been devised. The one described in this section is often called 'landmark registration', but we prefer to call it 'structural analysis' since it relies on a common structure underlying most growth curves. Here, we give an algorithmic description of how the alignment procedure works; a mathematically sound formalization can be found in Kneip and Gasser (1992). The procedure is as follows:

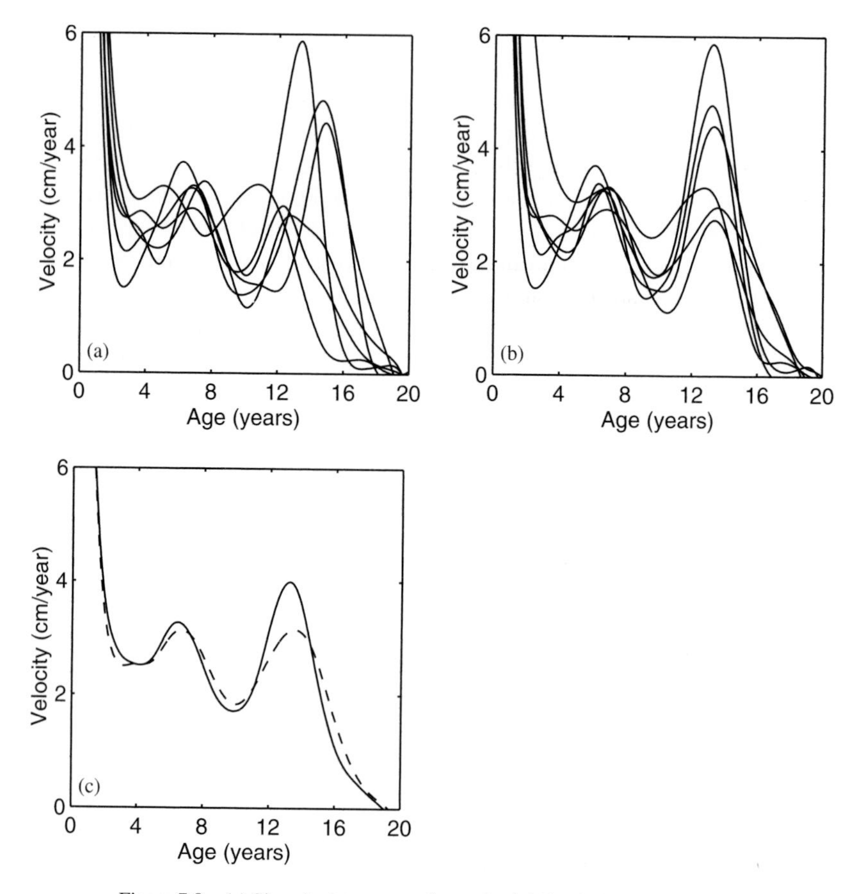

Figure 7.8 (a) Six velocity curves of growth of sitting height (kernel fitting). (b) The same, after alignment procedure. (c) Cross-sectional average curve (dashed), aligned or structural average curve (solid line).

1. Structural points $\tau_1, \ldots, \tau_p$ are defined which occur in all or most curves: these can be ages when a minimum or a maximum of velocity or acceleration occurs during the pubertal spurt and during the mid-growth spurt. A structural point could also be the age where a percentage decrease has been realized, e.g. for the velocity curve in infancy.
2. The structural points $\tau_1, \ldots, \tau_p$ are determined via kernel estimation from individual velocity or acceleration curves, rendering $\tau_{1i}, \ldots, \tau_{pi}$ for subject i. Usually, some plausible rules are needed in addition; for example, an age range is known where a peak is expected. When a structural point cannot be safely determined in a subject, it is put to missing. Often the bandwidth h can be chosen a bit larger than for the curve itself.

3. Based on $\tau_{1i}, \ldots, \tau_{pi}$ the alignment functions $h_i(t)$ are constructed in the following way:
   - $h(\bar{\tau}_s) = \tau_{si}$ for $s = 1, \ldots, p$, where $\bar{\tau}_s = \sum_{i=1}^{m} \tau_{si}/m$ (characterizing average tempo). This defines the transformation at structural points.
   - Between structural points, linear interpolation (or a more refined monotone interpolation as in Kneip and Gasser (1992)) is used. The resulting function $h_i(t)$ is smooth and monotone increasing (a normal requirement for time transformations).
4. Before aligning individual data to an average tempo, a continuous velocity curve $v_i(t)$ is constructed out of the raw velocities, for example by simple linear interpolation. The aligned individual velocity curve is then $v_i(h_i(t))$ and the aligned average curve (the 'structural average') is:

$$\bar{v}(t) = \frac{1}{m} \sum_{i=1}^{m} v_i[h_i(t)]$$

5. Due to averaging, $\bar{v}(t)$ will be much smoother than the interpolated and aligned individual raw velocities $v_i(t)$. If $\bar{v}(t)$ is not smooth enough, some smoothing by kernel estimation is appropriate.

At present (year 2002) the authors' software implementing this algorithm is not available on the World Wide Web but will be provided at request. Related software, that parallels the book by Ramsay and Silverman (1997), can be found at the home page of Jim Ramsay (see Ramsay, 2004).

The structural average has little or negligible bias compared to smoothed individual curves, since little smoothing (if at all) is involved. Due to the averaging process, it has also little variability, thus rendering a very accurate quantification of average growth. When determining for example the peak velocity of the pubertal spurt from the structural average $\bar{v}(t)$, the result is much more accurate than the mean of individually determined peak velocities (following e.g. individual smoothing; see p. 185). It can also be shown that the resulting structural average is closer to the underlying growth model. For mathematical details see Kneip and Gasser (1992). Since two types of variability are involved in this process (amplitude and time), it is mathematically not easy to give a standard deviation for the resulting structural average curve. One possibility is to take subsamples from the total sample, compute structural averages and assess variability visually.

Structural average curves are ideal for comparing growth of different groups, for example of boys and girls. They can also highlight, in a lucid way, differences in growth between various parts of the body (Gasser *et al.*, 1991b,c). For instance, clear-cut results for differences in the growth of trunk and legs

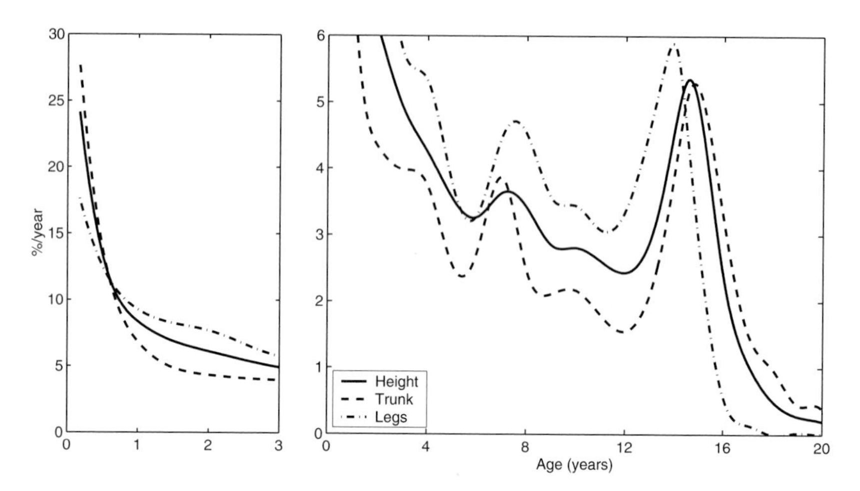

Figure 7.9 Structural average velocity curves of growth of height (solid line), trunk (dashed line) and legs (dash–dot line) for boys. Left: The first three years on a different scale.

could be obtained. While legs are relatively small at birth, they have a more intense growth after about 1.5 years until the advent of the pubertal spurt. The mid-growth spurt occurs earlier in the trunk than in the legs, while the pubertal spurt occurs later in the trunk. The pubertal spurt of the trunk is more intense compared to the legs and has also a different shape: while the legs stop growing rather abruptly, the spurt of the trunk fades away slowly and is thus responsible for late growth and further changes of body proportions. Similarly detailed results could be obtained for other somatic variables. Figure 7.9 gives an illustration for the growth of height, trunk and legs (in relative velocity, percentage increase per year). Legs show a low velocity in infancy, an elevated one thereafter until the pubertal spurt, and an early and not too impressive spurt, followed by a rapid stopping of growth. The trunk has a high velocity in infancy (continuing the intrauterine pattern) followed by a relatively low velocity until the pubertal spurt (which is relatively late and strong) and then slowly decreasing after the maximum. Height growth is sort of a compromise between leg and trunk growth. Interestingly, the mid-growth spurt is displaced for trunk and legs, and consequently it is weak for height. Biologically, then, it makes more sense to study trunk and legs separately, with their unique patterns.

Differences between groups were assessed by Schäfer *et al.* (1990) by comparing height growth in normal children with height growth in children with nephropathy. Ratcliffe *et al.* (1992) obtained clear-cut results about growth of normal men compared with men with an XYY chromosomal disposition. The higher stature of XYY men is due both to a higher prepubertal velocity level and to a more intense pubertal spurt.

Interesting results could also be obtained by comparing the structural average curves of subjects with small or large adult sizes, leading to insights into the biological mechanisms responsible for these differences (Gasser *et al.*, 2001b). For sitting height and biiliac width, an increased prepubertal velocity level is responsible for a large adult size; for legs this is also an important factor, but a delay in the pubertal spurt – and thus a longer growth period – is a further factor. For bihumeral width, the size of the pubertal peak is the decisive factor for adult size reached. The differences in growth of early and late maturers was also compared in this way (Gasser *et al.*, 2001c). Except for legs, adult size was very similar. The shorter growth period of early maturers was mainly compensated by a higher prepubertal velocity level; only for boys was a slightly higher pubertal peak found.

Concluding remarks

We have discussed in this chapter three statistical approaches which have proved to be useful for analysing longitudinal data of growth of skeletal variables: kernel estimation for non-parametric function fitting (including first and second derivatives), shape-invariant modelling for determining shape functions (i.e. the regression model) as well as individual parameters, and an alignment procedure to synchronize curves before further analysis (e.g. before averaging curves). All three methods are of interest beyond applications to growth data and are in fact applied in fields as diverse as econometrics and engineering. Closer to auxology, kernel estimators might be useful for analysing longitudinal hormonal data, while shape-invariant modelling has a potential for analysing circadian rhythms in a sample of subjects.

The growth of skeletal variables is almost exclusively genetically determined, at least in developed countries. For variables involving soft tissue, exogenous factors come into play, and cause a more variable shape of growth from subject to subject. In our experience, kernel estimation and structural analysis still work for weight, body mass index and circumferences. Determining milestones can become more problematic sometimes. We have not tried shape-invariant modelling for these variables. Skinfolds are different, since they are mainly influenced by exogenous factors. One can, of course, apply kernel estimation to individual data, but the curves show a highly variable pattern from subject to subject. This makes further analysis difficult. One approach that has been fruitful is the following: we used the alignment functions of height to align interpolated raw distances and velocities of skinfolds. The pattern of the resulting structural average curves was more relevant than the one obtained from the cross-sectional average. The biomedically relevant results can be found in a series of papers (Gasser *et al.*, 1993b, 1994a,b,c, 1995).

Among the three non-parametric methods discussed here, kernel estimation is the most mature one scientifically (and the same applies to alternative methods of non-parametric function fitting such as splines and local polynomials). In practice, it is advisable to start a data analysis with kernel estimators of the individual curves and to have a close look at the graphics obtained, in order to get an understanding of the process being studied. In a second step one might wish to extract features (meaningful parameters) from the individually estimated curves (be it distance, velocity or acceleration curves). They can be used for further statistical analysis. Often one will stop at this point. A natural next step would be the estimation of structural average curves to obtain a valid picture across somatic variables and for comparison of the two genders. Typically, these average curves are much more informative than the individual curves. Often the analysis of subgroups – e.g. subjects with a small or large size in some somatic variable, or early and late maturers – leads to substantial medical insight, not easy to obtain by other means. Shape-invariant modelling, on the other hand, is advisable at a later stage of research, and if sufficient statistical resources are available.

ACKNOWLEDGEMENT

This work was supported by the Swiss National Science Foundation (Projects Nos. 3200–064047.00/1 and 2000–063579.00/1).

REFERENCES

Bock, R., Wainer, H., Petersen, A., *et al.* (1973). A parametrization for individual human growth curves. *Annals of Human Biology*, **45**, 63–80.

Brockmann, M., Gasser, T. and Herrmann, E. (1993). Locally adaptive bandwidth choice for kernel regression estimators. *Journal of the American Statistical Association*, **88**, 1302–9.

Falkner, F. (ed.) (1960). *Child Development: An International Method of Study.* Basel, Switzerland: Karger.

Fan, J. and Gijbels, I. (1996). *Local Polynomial Modelling and Its Applications*. London: Chapman and Hall.

Gasser, T. and Müller, H. (1984). Estimating regression functions and their derivatives by the kernel method. *Scandinavian Journal of Statistics*, **11**, 171–85.

Gasser, T., Köhler, W., and Müller, H. (1984a). Velocity and acceleration of height growth using kernel estimation. *Annals of Human Biology*, **11**, 397–411.

Gasser, T., Müller, H., Kohler, W., Molinari, L. and Prader, A. (1984b). Nonparametric regression analysis of growth curves. *Annals of Statistics*, **12**, 210–29.

Gasser, T., Müller, H., Köhler, W., *et al.* (1985a). An analysis of the mid-growth spurt and the adolescent growth spurt of height based on acceleration. *Annals of Human Biology*, **12**, 129–48.

Gasser, T., Müller, H. and Mammitzsch, V. (1985b). Kernels for nonparametric curve estimation. *Journal of the Royal Statistical Society B*, **47**, 238–52.

Gasser, T., Sroka, L. and Jennen-Steinmetz, C. (1986). Residual variance and residual pattern in nonlinear regression. *Biometrika*, **73**, 625–33.

Gasser, T., Kneip, A., Ziegler, P., Largo, R. and Prader, A. (1990). A method for determining the dynamics and intensity of average growth., *Annals of Human Biology*, **17**, 459–74.

Gasser, T., Kneip, A., Binding, A., *et al.* (1991a). The dynamics of linear growth in distance, velocity and acceleration. *Annals of Human Biology*, **18**, 187–205.

Gasser, T., Kneip, A. and Köhler, W. (1991b). A flexible and fast method for automatic smoothing. *Journal of the American Statistical Association*, **86**, 643–52.

Gasser, T., Kneip, A., Ziegler, P., *et al.* (1991c). The dynamics of growth of width in distance, velocity and acceleration. *Annals of Human Biology*, **18**, 449–61.

Gasser, T., Engel, J. and Seifert, B. (1993a). Nonparametric function estimation. In *Handbook of Statistics*, ed. C. Rao, pp. 423–65. Amsterdam, The Netherlands: Elsevier Science.

Gasser, T., Ziegler, P., Kneip, A., *et al.* (1993b). The dynamics of growth of weight, circumferenccs and skinfolds in distance, velocity and acceleration. *Annals of Human Biology*, **20**, 239–259.

Gasser, T., Kneip, A., Ziegler, P., *et al.* (1994a). Development and outcome of indices of obesity in normal children. *Annals of Human Biology*, **21**, 275–86.

Gasser, T., Ziegler, P., Largo, R., *et al.* (1994b). A longitudinal study of lean and fat area at the arm. *Annal of Human Biology*, **21**, 303–14.

Gasser, T., Ziegler, P., Seifert, B., *et al.* (1994c). Measures of body mass and of obesity from infancy to adulthood and their appropriate transformation. *Annals of Human Biology*, **21**, 111–25.

Gasser, T., Ziegler, P., Seifert, B., *et al.* (1995). Prediction of adult skinfolds and body mass from infancy through adolescence. *Annals of Human Biology*, **22**, 217–33.

Gasser, T., Sheehy, A. and Largo, R. (2001a). Statistical characterization of the pubertal growth spurt. *Annals of Human Biology*, **28**, 395–402.

Gasser, T., Sheehy, A., Molinari, L. and Largo, R. (2001b). Growth processes leading to a large or small adult size. *Annals of Human Biology*, **28**, 319–27.

Gasser, T., Sheehy, A., Molinari, L. and Largo, R. (2001c). Growth of early and late maturers. *Annals of Human Biology*, **28**, 328–36.

Jolicoeur, P., Pontier, J., Pernin, M.-O. and Semp, M. (1988). A lifetime asymptotic growth curve for human height. *Biometrics*, **44**, 995–1003.

Karlberg, J. (1987). On the modelling of human growth. *Statistics in Medicine*, **6**, 185–92.

Kneip, A. and Gasser, T. (1988). Convergence and consistency results for self modeling nonlinear regression. *Annals of Statistics*, **16**, 82–112.

(1992). Statistical tools to analyze data representing a sample of curves. *Annals of Statistics*, **20**, 1266–1305.

Largo, R., Gasser, T., Prader, A., Stützle, W. and Huber, P. (1978). Analysis of the adolescent growth spurt using smoothing spline functions. *Annals of Human Biology*, **5**, 421–34.

Marubini, E., Resele, L., Tanner, J. M. and Whitehouse, R. (1972). The fit of Gompertz and logistic curves to longitudinal data during adolescence on height, sitting height and biacromial diameter in boys and girls of the Harpenden Growth Study. *Human Biology*, **44**, 511–24.

Prader, A., Largo, R., Molinari, L. and Issler, C. (1989). Physical growth of Swiss children from birth to 20 years of age. *Helvetica Paediatrica Acta (Supplement)* **52**, 1–125.

Preece, M. and Baines, M. (1978). A new family of mathematical models describing the human growth curve. *Annals of Human Biology*, **5**, 1–24.

Ramsay, J. (2004). Home page. http://www.psych.mcgill.ca/faculty/ramsay/ramsay.html

Ramsay, J. and Silverman, B. (1997). *Functional Data Analysis.* New York: Springer-Verlag.

Ratcliffe, R., Pan, H. and McKie, M. (1992). Growth during puberty in the XYY boy. *Annals of Human Biology*, **19**, 579–87.

Schäfer, F., Seidel, C., Binding, A., Gasser, T. and Sächrer, K. (1990). The pubertal growth spurt in chronic renal failure. *Pediatric Research*, **28**, 5–10.

Sheehy, A., Gasser, T., Molinari, L. and Largo, R. (1999). An analysis of variance of thc pubertal and midgrowth spurts for length and width. *Annals of Human Biology*, **26**, 309–331.

(2000). Contribution of growth phases to adult size. *Annals of Human Biology*, **27**, 281–98.

Shuttleworth, F. (1937). *Sexual Maturation and the Physical Growth of Girls Age Six to Nineteen,* vol. 2 of *Monographs of the Society for Research in Child Development*, Serial no. 12. Washington, DC: National Research Council.

Stützle, W., Gasser, T., Molinari, L., *et al.* (1980). Shape-invariant modelling of human growth. *Annals of Human Biology*, **7**, 507–28.

University of Zürich (2004). http://www.unizh.ch/biostat.

8 *Parametric models for postnatal growth*

Roland C. Hauspie
Free University of Brussels

and

Luciano Molinari
Kinderspital, Zürich

Why model growth data?

Growth can be considered as the process that makes children change in size and shape over time. The dynamics of growth is best understood from the analysis of longitudinal data, i.e. from serial measurements taken at regular intervals on the same subject. Table 8.1 gives an example of longitudinal growth data for height of a boy measured at birth and at each birthday thereafter up to the age of 18 years. Such data usually form the basis to estimate the underlying process of growth, which is supposed to be continuous. Recent analysis of frequent measurements of size (at daily or weekly intervals) with high-precision techniques (such as knemometry where measurement error is about 0.1 mm) has shown that the growth process is, at microlevel, not as smooth as we usually assume (Hermanussen, 1998; Lampl, 1999). However, we may readily assume that the growth process is continuous when we are dealing with measurements taken at yearly intervals, or even 3- to 6-monthly intervals, using classical anthropometric techniques. Various mathematical models have been proposed to estimate such a smooth growth curve on the basis of a set of discrete measurements of growth of the same subject over time (Marubini and Milani, 1986; Hauspie, 1989, 1998; Simondon *et al.*, 1992; Bogin, 1999).

The main goals of mathematical modelling of longitudinal growth data are:

Methods in Human Growth Research, eds. R. C. Hauspie, N. Cameron and L. Molinari.
Published by Cambridge University Press.

Table 8.1 *Attained height (in cm) and yearly increments in height (in cm/year) of a boy taken at birth and at each subsequent birth date up to the age of 18 years; these data are used in the examples of Figures 8.1 to 8.7*

Attained height		Yearly increments	
Age	Height	Age	Height
0	49.1	0.5	26.9
1	76.0	1.5	13.4
2	89.4	2.5	7.8
3	97.2	3.5	6.3
4	103.5	4.5	5.9
5	109.4	5.5	7.0
6	116.4	6.5	6.3
7	122.7	7.5	5.3
8	128.0	8.5	5.5
9	133.5	9.5	4.7
10	138.2	10.5	5.6
11	143.8	11.5	4.9
12	148.7	12.5	6.3
13	155.0	13.5	9.5
14	164.5	14.5	6.2
15	170.7	15.5	3.3
16	174.0	16.5	1.0
17	175.0	17.5	0.0
18	175.0		

- To estimate the continuous growth process from a set of discontinuous measures of growth in order to obtain a smooth graphical representation of the growth curve
- To estimate growth between measurement occasions (in that sense, curve fitting is an interpolation technique)
- To summarize the growth data by a limited number of constants or function parameters (therefore curve fitting is also a data reduction technique)
- To estimate particular milestones of the growth process (the so-called biological parameters) such as final size or age, size and velocity at take-off and at peak velocity, which characterize the shape of the growth curve and usually form the basis for further analysis
- To estimate a smooth velocity curve representing instantaneous velocity (i.e. by taking the mathematical first derivative of the fitted curve)
- To estimate the 'typical average' curve in the population, such as the mean-constant curve in the case of structural growth models.

Fitting a growth model consists of finding the set of function parameters that yield the *best-fitting curve*. The best-fitting curve is usually estimated by the least-squares method, i.e. the method that yields the curve with the smallest value for the sum of squared residuals (deviations of the observations from the fitted curve). Other parameter estimation methods may be envisaged. A thorough discussion of this topic can be found in Chapter 9.

Non-structural versus structural models

Broadly speaking, we can subdivide growth models into structural (or parametric) and non-structural models (Bock and Thissen, 1980). Non-structural models do not postulate a particular form of the growth curve. They provide smoothing techniques suppressing measurement error and short-term variation. Typical examples are polynomials and cubic splines (Largo *et al.*, 1978). Non-structural models:

- Do not postulate a particular form of the growth curve
- Usually have a large number of parameters with no biological interpretation
- Do not tend to an asymptotic value
- Are usually unstable in the extremities of the data range
- Are easy to fit.

Figure 8.1 shows the example of the fit of a 4th-degree (Figure 8.1a) and 9th-degree polynomial (Figure 8.1b) to the yearly increments in height of the boy shown in Table 8.1. The dashed line shows the subject's yearly increments in height (i.e. the differences between height measurements, one year apart) compared to the polynomial fits. It is obvious that the 4th-degree polynomial (with five parameters) does not adequately describe the yearly increments in height. The 9th-degree polynomial (with ten parameters) performs much better, but still cannot correctly describe the maximum increment in height. Polynomials can be considered as inadequate to fit growth data over wide age ranges, but can be used to fit growth over reasonably short-term intervals (a few years).

The flexibility of polynomials can be much improved by approaches like smoothing splines, which consist of series of lower-order polynomials (3rd-degree or cubic, for example) that are fitted over only a small range of ages, and which are connected by constraints of continuity, i.e. equality of the first and second derivative at the points of transition or 'knots'. These models give considerable better fits than higher-order polynomials and are better able to model local variations in the growth pattern such as the mid-growth spurt or the decrease in velocity prior to the adolescent growth spurt (Goldstein, 1984). The subjective element in this approach is the determination of the number and

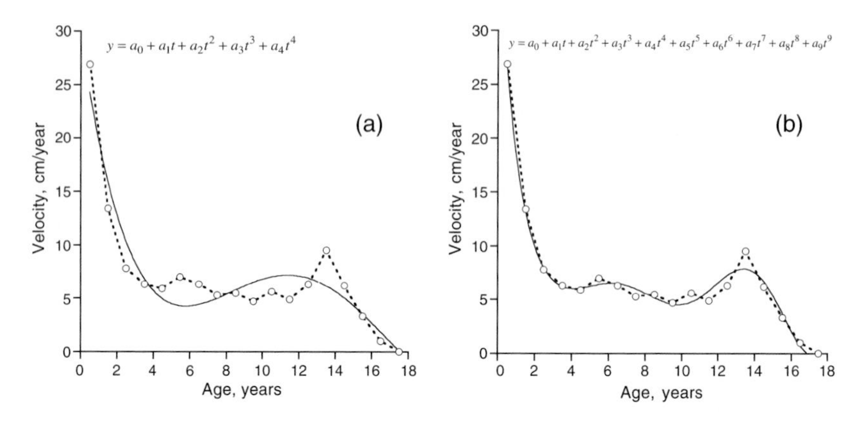

Figure 8.1 Yearly increments in height of the boy whose growth data are shown in Table 8.1. (a) Fit of 4th degree polynomial, (b) fit of 9th degree polynomial. y = increment, t = age in years, a_0 to a_k the function parameters. (After Hauspie and Chrzastek-Spruch, 1999.)

position of the knots in the fitting procedure. Other most useful developments of non-structural approaches in modelling growth are discussed in Chapter 7.

Structural or parametric models:

- Imply a basic functional form of the growth model
- Usually have fewer parameters that allow some functional/biological interpretation
- Usually tend to an upper asymptote (final size).

Structural or parametric models sometimes impose too rigid a shape on the growth data, which may result in slight, but systematic, bias. They also require more sophisticated curve-fitting techniques, but most statistical and several graphical software packages nowadays offer the possibility of non-linear regression analysis of user-defined functions. The estimation algorithms may vary from one type of software to another, but they are all based on iterative numeric minimization techniques and require more or less rough guesses of the values of the function parameters to be estimated, the so-called starting values. Good starting values will often allow an iterative technique to converge to a solution more quickly.

Reaching convergence means that the numeric minimization procedure has found a minimum in the multidimensional plane of the sum of squared deviations. However, it may occur that the process has encountered a local minimum but has not yet reached the absolute true minimum, which then leads to a misfit of the data. The occurrence of local minima in the estimation procedures is intrinsic to non-linear regression and does not depend on the minimization algorithm, but rather on the functional form of the growth model. The risk of reaching convergence at a local minimum is greater if the starting values are badly chosen, the scatter in the data is large (usually not a problem in longitudinal data except for outliers due to erroneous measurements), and the range of the data is insufficient for the model at hand. A robust model is one that has few or no local minima near the real minimum and hence is not too sensitive towards the choice of starting values. To test for the presence of a false minimum, one can run the curve-fitting procedure with different sets of starting values of the parameters. If they all converge to nearly the same solution, then it is very likely that you have found the true minimum. If one set of starting values results in a substantially lower sum of squares, then you should keep them as the new starting values and repeat the procedure since it is likely that you are now nearer to the true minimum (see Chapter 9 for a more extensive discussion of numerical minimization techniques).

Usually, the population means for the function parameters can serve as starting values for these numerical minimization techniques although individual adjustments are sometimes required. When studying a new population, one can utilize starting values taken from the literature. Tables 8.2 and 8.3 provide sets of starting values for the models discussed in this chapter. They will not necessarily be suitable to fit all growth curves in a specific population, but they can be used in a preliminary analysis of specific data. A more optimal set of starting values can then be obtained from the means of the successful fits.

Most structural growth models are monotonously increasing functions and are therefore in the first place designed to describe growth of skeletal dimensions for which, strictly speaking, we have only positive growth. For this reason, structural models are not suitable for traits such as body weight, body mass index and skinfolds, for instance. The latter traits may show negative as well as positive growth, and the general shape of the growth pattern of those traits usually does not match the functional form of the models. Non-structural approaches are more apt to fit those traits. Most structural models designed to describe adolescent growth tend to an upper asymptote (final size) towards the end of the growth phase, and also allow for an adolescent spurt. Therefore, they are suitable for postcranial skeletal dimensions (length and width measurements of the body), but perform badly for measurements of the head and face, which have virtually no adolescent growth spurt.

Growth in infancy and childhood

A long time ago, Jenss and Bayley (1937) proposed a four-parameter nonlinear model which fits satisfactorily growth data from birth to 8 years. The formulation of the Jenss curve is as follows:

$$y = a + bt - \mathrm{e}^{c+dt}$$

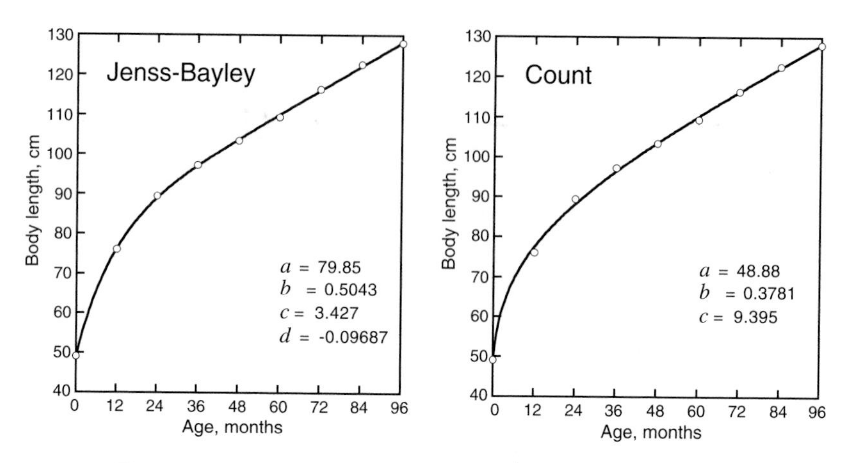

Figure 8.2 Fit of Jenss–Bayley model and Count model to the data of growth in height from birth to 8 years of age (Table 8.1).

where y is size, t is age, and a, b, c and d are the four function parameters. The model has a linear component $(a + bt)$ in which the parameter b determines the childhood growth velocity and an exponential component (e^{c+dt}), determining the decreasing growth rate shortly after birth.

The Jenss curve has been successfully applied by Deming and Washburn (1963), Manwani and Agarwal (1973), Berkey (1982), and several others. The model is suitable for describing growth of body length and of various dimensions of the head (typical head circumference) during infancy and early childhood. It has often been used to fit weight data as well, despite the problems that may arise when growth in weight is not monotonously increasing or has an irregular pattern, which often occurs shortly after birth.

Another model that fits early childhood data fairly well is the three-parameter model proposed by Count (1942, 1943), slightly modified by Livshits *et al.* (2000) in order to allow the inclusion of birth data:

$$y = a + bt + c\ln(t + 1)$$

where y is size, t is age, and a, b and c are the three function parameters.

Figure 8.2 shows the height data of Table 8.1 for ages 0 to 8 years with Jenss–Bayley and Count curve fittings. For the purpose of fitting those two models, it is better to express the ages in months. The estimates of the parameters for the respective fits are shown in the figures. At first glance, both models seem to describe adequately body length during the first 8 years of postnatal life, although visual inspection of the graphs shows a slightly better fit of the Jenss–Bayley curve. The residual standard deviation (RSD) is 0.48 cm for the

Table 8.2 *Starting values for fitting the Jenss–Bayley model and the Count model to growth of body length, weight and head circumference with numeric minimization algorithms (when age is expressed in months)*

	Jenss–Bayley			Count		
Parameter	Length (cm)	Weight (kg)	Head circumference (cm)	Length (cm)	Weight (kg)	Head circumference (cm)
a	83	8.5	50	50	3	35
b	0.5	0.2	0.03	0.4	0.1	−0.03
c	3.5	1.7	2.6	9.5	2	4.5
d	−0.08	−0.3	−0.1			

Jenss–Bayley curve and 0.93 cm for the Count curve. Both models are fairly robust towards the choice of starting values. Table 8.2 gives a set of starting values for body length (in cm), body weight (in kg) and head circumference (in cm) when the age is expressed in months. The sex differences in growth are much smaller than the normal variations in growth during infancy so that a single set of starting values suffices for both genders.

Berkey (1982) compared the reliability, efficiency, precision and goodness-of-fit of the Count and Jenss–Bayley models and concluded that the latter model fitted the growth data better than Count's model, especially prior to 1 year of age. Berkey and Reed (1973) have greatly enhanced the flexibility of the Count function by adding one or more deceleration terms. They proposed the following two functions:

Reed 1st order $$y = a + bt + c\ln(t) + \frac{d}{t}$$

Reed 2nd order $$y = a + bt + c\ln(t) + \frac{d_1}{t} + \frac{d_2}{t^2}$$

where y is size, t is age, and a, b, c, d and a, b, c, d_1, d_2 are the parameters.

The Reed models can accommodate one or more inflection points (depending on the number of reciprocal terms), allowing the description of one or more periods of growth acceleration and thus fitting a wider variety of both normal and abnormal growth patterns in early childhood. However, if birth is included, then chronological age since birth cannot be used, and an alternative age scale has to be chosen. Berkey and Reed (1973) suggest the age transformation $t =$ (months since birth + 9)/9, which assigns $t = 0$ at conception and $t = 1$ at birth. They showed that the four-parameter Reed model provided significantly better overall fits than the Jenss–Bayley model, which has also four parameters. Moreover, by the fact that the Reed models are linear in their constants, they can be fitted by simpler statistical methods than the non-linear Jenss–Bayley curve. Simondon *et al.* (1992) made an interesting comparison of five growth models to fit weight data between birth and 13 months of age.

Growth at adolescence

Logistic and Gompertz functions

The first attempts to fit the adolescent growth cycle were made by using the logistic and the Gompertz function. These models are special cases of the generalized logistic model (Nelder, 1961) of which the differential equation integrates to:

$$y = K(1 + ce^{-bt})^{1/(1-m)} \quad \text{for } m > 1$$

For $m > 1$, this curve has an S-shape with a lower and upper asymptote, equal to zero and K, and one point of inflection. Parameter b is a rate constant, determining the spread of the curve along the time axis, while parameter c is an integration constant. For the purpose of fitting the adolescent growth cycle, the lower asymptote is set different from zero by adding a constant P.

For $m = 2$, the generalized logistic leads to the autocatalytic or logistic curve, which can, after reparameterization, be written in the form:

$$y = P + \frac{K}{1 + e^{a-bt}}$$

where y is size, t is age, and with P, K, $a = \log(c)$ and b as stated above. In the logistic model, relative growth rate (growth velocity divided by size) declines linearly with size. Hence, the curve is symmetrical around its inflection point $y_I = P + K/2$ at $t_I = a/b$ (age at peak velocity) with maximal peak velocity given by $bK/4$.

For $m = 1$, the generalized logistic equation breaks down, but it can be shown that for $m \to 1$, the model leads to the Gompertz curve, in which relative growth rate declines exponentially with size:

$$y = P + Ke^{-e^{a-bt}}$$

where y is size and t is age. The Gompertz curve is asymmetrical around its point of inflection: $y_I = P + K/e \approx P + 0.37K$, at $t_I = a/b$ (age at maximal velocity) with maximal velocity given by bK/e. In both, the logistic and Gompertz function, the inflection point is functionally related to the amount of adolescent growth (respectively 50% and $\pm$37%).

The logistic and Gompertz functions were used to fit the adolescent growth data of several body dimensions (Deming, 1957; Marubini *et al.*, 1971, 1972; Tanner *et al.*, 1976). In a longitudinal study of 35 Belgian girls (Hauspie *et al.*, 1980), it was shown that both models fit adolescent data well with pooled residual variances of 0.45 cm^2 for the logistic and 0.61 cm^2 for the Gompertz function (total number of degrees of freedom 110). Nevertheless, Wilcoxon's signed rank test revealed significantly better fits with the logistic than with the Gompertz function ($P < 0.05$). However, none of the derived biological variables differed significantly between the two models.

The major drawback of both models is that the lower age bound of the data to be fitted (i.e. the cut-off point between the prepubertal and adolescent growth cycle) has to be determined arbitrarily for each individual. This cut-off point is usually taken as the age at minimal prepubertal growth velocity (age at take-off), obtained through a graphical inspection of a plot of the yearly increments in function of age. This procedure is, in practice, not always so easy and may lead to subjective decisions. Nevertheless, errors in assessing the take-off point of less than 1 year should not greatly affect the estimates of age and velocity at the peak of the growth spurt (Marubini, 1978; Hauspie, 1981). Due to these drawbacks and inconveniences both the logistic and Gompertz functions have been abandoned as ways to fit adolescent growth, but they have proved to be still very useful as elements of more complex models describing wider age ranges and for which the graphical determination of the age at take-off is not required.

Preece–Baines model 1 (PB1)

Preece and Baines (1978) proposed the following multiplicative exponential-logistic model:

$$y = h_1 - \frac{2(h_1 - h_\theta)}{e^{s_0(t-\theta)} + e^{s_1(t-\theta)}}$$

where y is size, t is age, and h_1, h_θ, s_0, s_1 and θ are the five function parameters. Adult size is given by parameter h_1. Parameter θ locates the adolescent growth spurt along the time axis. This parameter is highly correlated with age at peak velocity. h_θ is the size at age θ. The parameters s_0 and s_1 are growth-rate constants, related to prepubertal and pubertal velocity.

This five-parameter model is designed to fit the adolescent growth cycle starting from childhood. The success with which the model describes adolescent growth depends, among other things, on how much childhood data is included in the fitting procedure. It was shown that the lower limit of the age range

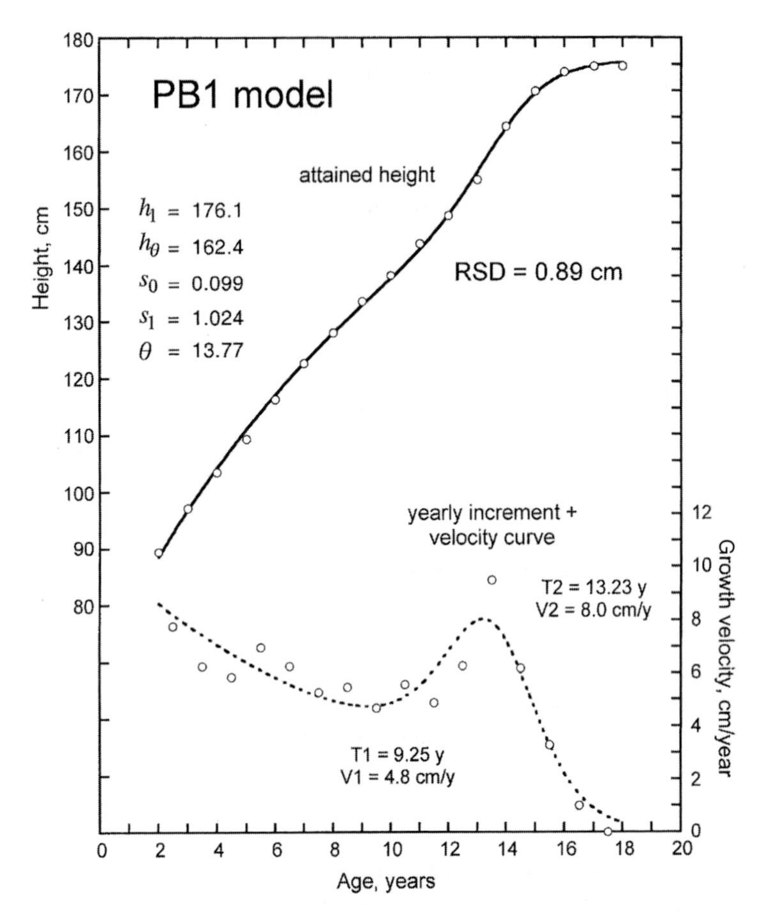

Figure 8.3 Plot of Preece–Baines model 1 (PB1) for the longitudinal data of the boy whose growth data are given in Table 8.1. RSD, residual standard deviation. For definitions of parameters, see text.

should not be under 2 years of age, and that the fit of the adolescent growth cycle is substantially better if the age range includes data from not more than a few years before the age at take-off (Hauspie *et al.*, 1980). Therefore, the PB1 model should essentially be considered as a model to fit the adolescent growth cycle from before take-off up to adulthood.

Figure 8.3 shows the results of the PB1 function, fitted to the data of Table 8.1 including measurements from age 2 years onwards. The values of the function parameters for this particular fit are shown in the figure. The upper part of the graph shows the PB1 fit plotted onto the raw measurements for

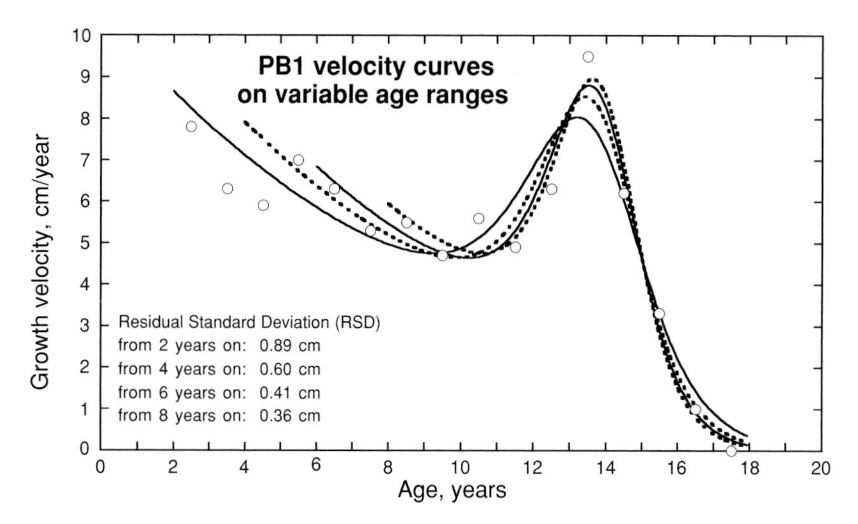

Figure 8.4 Velocity curves obtained by fitting Preece–Baines model 1 (PB1) to variable age ranges of the boy whose growth data are given in Table 8.1.

attained height. The lower part shows the yearly increments in height together with the instantaneous velocity curve obtained as the mathematical first derivative of the PB1 model. The values for age at take-off (T_1), velocity at take-off (V_1), and the corresponding parameters at peak velocity, T_2 and V_2, are also shown in the graph. The graph of the fitted curve seems to match the raw data reasonably well, but we should note that the RSD of 0.89 cm is nevertheless relatively high and well above the measurement error for height. Also the PB1 model underestimates peak velocity when compared to the yearly increments. The boy in this example has a particularly 'peaked' pubertal spurt, and this model generally copes better with less sharp pubertal spurts.

However, when gradually including fewer data from childhood in the fitting procedure, we obtain better fits of the adolescent growth cycle. Figure 8.4 shows the growth velocity curves obtained by fitting the PB1 model on variable age ranges of the height-for-age data of the same subject. The velocity curves are obtained by feeding in the respective values of the function parameters into the first derivative (velocity function) of the PB1 model:

$$y' = \frac{2(h_1 - h_\theta)\left(s_0 e^{s_0(t-\theta)} + s_1 e^{s_1(t-\theta)}\right)}{\left(e^{s_0(t-\theta)} + e^{s_1(t-\theta)}\right)^2}$$

Note that the first derivatives of growth models may take fairly complex forms. However, velocity curves can be easily plotted as the numeric derivative of the fitted distance curve. If the time intervals over which the numeric derivatives are calculated are small (0.05 year, for instance), we obtain a sufficiently smooth representation of the growth velocity.

It is clear that the results differ, depending on the amount of data included in the fitting procedure, and that the best fit is obtained with the data starting at 8 years of age (RSD = 0.36 cm). Therefore, when using the PB1 model, it is advisable to optimize the age range of the data in order to reduce bias in estimating the adolescent growth spurt by reducing the amount of preadolescent data offered to the fitting procedure, but, at the same time being careful that take-off of the adolescent growth spurt remains estimated within the age range of the observations. The fitted curve should include at least a few data points prior to age at take-off.

The above example clearly illustrates that curve fitting should not be applied blindly, but requires graphical inspection of the results for each and every subject in the sample. Plots of the velocity curve onto the yearly increments are a very useful tool in this respect although one should bear in mind that increments depict the *average* velocity in the interval over which the increment was calculated, whereas instantaneous velocity is the velocity at a precise age. Therefore, the peak of an instantaneous velocity curve can be higher than the maximum of the yearly increments in height.

Optimal selection of the age range is just one aspect in the search for a 'best' fit; the inclusion/exclusion of certain data points that are 'out of line' (because of measurement error, transcription error, etc.) and which may sometimes grossly disturb the curve fitting, is another. Therefore, curve fitting does not offer an absolutely unequivocal solution to the problem of estimating a smooth growth curve, but requires a good deal of common sense in deciding which data to offer to the curve fitting procedure and in deciding which curve gives an acceptable description of the data. The fitting of non-linear models is non-trivial, and the curves obtained must be carefully checked to avoid silly results, especially if the range of the data is rather limited or if the density of the data is scarce in crucial periods, such as the age at peak velocity, for instance.

The starting values for the fit of PB1 to height data by numerical estimation algorithms can be defined as follows. Parameter h_1 can be set as the sex-specific mean value for adult height in the considered population. A workable starting value for h_θ usually is $0.9 \times h_1$. The starting value for parameter s_0 can be set equal to 0.1 and for parameter s_1 equal to 1.2. A rough estimate of mean age at peak velocity in the considered population will usually do as starting value for θ, i.e. approximately 12 for girls and 14 for boys.

Growth from birth/early childhood to adulthood

The BTT model

Among the first attempts to describe the human growth curve over a wider age range, we should note the double-logistic function, a six-parameter model resulting from a combination of two logistic functions (Bock *et al.*, 1973). The model can be reduced to a five-parameter model if adult stature is not estimated but taken from the data, and given in as a constant. Later, Thissen *et al.* (1976) also tested other two-component combinations of logistic and Gompertz functions. They found that the linear summation of two logistic functions was superior to the other three combinations. Hauspie *et al.* (1980) found that the double-logistic function was also fairly sensitive with respect to variations in the lower age bound of the data range. In 1980, Bock and Thissen developed the triple logistic function (comprising nine parameters). This model was based on the concept that mature size is a summation of three processes, each of which can be described by a logistic function:

$$y = a_1\left[\frac{1-p}{1+e^{-b_1(t-c_1)}} + \frac{p}{1+e^{-b_2(t-c_2)}}\right] + \frac{a_2}{1+e^{-b_3(t-c_3)}}$$

where y is size, t is age, and a_1, b_1, c_1, a_2, b_2, c_2, b_3, c_3 and p are the nine parameters.

The authors suggest a possible physiological interpretation of their model by attributing the first component to the period of infancy of declining foetal adrenal functioning, the second to the onset of mature adrenal functioning in the mid-childhood period, and the third to the onset of mature gonadal functioning in adolescence. Bock *et al.* (1994) improved the triple logistic model by replacing the logistic function in each of the three components with a generalized logistic function, the BTT model with eight parameters:

$$y = \frac{a_1}{(1+e^{-b_1t-c_1})^{d_1}} + \frac{a_2}{(1+e^{-b_2t-c_2})^{d_2}} + \frac{a_3}{(1+e^{-b_3t-c_3})^{d_3}}$$

where y is size, t is age, and a_1, b_1, a_2, b_2, c_2, a_3, b_3, c_3 are the eight function parameters, and with the following fixed constants (Bock, 1995):

$c_1 = 0.25$

	d_1	d_2	d_3
male	0.815	0.70	0.615
female	0.85	0.63	0.60

Figure 8.5 shows a plot of the BTT fit to the growth data of Table 8.1 including measurements from age 1 year onwards. The model is designed to fit growth data from birth up to adulthood, but performed better in this particular case

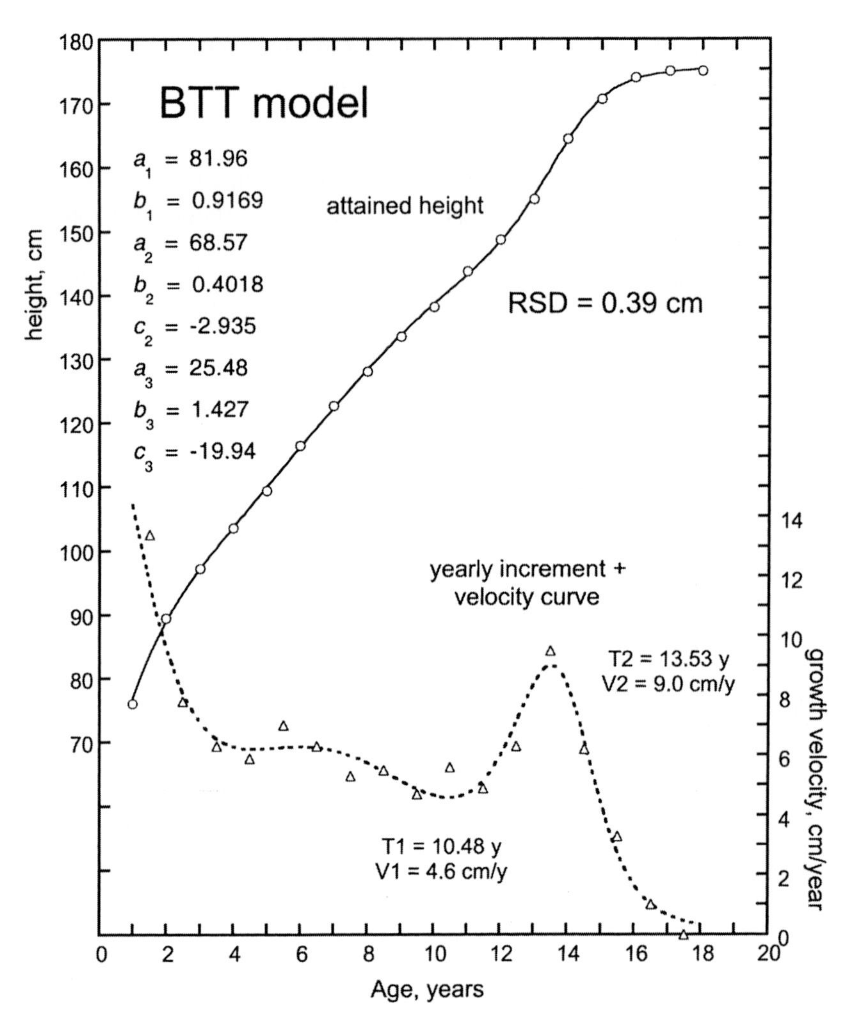

Figure 8.5 Plot of BTT model to the longitudinal data of the boy whose growth data are given in Table 8.1.

when omitting length at birth. In the present example, the fit yielded an RSD of 0.52 cm, which is close to measurement error for height. Age at peak velocity is now estimated by the BTT model at 13.53 years with a peak velocity of 9.0 cm/year, which is very close to the values obtained by the best fit with PB1 on the data from 8 years onwards, namely 13.60 years and 9.0 cm/year. The BTT model allows for a small rise in growth velocity during childhood, which the authors claim to represent the mid-growth spurt. However, our experience is

Table 8.3 *Starting values for fitting the BTT, Count–Gompertz and JPA-2 model to height data with iterative numerical minimization algorithms*

	BTT[a] (age in years)			Count–Gompertz[b] (age in months)			JPA-2[c] (age in years)	
	Boys	Girls		Boys	Girls		Boys	Girls
a_1	81	66	A	12	9	a_1	175	162
b_1	0.85	1.2	B	0.07	0.06	b_1	0.35	0.43
a_2	75	76	C	48	47	b_2	0.12	0.13
b_2	0.38	0.43	D	0.38	0.39	b_3	0.076	0.089
c_2	−2.8	−2.4	E	9.2	8.7	c_1	0.43	0.48
a_3	25	25	U	175	160	c_2	2.8	2.9
b_3	1.4	1.2				c_3	20	16
c_3	−19	−14				e	0.34	0.40

[a] After Bock (1995).
[b] After Jolicoeur *et al.* (1988).
[c] After Jolicoeur *et al.* (1992).

that mid-growth spurts are much sharper and occur over a shorter period of time than can actually be represented by any structural model. The small bump in the growth velocity of the BTT curve as shown in Figure 8.5 does not really match the shape of a mid-childhood spurt. It is even doubtful whether the boy shown in this figure really has a prepubertal spurt at all. The problem is even more complex since some children do not show a detectable prepubertal growth spurt at all, while others may show more than one such spurt (Hauspie and Chrzastek-Spruch, 1993, 1999). In any case, the analysis of the mid-childhood spurt(s) requires frequent measures of height (three-monthly at least) taken with high precision.

Table 8.3 shows a set of starting values for fitting the BTT model (after Bock, 1995).

The Shohoji–Sasaki or Count–Gompertz function

Shohoji and Sasaki (1987) have experimented with a combination of the Count and Gompertz function and proposed the following six-parameter Count–Gompertz growth model (Shohoji and Sumiya, 2001; Sumiya *et al.*, 2001):

$$y = [C + Dt + E\ln(1+t)]\left(1 - \mathrm{e}^{-\mathrm{e}^{A-Bt}}\right) + U\mathrm{e}^{-\mathrm{e}^{A-Bt}}$$

where y is size, t is age, and A, B, C, D, E and U are the function parameters.

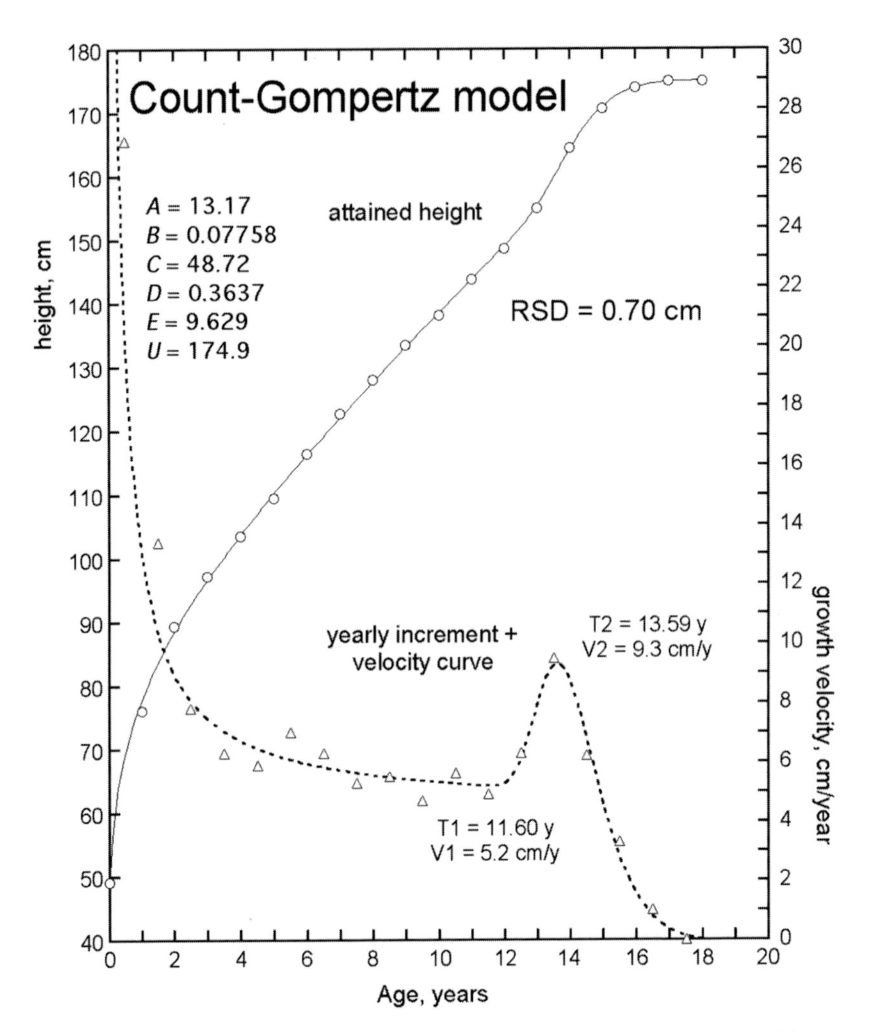

Figure 8.6 Plot of the Count-Gompertz model fitted to the longitudinal data of the boy whose growth data are given in Table 8.1. RSD, residual standard deviation.

The model has a childhood component described by the Count function and an adolescent component accounted for by the Gompertz function. Figure 8.6 shows an example of the Count–Gompertz curve fitted to the data from Table 8.1 including measurements from age birth onwards. The Count–Gompertz function fits better when the ages are expressed in months. For the sake of comparison, we have converted the ages into years in the graphs of Figure 8.6. However, the starting values given in Table 8.3 for the Count–Gompertz model are for when the ages are in months. Parameter U can best be set equal to the value of the observed final height of each individual curve. If problems of convergence occur in some individuals one may follow the procedure explained by Shohoji *et al.* (1989) to find optimal initial values.

The model fits the data reasonably well in the given example with a residual standard deviation of 0.70 cm. Peak velocity obtained as the first derivative of the fitted curve matches closely the value of the maximum yearly increment. This model, which has only six parameters, seems to cope very well with the fairly sharp adolescent growth spurt of this particular case, and age and velocity at take-off seem to be estimated correctly. Indeed, graphical inspection of the lower part of Figure 8.6 shows that the velocity curve of the Count–Gompertz model matches the take-off and peak velocity of the incremental data very well.

Shohoji and Sumiya (2001) have recently proposed an extended eight-parameter version of the Count–Gompertz model, allowing for a mid-childhood spurt, accounted for by an extra Gompertz function. However, the extended Count–Gompertz model suffers from the same limitations as the BTT model when it comes to describe prepubertal spurt(s). The authors rightly conclude that we need more detailed studies on the phenomenon of mid-childhood growth.

Starting values for fitting the Count–Gompertz model when the ages are expressed in months are given in Table 8.3.

The JPA-2 model

Jolicoeur and co-workers have developed three asymptotic models for studying longitudinal growth in stature over a wide age range. These are: the JPPS model with seven parameters (Jolicoeur *et al.*, 1988; Pontier *et al.*, 1988a,b), the JPA-1 and JPA-2 models with respectively seven and eight parameters (Jolicoeur *et al.*, 1992). While most models use postnatal age (measured from the day of birth), the JPPS and JPA-1 models utilize total age (measured from the day of fertilization), which is obtained by correcting postnatal age with the average duration of pregnancy (0.75 years). Models passing through the origin with respect to total age are defined before birth as well as after birth and, the authors claim, may be particularly suitable if prenatal data are to be included in the analysis or if prenatal extrapolations are desired. We will here only discuss the eight-parameter JPA-2 model (Jolicoeur *et al.*, 1992) designed to fit postnatal growth from birth to full maturity utilizing postnatal age because this model has proven to fit postnatal data from birth onwards very well indeed:

$$y = a\left\{1 - \frac{1}{1 + [b_1(t+e)]^{c_1} + [b_2(t+e)]^{c_2} + [b_3(t+e)]^{c_3}}\right\}$$

where y is size, t is age and a, b_1, b_2, b_3, c_1, c_2, c_3 and e are the function parameters.

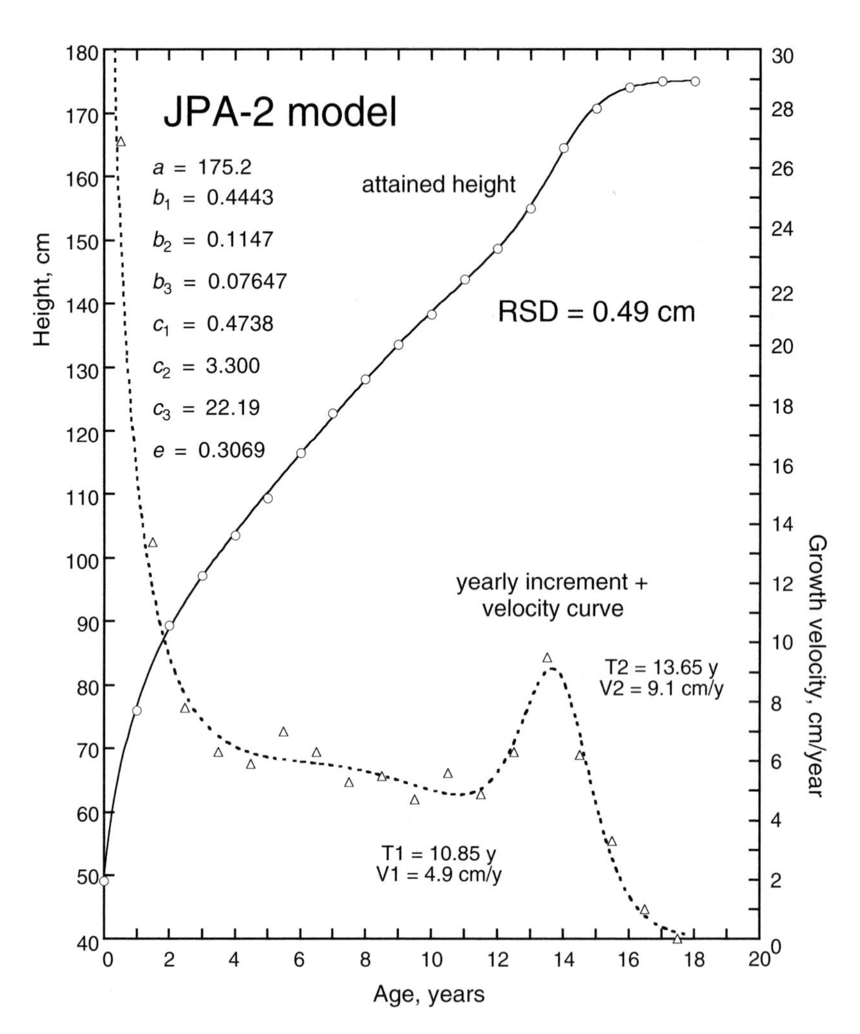

Figure 8.7 Plot of the JPA-2 model fitted to the longitudinal data of the boy whose growth data are given in Table 8.1. RSD, residual standard deviation.

Figure 8.7 shows an example of the fit of the JPA-2 model to the growth data of Table 8.1 including all measurements since birth. The curve fit is very satisfactory with an RSD of 0.49 cm.

Biological parameters and structural average curve

A major goal of growth modelling is to estimate biological parameters from the fitted curve. These biological parameters are features that characterize the

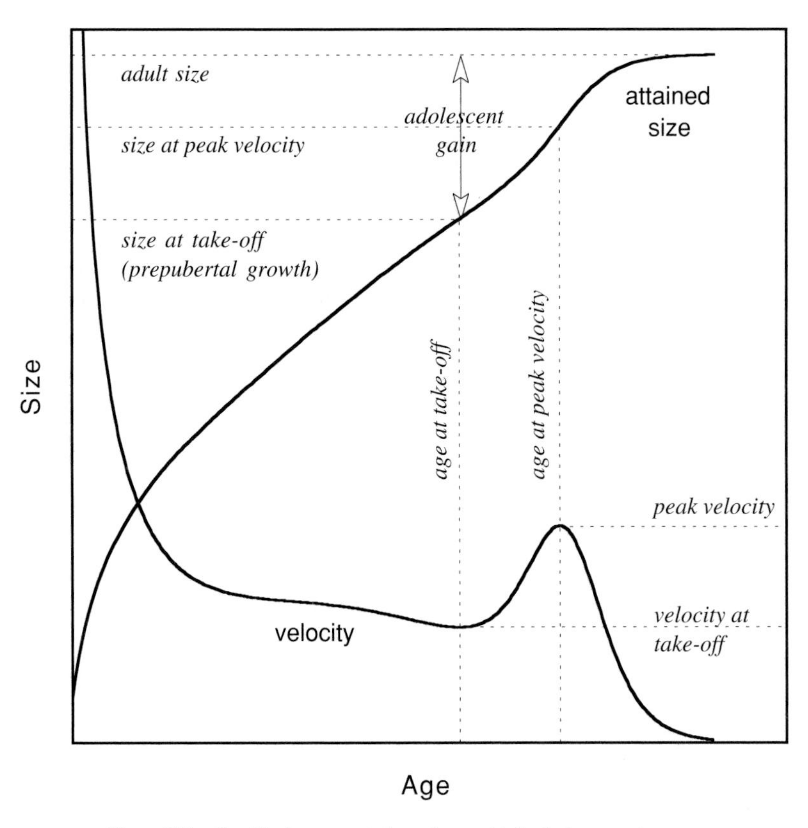

Figure 8.8 Graphical representation of some biological parameters.

shape of the growth curve such as age, size and velocity at the start of the adolescent growth spurt (at take-off) or at the time of maximal velocity (age at peak velocity). Size at take-off can be considered as a measure of prepubertal growth, while the difference between adult size and size at take-off is a measure of adolescent growth (adolescent gain). Figure 8.8 gives a graphical representation of a number of biological parameters.

In the case of the PB1 model, we can estimate the age at take-off (T_1) and at peak velocity (T_2) in the following way:

$$T_1 = \frac{\ln[(-A + B)/2]}{s_0 - s_1} + \theta \qquad T_2 = \frac{\ln[(-A - B)/2]}{s_0 - s_1} + \theta$$

$$\text{with } A = 4\frac{s_1}{s_0} - 1 - \left(\frac{s_1}{s_0}\right)^2 \quad \text{and} \quad B = \sqrt{A^2 - 4\left(\frac{s_1}{s_0}\right)^2}$$

By evaluating the PB1 function and its first derivative with the estimated function parameters at ages T_1 and T_2 we obtain size and velocity at take-off and at peak velocity. The other parameters can then be easily computed. Remember that adult size is equal to h_1, the upper asymptote of the function.

For models with more parameters, such as the BTT, the Count–Gompertz and the JPA-2 models, the mathematical expression of the first derivative or velocity curve becomes very large and complex. In these cases, biological parameters are usually estimated numerically.

Structural models that tend to an upper asymptote yield an acceptable estimate of final size on the condition that the data give a clear indication that growth has stopped or nearly stopped. If this is not the case, the estimation of the parameter that determines final size is not under control of the data, and hence may take quite awkward values, which are most of the time biologically meaningless. In addition, derived biological variables that depend on final size, such as adolescent gain for instance, will then also be biased. Whether growth has reached an end, or is nearly ended, is a matter of subjectivity. In clinical studies it is often accepted that growth has reached an end when growth velocity falls below 1 cm/year. For the purpose of curve fitting, one should, as a rule of thumb, suspect the estimate of final size by a structural model when the yearly increment calculated over the last two measurement occasions exceeds 2 cm, or if final size estimated by the model exceeds the last measurement by more than 2 cm (at least when we are dealing with growth in height). A graphical inspection of a plot of the fitted curve onto the raw data, and a good deal of common sense in interpreting the results, should help to decide whether the curve fit is eventually acceptable or not. Again this emphasizes the fact that curve fitting by a least-squares method is merely a descriptive rather than a predictive method and does not allow making extrapolations beyond the range of the data, unless one uses Bayesian estimation techniques (see Chapter 9).

A further interesting aspect of using non-linear growth models is that they allow to produce mean-constant (or structural average) curves. A mean-constant curve is obtained as the plot of the growth model with the mean values of the respective function parameters obtained for a group of subjects. The curve obtained thus shows the 'typical average' growth pattern for the group, and differs from the mean growth curve obtained by taking the cross-sectional average of the size of the subjects at each age. This is a particular property of non-linear models, i.e. models in which there are non-linear relationships between the parameters such as exponential and logarithmic functions. For linear models (such as polynomials, for instance) the mean-constant curve is identical to the cross-sectional mean of the individual curves and, hence, does not show the 'typical average' growth pattern for the group.

As an example, Figure 8.9 shows the velocity curves of two boys (obtained as the first derivatives of JPA-2 fits) together with the mean-constant velocity curve

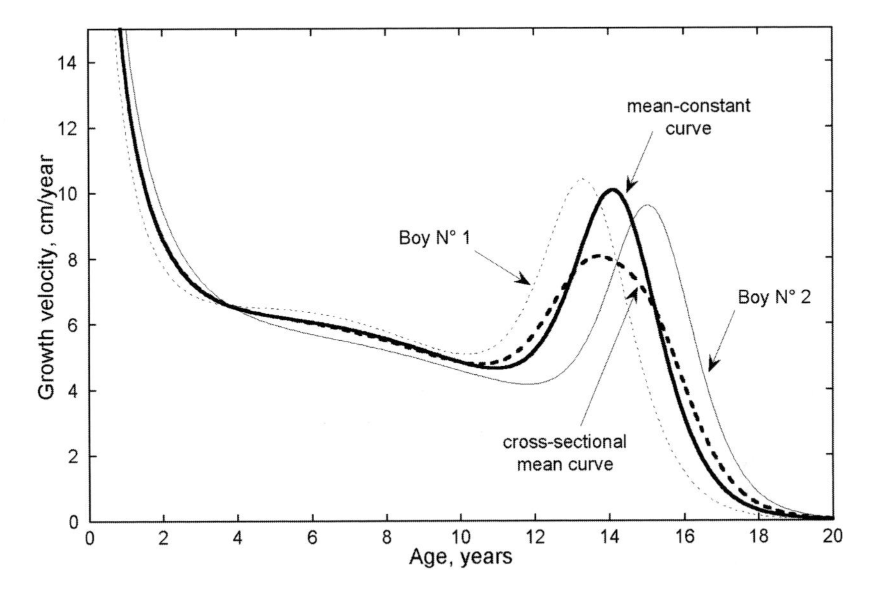

Figure 8.9 Instantaneous velocity curves (obtained as first derivatives of the JPA-2 model fitted to the distance data) of two boys together with the mean-constant curve and the cross-sectional mean velocity curve.

(obtained as the first derivative of the distance mean-constant curve), as well as the cross-sectional mean curve obtained by averaging the velocities of both individuals at each age.

From this example, it is clear that the mean-constant curve has a shape that is 'average' for both subjects, while the adolescent growth spurt shown by the cross-sectional mean curve is flattened out due to the fact that both subjects have a different timing of their spurt, the so-called 'phase-difference' effect (Tanner *et al.*, 1966a,b). The cross-sectional mean curve is not representative for any of the two individual curves. The phenomenon is best illustrated for velocities, but it equally holds of course for the distance curves where the cross-sectional mean curve is less steep than any of the individual curves while the distance mean-constant curve has a slope that is 'typical average' for the individuals in the sample. Therefore, cross-sectional analysis of longitudinal data, and obviously also of pure cross-sectional data, yield mean curves that give a distorted picture of the adolescent growth spurt. If the distribution of the function parameters of the growth models is very much skewed, one may envisage using the median values of the function parameters, rather than the mean values.

Although the use of mean-constant curves forms an elegant and interesting application of curve fitting, there may also be some pitfalls in using them, particularly if the subjects in the sample have quite variable age ranges. Indeed, the parameters controlling for the prepubertal growth are not much under control of the data in the curves where prepubertal growth data is scarce, and hence they may take quite awkward values. However, if those values are included in subsequent computation of the parameter means, the resulting mean-constant curve may become distorted and not representative for the prepubertal period. Again, it has all to do with the fact that least-squares estimation is not suitable for extrapolation beyond the range of observations. Bayesian estimations techniques (see Chapter 9) or 'structural analysis' (see Chapter 7) may in such cases be better approaches.

Tanner *et al.* (1966a,b) used a graphical procedure to correct for the phase-difference effect when producing their famous growth standards for growth and growth velocity in British children. Their technique consisted in centring each individual's longitudinal growth data at the age at peak velocity before averaging them. The resulting mean curves (and centile distributions) were described as tempo-conditioned and also represented the average shape of the growth in the population. The use of mean-constant curves is to some extent equivalent to Tanner's graphical approach, but has the advantage of not distorting the shape of preadolescent growth which may occur in curves centred on peak height velocity. Later Tanner and Davies (1985) used the same graphical procedures to produce clinical longitudinal standards for height and height velocity in North American children. Wachholder and Hauspie (1986) and Hauspie and Wachholder (1986) produced tempo-conditioned standards for growth and growth velocity using mean-constant curves as the reference for typical average growth. Moreover, they also produced reference curves for 'typical early' and 'typical late' maturing children. These curves were obtained from regression analysis of the function parameters on age at peak velocity and using the set of values corresponding to mean age at peak velocity $\pm$ 2 standard deviations (Hauspie *et al.*, 1993). Another approach to get rid of the differences in tempo of growth without having to align individual growth curves along the time axis is given by 'structural analysis' (see Chapter 7).

Assumptions and goodness of fit

Assumptions of non-linear regression

Curve fitting is actually an example of non-linear regression and, just as simple linear regression, requires that a number of assumptions be met. We will summarize the most important ones and briefly explain how one can test for violations of these assumptions.

Sensible model

It is up to the researcher to choose the model appropriate for the data at hand. Non-linear regression provides estimates of the function parameters of the model through least-squares techniques, but does not provide clues to find a better model. The previous sections should be helpful to decide about which model is most appropriate for the data.

Gaussian distribution of the residuals

Non-linear regression assumes that the scatter of data points around the fitted curve is Gaussian. This can be tested by calculating skewness and kurtosis, by a variety of statistical tests such as Kolmogorov–Smirnov or Shapiro–Wilk, or, graphically, by inspecting normal probability plots of the residuals (Royston and Wright, 2000). The latter technique not only allows to show deviations from Gaussianity in terms of skewness and kurtosis, but also to easily detect outliers. Significant deviation from Gaussianity for the error term may lead to a decreased efficiency of the parameter estimation and incorrect calculation of the standard errors of the parameters. In such cases, one should resort to formally robust methods (Hampel *et al.*, 1986; Marazzi *et al.*, 1993).

Homoscedasticity

Regression analysis assumes that the scatter of points around the best-fit curve has the same standard deviation all along the curve. The assumption is violated if the points with higher (or lower) y-values also tend to lie further away from the best-fit curve. This assumption can be verified graphically by the inspection of a graph of the residuals versus the predicted (fitted values), but is usually not a problem in longitudinal growth data.

Independence of residuals

The distribution of the residuals (the deviations of the observations from the fitted curve, or error term) should be random, i.e. without a systematic pattern over the time axis. However, with models fitted to longitudinal growth data, there is a substantial risk of such systematic bias. Within-subject systematic residual patterns may give the impression of correlated residuals, although, as we indicate in the following there is little reason for this. We may in fact think of several conceptually different sources of variation in the residuals and causes of systematic bias, none of which would cause residuals several months apart to be correlated, with the possible exception of catch-up growth.

- Measurement error is likely to create random variation of the data points around the fitted curve without causing systematic bias in the fit.
- Short-term variations in growth velocity, such as the ones described by Hermanussen (1998) and Lampl (1999), have a too small amplitude and too short phase to contribute to the random variation or systematic bias in the residuals. Also, classical anthropometric techniques to measure height are not precise enough, and measurement frequency in longitudinal surveys is usually much too small for detecting this level of short-term variation in growth rate.
- On the contrary seasonal variation in growth rate may be responsible for a cyclic pattern in the residuals and cause minor systematic bias in the curve fit, but only if the data is taken at sufficiently frequent intervals (say three-monthly) with a high level of precision. Growth models are designed to describe the growth curve over longer periods and hence cannot cope with this kind of fluctuations. The effect of seasonal variation can be avoided when measurements are taken at yearly intervals.
- Disease or adverse environmental conditions possibly followed by a period of catch-up growth may also create unusual patterns of growth that result in systematic bias in curve fits.
- A phenomenon that may cause systematic bias is the occurrence of one or several prepubertal spurts. These spurts last over a few months to a few years and can cause a non-random distribution of the residuals around the fitted curve in the preadolescent period (Hauspie and Chrzastek-Spruch, 1999). Parametric growth models are simply not designed to describe this phenomenon adequately.
- While growth models are supposed to estimate the overall shape of the growth curve in a particular age range, systematic bias can occur as a consequence of the lack of flexibility of the model in coping with the normal variation in individual growth patterns. Figure 8.10 shows the systematic bias in the pattern of residuals centred on age at peak velocity of the PB1 curve of 112 girls taken from the First Zürich Longitudinal Study (Prader *et al.*, 1989). The graph shows that PB1 slightly overestimates the attained stature at age of peak velocity and that it underestimates it by an even larger amount at about three years prior to peak velocity. This bias is introduced by the PB1 model; other models such as JPA-2 or Count–Gompertz will yield different residual patterns.

Goodness of fit

Measures of goodness of fit indicate how close the curve matches the observations. Examination of the deviations from the fitted curve (i.e. the residuals)

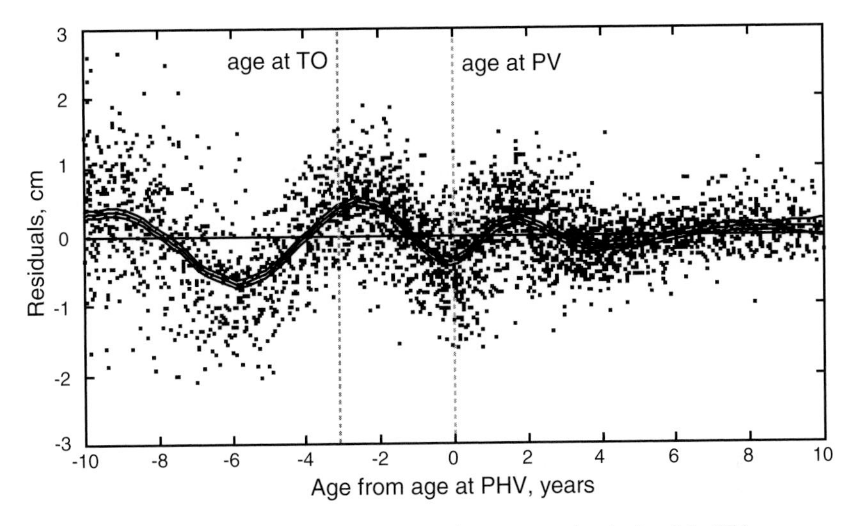

Figure 8.10 Pattern of residuals centred on age at peak velocity of the PB1 curve fitted to the height data of 112 girls from the First Zürich Longitudinal Study. Central tendency and confidence intervals are shown as well. The vertical lines show age at take-off (TO) and at peak velocity (PV).

may detect outliers in the data (see also the normal probability plots above). Removal of such outliers may considerably improve the fit.

A useful and simple statistical tool to quantify the overall goodness of fit is the residual standard deviation, also named standard error of estimate (SEE), or residual root mean square (RMSQ):

$$\text{residual standard deviation (RSD)} = \sqrt{\frac{\sum_{i=1}^{N}(y_i - \hat{y}_i)^2}{N-k}}$$

where y_i is the measurement at age x_i, $\hat{y}_i$ the value of the fitted curve at age x_i, N the number of observations, and k the number of parameters in the model. RSD is the standard deviation of the spread of the data points around the fitted curve. It is generally accepted that a fitted curve adequately describes the growth data if the RSD is of the same order of magnitude as the measurement error of the trait under consideration (typically 0.5 cm for stature in childhood and adolescence for instance). In growth curve analysis the residual standard deviation is preferred over the coefficient of determination R^2 since this value is usually very high in all curve fittings and relates less well to the notion of spread of the data than the RSD. For a group of subjects it is often useful to calculate the *pooled* residual variance, i.e. sum of squared deviations over all subjects divided by the total number of degrees of freedom. The square root of this value is the *pooled* RSD for the group under study.

Lack of flexibility of all parametric models to describe subtle, small growth patterns, such as prepubertal spurts, will leave systematic biases in the residuals. This minor bias usually does not affect the overall fit to an extent that the obtained curve is not a useful description of the general growth pattern. In a way, these patterns of non-random variation in the residuals can be considered as a proxy for short-term variation in growth not accounted for by the growth model and has, as such, been exploited to study genetic variance in patterns of these short-term variation in growth among twins (Hauspie *et al.*, 1994), or in a study of the similarities in patterns of growth among various body dimensions (Hauspie and Das, 1995), for instance.

Large values of the residual standard deviation and systematic bias may, among others, be due to:

- Low-precision data (large measurement error, inexperienced measurers, large inter-measurer variation, inadequate measuring devices, etc.).
- Inappropriate age range of the growth data for the chosen model (each model is designed to fit growth data in a particular age range; including data beyond that age range results in bad curve fittings).
- Inappropriate models for the type of variable (models designed for postcranial skeletal dimensions during the adolescent period are inappropriate for fitting head dimensions or weight, and for growth data in early childhood).
- The presence of particular features in the growth data that cannot be described by the growth model (prepubertal growth spurts, unusual variations of growth rate caused by severe changes in environmental conditions or disease).

REFERENCES

Berkey C. S. (1982). Comparison of two longitudinal growth models for preschool children. *Biometrics*, **38**, 221–34.

Berkey, C. S. and Reed B. (1973). A model for describing normal and abnormal growth in early childhood. *Human Biology*, **59**, 973–87.

Bock, R. D. (1995). Growth models. In *Essays on Auxology presented to James Mourilyan Tanner*, eds. R. C. Hauspie, G. Lindgren and F. Falkner, pp. 28–38. Welwyn Garden City, UK: Castlemead Publications.

Bock, R. D. and Thissen, D. M. (1980). Statistical problems of fitting individual growth curves. In *Human Physical Growth and Maturation*, eds. F. E. Johnston, A. F. Roche and C. Susanne, pp. 265–90. Plenum Press: New York.

Bock, R. D., Wainer, H., Petersen, A., *et al.* (1973). A parameterization for individual human growth curves. *Human Biology*, **45**, 63–80.

Bock, R. D., du Toit, S. H. C. and Thissen, D. (1994). *AUXAL: Auxological Analysis of Longitudinal Measurements of Human Stature*. Chicago, IL: Scientific Software International.

Bogin, B. (1999). *Patterns of Human Growth*, 2nd edn. Cambridge, UK: Cambridge University Press.

Count, E. W. (1942). A quantitative analysis of growth in certain human skull dimensions. *Human Biology*, **14**, 143–65.

(1943). Growth patterns of the human physique: an approach to kinetic anthropometry. *Human Biology*, **15**, 1–43.

Deming, J. (1957). Application of the Gompertz curve to the observed pattern of growth in length of 48 individual boys and girls during the adolescent cycle of growth. *Human Biology*, **29**, 83–122.

Deming, J. and Washburn, A. H. (1963). Application of the Jenss curve to the observed pattern of growth during the first eight years of life in forty boys and forty girls. *Human Biology*, **35**, 484–506.

Goldstein, H. (1984). Current developments in the design and analysis of growth studies. In *Human Growth and Development*, eds. J. Borms, R. C. Hauspie, C. Susanne and M. Hebbelink, pp. 733–52. New York: Plenum Press.

Hampel, F. R., Ronchett, E. M., Rousseeuw, P. J. and Stahel, W. A. (1986). *Robust Statistics: The Approach Based on Influence Functions*. New York: John Wiley.

Hauspie, R. C. (1981). L'Ajustement de modèles mathématiques aux données longitudinales. *Bulletin de la Société royale belge d'Anthropologie*, **92**, 157–165.

(1989). Mathematical models for the study of individual growth patterns. *Revue d'Epidémiologie et de Santé Publique*, **37**, 461–76.

(1998). Curve fitting. In *The Cambridge Encyclopedia of Human Growth and Development*, eds. S. J. Ulijaszek, F. E. Johnston and M. A. Preece, pp. 114–15. Cambridge, UK: Cambridge University Press.

Hauspie, R. C. and Chrzastek-Spruch H. (1993). The analysis of individual and average growth curves: some methodological aspects. In *Kinanthropometry*, vol. 4, eds. W. Duquet and J. A. P. Day, pp. 68–83. London: E.&F. Spon.

(1999). Growth models: possibilities and limitations. In *Human Growth in Context*, eds. F. E. Johnston, B. Zemel and P. B. Eveleth, pp. 15–24. London: Smith-Gordon.

Hauspie, R. C. and Das, S. R. (1995). Short-term variations in growth rate of height, sitting height and biacromial diameter in Bengali children. In *Essays on Auxology presented to James Mourilyan Tanner*, eds. R. C. Hauspie, G. Lindgren and F. Falkner, pp. 260–8. Welwyn Garden City, UK: Castlemead Publications.

Hauspie, R. C. and Wachholder, A. (1986). Clinical standards for growth velocity is height of Belgian boys and girls, aged 2 to 18 years. *International Journal of Anthropology*, **1**, 339–48.

Hauspie, R. C., Wachholder, A., Baron, G., *et al.* (1980). A comparative study of the fit of four different functions to longitudinal data of growth in height of Belgian girls. *Annals of Human Biology*, **7**, 347–58.

Hauspie, R. C., Wachholder, A. and Vercauteren, M. (1993). Normes de croissance staturale et pondérale et de vitesse de croissance de garçons et de filles belges de 3 à 18 ans. *Archives françaises de Pédiatrie*, **50**, 763–9.

Hauspie, R. C., Bergman, P., Bielicki, T. and Susanne, C. (1994). Genetic variance in the pattern of the growth curve for height: a longitudinal analysis of male twins. *Annals of Human Biology*, **21**, 347–62.

Hermanussen, M. (1998). The analysis of short-term growth. *Hormone Research*, **49**, 53–64.

Jenss, R. M. and Bayley, N. (1937). A mathematical method for studying the growth of a child. *Human Biology*, **9**, 556–63.

Jolicoeur, P., Pontier, J., Pernin M.-O. and Sempé, M. (1988). A lifetime asymptotic growth curve for human height. *Biometrics*, **44**, 995–1003.

Jolicoeur, P., Pontier J. and Abidi, H. (1992). Asymptotic models for the longitudinal growth of human stature. *American Journal of Human Biology*, **4**, 461–8.

Kneip, A. and Gasser, T. (1992). Statistical tools to analyze data representing a sample of curves. *Annals of Statistics*, **20**, 1266–305.

Lampl, M. (1999). *Saltation and Stasis in Human Growth: Evidence, Methods and Theory*. London: Smith-Gordon.

Largo, R. H., Gasser, T., Prader, A., Stuetzle, W. and Huber, P. J. (1978). Analysis of the adolescent growth spurt using smoothing spline functions. *Annals of Human Biology*, **5**, 421–34.

Livshits, G., Peter, I., Vainder, M. and Hauspie, R. C. (2000). Genetic analysis of growth curve parameters of body weight, height and head circumference. *Annals of Human Biology*, **27**, 299–312.

Manwani, A. H. and Agarwal, K. N. (1973). The growth pattern of Indian infants during the first year of life. *Human Biology*, **59**, 973–87.

Marazzi, A., Joss, J. and Randriamiharisoa, A. (1993). *Algorithms, Routines, and S Functions for Robust Statistics: The FORTRAN Library ROBETH with an Interface to S-Plus*. London: Chapman and Hall.

Marubini, E. (1978). The fitting of longitudinal growth data of man. In *Auxology: Human Growth in Health and Disorders*, eds. L. Gedda and L. Parisi, pp. 123–32. London: Academic Press.

Marubini, E. and Milani, S. (1986). Approaches to the analysis of longitudinal data. In *Human Growth: A Comprehensive Treatise*, 2nd edn, eds. F. Falkner and J. M. Tanner, pp. 79–94. New York: Plenum Press.

Marubini, E., Resele, L. F. and Barghini, G. (1971). A comparative fit of the Gompertz and logistic function to longitudinal height data during adolescence in girls. *Human Biology*, **43**, 237–52.

Marubini, E., Resele, L. F., Tanner, J. M. and Whitehouse, R. H. (1972). The fit of the Gompertz and logistic curves to longitudinal data during adolescence on height, sitting height, and biacromial diameter in boys and girls of the Harpenden Growth Study. *Human Biology*, **44**, 511–24.

Nelder, J. A. (1961). The fitting of a generalization of the logistic curve. *Biometrics*, **17**, 89–110.

Pontier, J., Jolicoeur, P., Abidi, H. and Sempé, M. (1988a). Croissance staturale chez l'enfant: le modèle JPPS. *Biométrie et Praximétrie*, **28**, 27–44.

Pontier, J., Jolicoeur, P., Pernin, M.-O., Abidi, H. and Sempé, M. (1988b). Modelisation de la courbe staturale chez l'enfant: le modèle JPPS. *Cahiers d'Anthropologie et Biométrie Humaine (Paris)*, **6**, 71–85.

Prader, A., Largo, R. H., Molinari, L. and Issler, C. (1989). Physical growth of Swiss children from birth to 20 years of age. *Helvetica Paediatrica Acta* (*Supplement*), **52**.

Preece, M. A. and Baines, M. K. (1978). A new family of mathematical models describing the human growth curve. *Annals of Human Biology*, **5**, 1–24.

Royston, P. and Wright, E. M. (2000). Goodness-of-fit statistics for age-specific reference intervals. *Statistics in Medicine*, **19**, 2943–62.

Shohoji, T. and Sasaki, T. (1987). Individual growth of Japanese. *Growth*, **51**, 432–50.

Shohoji, T. and Sumiya, T. (2001). Individual physical growth models and biological parameters of Japanese. In *Perspectives in Human Growth, Development and Maturation*, eds. P. Dasupta and R. C. Hauspie, pp. 17–32. Dordrecht, The Netherlands: Kluwer Academic Publishers.

Shohoji, T., Kanefuji, K., Kuwabara, M. and Sumiya, T. (1989). Growth of height of Japanese girls. *Acta Medica Auxologica*, **21**, 115–27.

Simondon, K. B., Simondon, F., Delpeuch, F. and Cornu, A. (1992). Comparative study of five growth models applied to weight data from Congolese infants between birth and 13 months of age. *American Journal of Human Biology*, **4**, 327–35.

Sumiya, T., Nakahara, H. and Shohoji, T. (2001). Relationships among biological growth parameters for body weight in Japanese children. *Growth, Development and Aging*, **64**, 91–112.

Tanner, J. M. and Davies, P. S. W. (1985). Clinical longitudinal growth standards for height and height velocity for North American children. *Journal of Pediatrics*, **107**, 317–29.

Tanner, J. M., Whitehouse, R. H. and Takaishi, M. (1966a). Standards from birth to maturity for height, weight, height velocity, and weight velocity: British children, 1965. Part I. *Archives of Disease in Childhood*, **41**, 454–71.

(1966b). Standards from birth to maturity for height, weight, height velocity, and weight velocity: British children, 1965. Part II. *Archives of Disease in Childhood*, **41**, 613–35.

Tanner, J. M., Whitehouse, R. H., Marubini, E. and Resele, L. (1976). The adolescent growth spurt of boys and girls of the Harpenden Growth Study. *Annals of Human Biology*, **3**, 109–26.

Thissen, D., Bock, R. D., Wainer, H. and Roche, A. F. (1976). Individual growth in stature: a comparison of four growth studies in the USA. *Annals of Human Biology*, **3**, 529–42.

Wachholder, A. and Hauspie, R. C. (1986). Clinical standards for growth in height of Belgian boys and girls, aged 2 to 18 years. *International Journal of Anthropology*, **1**, 327–38.

9 *Parameter estimation in the context of non-linear longitudinal growth models*

R. Darrell Bock
University of Chicago

and

Stephen H. C. du Toit
Scientific Software International, Chicago

Longitudinal measurements of human growth present special difficulties for statistical analysis, both in the fitting of parametric models for individual growth and when comparing growth in various populations. The main objective of parametric analysis is to describe or predict growth, or differences in growth, as a function of chronological age. Although it would be convenient in this connection to assume complete records of measurements at the same time points for all cases, that condition is almost impossible to fulfil even in carefully conducted longitudinal studies. For this reason, conventional multivariate statistical methods that assume measurement records of fixed dimensionality do not apply.

Further difficulties arise from the fact that growth is not a simple deterministic process that can be represented by a continuous function of time with a manageable number of parameters. Even well-fitting functional models will exhibit residual variation attributable to so-called 'equation error'. To the extent that the residuals are unbiased (i.e. tend to sum to zero), the fitted growth curve will pass through the data points in such a way that the points in successive intervals will lie on one side of the curve and then on the other. This introduces autocorrelation among the residuals and violates the assumption of independent error required in most curve-fitting procedures – although it does not bias the shape of curves appreciably if the growth measurements are regularly spaced in time. If the measurements are more dense at some age intervals than others, however, ignoring autocorrelation will bias the curve towards better fit in those

Methods in Human Growth Research, eds. R. C. Hauspie, N. Cameron and L. Molinari. Published by Cambridge University Press.

intervals at the expense of poorer fit elsewhere. In effect, autocorrelation reduces the information about the shape of the curve when successive points are closer together. Even when shape is not affected, ignoring autocorrelation can still lead to underestimation of the standard errors of estimated model parameters.

In addition to equation error, there are also sources of error arising from quotidian variability depending on the time of day and the activity of the subject prior to measurement. Technical errors in the measurement procedure will also inevitably be present. These types of error can safely be assumed independent over intervals of a day or more, but their variance may not be homogeneous at different age levels.

Dealing with autocorrelation and non-homogeneous error are statistical issues that apply to all growth models. There are other issues, however, connected with the functional forms of the models themselves. In any given data, some models are better conditioned for estimation than others: good conditioning leads to smaller standard errors of the parameter estimators and lower correlation between parameters in the population of subjects; poor conditioning makes iterative numerical solutions of estimation equations sensitive to starting values and may prevent convergence. In some cases, conditioning can be improved by suitable reparameterization – that is, by non-singular transformation of the model parameters and suitable alteration of the functional form of the model. In this chapter we examine these issues and outline some solutions to the problems they present for data analysis. These include: conditioned maximum likelihood estimation, empirical Bayes estimation, and non-homogeneous and autocorrelated residuals. We illustrate the various results with two well-known growth models applied to data from the Fels Longitudinal Growth Study.

Maximum likelihood estimation

Maximum likelihood (ML) estimation has certain good properties that recommend it in fitting non-linear longitudinal growth models. The ML estimator is in general *consistent*, which means that it attains the true values of the model parameters as the number of measurements increases without limit; it is *efficient*, which means that its error variance is the minimum attainable; and it has *parameter invariance* with respect to any one-to-one transformation, which means that the corresponding transformation of the estimate is the ML estimate of the transform.

ML estimation of growth-model parameters requires, however, an assumed distribution of the residual variation conditional on age. Fortunately, the usual assumption of normally distributed residuals is reasonable in the context of growth in linear dimensions (although possibly not for measures of area or volume, or derived indices such as body mass index). The stronger assumptions that the conditional distributions have the same variance at different ages and are independent may not be reasonable; they are relaxed below. Initially, however, let us assume that at age x_j, $j = 1, 2, \ldots, n$, the deviation of the observed growth measurement y_j from the corresponding value of an m-parameter growth model $f(x_j)$

$$u_j = y_j - f(x_j) \tag{9.1}$$

is independent normal with mean zero and variance σ_u^2. For ML estimation n must be greater than the number of free parameters in the model, preferably several times greater. (As discussed below, this is not necessary in Bayes estimation.) For the n measurements, the likelihood to be maximized with respect to ν_k, $k = 1, 2, \ldots, m$, is defined as

$$L(\nu_k) = \prod_{j=1}^{n} \frac{1}{\sigma_u \sqrt{2\pi}} \exp\left(-u_j^2/2\sigma_u^2\right) \tag{9.2}$$

Taking the logarithm of the likelihood and differentiating, we obtain the corresponding likelihood equations, say,

$$G(\nu_k) = \frac{1}{\sigma_u^2} \sum_{j=1}^{n} u_j \frac{\partial f(x_j)}{\partial \nu_k} = 0 \tag{9.3}$$

Let $\hat{\nu}_k$, $k = 1, 2, \ldots, m$, be a solution of these simultaneous non-linear equations for a model with m free parameters. Provided the log likelihood is continuous in its first two derivatives, a solution can in general be approximated numerically with sufficient accuracy by an iterative Newton method. This requires the second derivatives of the log likelihood with respect to ν_k and ν_ℓ, $k = 1, 2, \ldots, m$; $\ell = 1, 2, \ldots, m$;

$$H(\nu_k, \nu_\ell) = -\sum_{j}^{n} \left[\frac{\partial f(x)}{\partial \nu_k} \cdot \frac{\partial f(x)}{\partial \nu_\ell} + u_j \frac{\partial^2 f(x_j)}{\partial \nu_k \partial \nu_\ell} \right] \tag{9.4}$$

Approximating Eqn 9.3 by the first two terms of a Taylor series for the provisional values of the parameter estimate at iteration $t + 1$, about the values at iteration t, we have the m linear equations

$$G\left(\nu_k^{(t)}\right) + \sum_{\ell}^{m} H\left(\nu_k^{(t)}, \nu_\ell^{(t)}\right)\left(\nu_\ell^{(t+1)} - \nu_\ell^{(t)}\right) = 0 \tag{9.5}$$

To express the simultaneous solution of these equations, it is convenient to adopt matrix notation and write

$$G_t + H_t(\nu_{t+1} - \nu_t) = 0 \tag{9.6}$$

where the $m \times 1$ matrix G is called the *gradient* vector, the $m \times m$ symmetric matrix H is called the *Hessian* matrix, and the $m \times 1$ matrix 0 is called the *null* vector. A necessary condition on the solution is that H be *positive-definite*; i.e. the so called *pivots* of the solution of Eqn 9.6 by Gaussian elimination must be positive. This insures that the solution corresponds to a maximum of the m-dimensional likelihood.

The computations of the solution may be programmed in matrix operations subroutines: first calculate H^{-1}, the inverse or reciprocal matrix of H, then the matrix product of the term in Eqn 9.6 and rearranged as

$$\nu_{t+1} = \nu_t - H_t^{-1} G_t \tag{9.7}$$

If at the first iteration all elements of ν_0 are sufficiently close to the solution point, the iterations will converge quadratically; that is, the magnitude of the elements in $\nu_{t+1} - \nu_t$ will square at each iteration and the number of correct decimal places in the solution will double. As a scale-free criterion of the approach to convergence, one may check the closeness of the gradient to zero by dividing its elements by the square roots of the corresponding diagonal elements of H, or alternatively, the maximum magnitude of the ratio corrections (i.e. the change in the estimated values between iterations) to the current value of the corresponding parameter.

When a normal distribution of the residuals is assumed, ML estimation is identical to the method of least squares. Gauss pointed this out in 1809 and observed that the expected value of the second term in the Hessian matrix goes to zero with an increasing number of observations and can generally be neglected in practical computation. R. A. Fisher adopted this technique for numerical solution of non-linear equations in ML estimation and defined the so-called *Fisher information matrix*, the k, ℓ element of which is

$$I(\nu_k, \nu_\ell) = -\frac{1}{\sigma_u^2} \sum_{j=1}^{n} \frac{\partial f(x_j)}{\partial \nu_k} \cdot \frac{\partial f(x_j)}{\partial \nu_\ell} \tag{9.8}$$

In this form, the iterative solution of the likelihood equations may be expressed as

$$\nu_{t+1} = \nu_t + I_t^{-1} G_t \tag{9.9}$$

sometimes referred to as the *Fisher scoring* solution. This method has the important advantage of depending only on the functional form of the model and not on the data. For any model with finite first derivatives with respect to the parameters, the information matrix is almost always positive-definite if the number of data points equals or exceeds the number of free parameters. This may not be the case with the full Hessian matrix if there are outlying negative residuals and the number of measurements is small relative to the number of free parameters. For this reason, we will confine our attention to Fisher scoring solutions in this chapter. Note that, when the model is linear, the ML solution becomes identical to the non-iterative least-squares solutions in multiple regression analysis.

A convenient property of the Fisher scoring solution is that the inverse information matrix in Eqn 9.9 is the large-sample estimator of the variance covariance matrix of the parameter estimates on the assumption that the residuals are independently distributed with mean zero and variance σ_u^2. The variance of the residuals may be estimated in the usual way (note the residuals are not restricted to sum to zero in non-linear estimation):

$$\hat{\sigma}_u^2 = \frac{1}{n-1} \sum_{j=1}^{n} \hat{u}_j^2 \tag{9.10}$$

where $\hat{u}_j$ is the residual corresponding to x_j when the fitted model is evaluated. In published reports, it is common practice to show only the standard errors of the estimated parameters, i.e. the square roots of the diagonal elements of the inverse information matrix. Especially with heavily parameterized models, however, it is advisable to show also the correlation matrix corresponding to the inverse matrix. High correlations in this matrix are another indication that the model is poorly conditioned for estimation.

Ill-conditioned models pose problems for Newton solutions of the likelihood equations because they require the starting parameter values to be very close to the solution point. If one is working with a well-known growth model there may be studies in the literature that give typical values for the parameters among subjects of various ages; these may provide satisfactory initial values. If no such information is available or the model has not previously been investigated, trial-and-error may be the only resort. Exploratory studies of how the parameters affect the shape of the growth curve may suggest plausible starting values.

In most cases, poor starting values are evident in the first iteration because one or more of the corrections, $\nu_k^{(t+1)} - \nu_k^{(t)}$ are of the same order of magnitude as $\nu_k^{(t)}$. If the initial value is in a plausible neighbourhood of the solution, e.g. mean value in the population, the solution can still be achieved by so called *ridge regression*. In this technique the diagonal elements of the information matrix are multiplied by a positive constant large enough to bring the magnitude of the initial corrections down to a small fraction of the trial value of the parameter. As the solution begins to converge with that constant, its magnitude can be reduced in succeeding iterations until it is 1.0 at the solution point. This procedure is optimized in so-called *conditioned Newton–Raphson* computer procedures, but in working with growth models this simpler version of ridge regression is quite adequate. As we shall see in the next sections, Bayesian methods of fitting the model eliminate the starting value problem entirely.

Example 1: Fitting the Jenss–Bayley model

The four-parameter Jenss–Bayley (Jenss and Bayley, 1937) model

$$y = a + bx - \exp(c - dx)$$

provides a reasonably good account of a child's linear growth from 6 months to about 6 years of age. It is not intended to fit later growth. The model is linear in parameters a and b, and non-linear in c and d.

The elements of the gradient vector are

$$G_a = n$$

$$G_b = \sum_j^n x_j$$

$$G_c = -\sum_j^n \exp(c - dx_j)$$

$$G_d = \sum_j^n x_j \exp(c - dx_j)$$

As a computing example of ML estimation, we fit this model to recumbent length measurements of a boy and girl selected from the Fels Institute data files. The ML estimation required an initial ridge constant of 3.0. The parameter estimates, standard errors, final corrections and error correlations are shown in Table 9.1. Observed and fitted quantities are shown in Table 9.2.

The standard errors and error correlations in Table 9.1 are computed from the information matrix without the ridge (i.e. ridge constant equals 1.0). The high error correlations shown in Table 9.1 show the Jenss–Bayley model to be very ill-conditioned for estimation. Nevertheless, the fitted heights in Table 9.2 are a good representation of the main trend during this period of rapid growth. The appearance of isolated large residuals suggests measurement errors. In their presence, the velocities derived from the model for height are much more dependable indicators of rate of growth than the observed velocities.

Maximum a posteriori estimation

Empirical Bayes fitting of longitudinal growth models assumes that the measurements are drawn from some population in which the distribution of the model parameters, called the *prior* distribution, is known. Estimates of prior

Table 9.1 *Maximum likelihood estimates, standard errors and solution characteristics ridges: 3.0, 1.5, 1.0; convergence criterion: 0.0006; and residuals assumed uncorrelated*

	Parameter			
	a	*b*	*c*	*d*
Case no. 599 male		*Number of iterations* = 74		
Starting values	85.3640	5.5190	3.3060	0.6210
Estimate	84.9325	5.4704	3.3516	0.7496
Last correction	−0.0082	0.0013	−0.0001	0.0005
Standard error	3.6540	0.6126	0.1046	0.1557
Error correlation	1.0000			
	0.9950	1.0000		
	0.9198	−0.9343	1.0000	
	−0.9588	0.9323	−0.7859	1.0000
Case no. 616 female		*Number of iterations* = 47		
Starting values	83.2680	5.6380	3.3290	0.6700
Estimate	82.8930	5.4512	3.2197	0.6556
Last correction	−0.0013	0.0003	−0.0001	−0.0001
Standard error	4.7582	0.7448	0.1571	0.1825
Error correlation	1.0000			
	−0.9966	1.0000		
	0.9628	−0.9724	1.0000	
	−0.9733	0.9543	−0.8828	1.0000

distributions have to come from longitudinal growth studies of children sampled from the population. Fortunately, such studies are available in many parts of the world (Eveleth and Tanner, 1976). Computationally, it is convenient to assume that the prior distribution is multivariate normal as specified by its mean and covariance matrix. In the following section, we discuss estimation of these quantities. If necessary, the model parameters may be transformed to better approximate normality.

Before any measurement of the case is in hand, the Bayes estimates of the model parameters are simply the population means. As soon as at least one measurement is available, however, the Bayes estimates move away from the prior mean towards the so-called *posterior* mean of the case in question; similarly, the estimate of the covariance matrix moves from its prior value toward the posterior covariance matrix of the case. The extent of the movement depends upon the amount of information in the measurements relative to that of the prior. As the number of informative measurements increases, the Bayes estimate approaches the ML estimate. At some point, one is in the position of so-called 'strong inference', where the data dominate the prior to such an extent that exact specification of the prior is not necessary for dependable estimation. Bayes

Table 9.2 *Maximum likelihood fit of the Jenss–Bayley model to measurement of two cases from the Fels Longitudinal Growth Study*

	Case no. 599 male					Case no. 616 female				
Age	Height (cm)		Residual	Velocity (cm/yr)		Height (cm)		Residual	Velocity (cm/yr)	
(yrs)	Observed	Fitted	(cm)	Fitted	Observed	Observed	Fitted	(cm)	Fitted	Observed
0.50	68.0	68.0	−0.04			66.7	67.6	−0.89		
				16.59	18.11				14.71	18.61
0.75	72.4	72.8	−0.36			72.7	71.8	0.94		
				13.83	18.62				21.85	13.25
1.00	77.0	76.9	0.13			75.8	75.2	0.60		
				11.22	14.01					
1.50	84.2	84.0	0.20							
				9.44	13.43					
2.00	90.7	89.4	1.25			85.7	86.9	−1.24		
				8.20	7.33				8.15	12.65
2.50	94.4	94.2	0.17			92.2	91.7	0.51		
				7.35	5.08				7.40	7.00
3.00	96.9	98.3	−1.37			95.7	95.8	−0.07		
				6.75	9.20				6.85	7.00
3.50	101.5	102.0	−0.45			99.1	99.4	−0.27		
				6.36	5.69				6.46	8.35
4.00	104.5	105.5	−1.01			103.5	103.0	0.51		
				6.08	9.96				6.19	5.32
4.50	109.1	108.5	0.65			106.0	106.0	−0.03		
				5.90	6.42				5.99	6.98
5.00	112.4	111.6	0.82			109.4	109.0	0.36		

estimators have the same good properties as ML with respect to an increasing number of measurements. But they do not have invariance under transformation of the parameters: the transformed parameters have to be re-estimated from the original data.

Maximum a posteriori (MAP) is a form of Bayes estimation in which the mode of the posterior distribution of the model parameters, given the data, is estimated rather than the mean. The term 'Bayes estimate' usually refers to the latter. Although the MAP estimate generally has larger sampling variance than the Bayes estimate, it is much easier to compute for multiparameter non-linear growth models, where a closed form of the posterior distribution does not exist and high-dimensional numerical integration becomes necessary (see pp. 000–00).

To express the probability density function of the posterior distribution for fitting an m-parameter model, let ν represent the $m \times 1$ vector of parameters, and let $L(\nu)$ be its likelihood as in Eqn 9.2; let the $n \times 1$ vector y represent the n measurements of the child in question. Suppose $g(\nu)$ is a prior probability density function that integrates to 1. Then the posterior distribution of ν is, say,

$$p(\nu|y) = L(\nu)g(\nu) \tag{9.11}$$

Assume $g(\nu)$ is multivariate normal with mean μ and covariance matrix Σ:

$$g(\nu) = \frac{|\Sigma|^{-\frac{1}{2}}}{(2\pi)^{m/2}} \exp\left[-\frac{1}{2}\sum_{k}^{m}\sum_{\ell}^{m} \Sigma^{(k\ell)}(\nu_k - \mu_k)(\nu_\ell - \mu_\ell)\right] \tag{9.12}$$

where $\Sigma^{k\ell}$ is the $k\ell$ element of the inverse matrix, Σ^{-1}.

Then the log posterior density is

$$\log p(\nu|y) = \log L(\nu) + \log g(\nu)$$

and in terms of the residuals, $u_j = y_j - f(x_j)$, the MAP equation for parameter ν_k, the solution of which is, say, $\bar{\nu}_k$, is

$$G(\nu_k) = \frac{1}{\sigma_u^2}\sum_{j=1}^{n} u_j \cdot \frac{\partial f(x_j)}{\partial \nu_k} - \sum_{\ell}^{m} \Sigma^{(k\ell)}(\nu_\ell - \mu_\ell) = 0 \tag{9.13}$$

The k, ℓ element of the Hessian matrix is

$$H(\nu_k, \nu_\ell) = -\frac{1}{\sigma_u^2}\sum_{j=1}^{n}\left[\frac{\partial f(x_j)}{\partial \nu_k} \cdot \frac{\partial f(x_j)}{\partial \nu_\ell} + u_j \frac{\partial^2 f(x_j)}{\partial \nu_k \partial \nu_\ell}\right] - \Sigma^{(k\ell)} \tag{9.14}$$

and the corresponding element of the posterior information matrix is

$$I(\nu_k, \nu_\ell) = \frac{1}{\sigma_u^2}\sum_{j=1}^{n} \frac{\partial f(x_j)}{\partial \nu_k} \cdot \frac{\partial f(x_j)}{\partial \nu_\ell} + \Sigma^{k\ell} \tag{9.15}$$

The population covariance matrix must be positive-definite, which assures that its determinant is positive and its inverse is finite. The MAP equation differs from the likelihood equation only by the presence of the so-called 'penalty' term, $\Sigma^{-1}(\nu - \mu)$, which stochastically constrains the estimate to a plausible distance from the population mean. The Hessian and information matrix now includes effectively the 'ridge', Σ^{-1}, which reduces the size of the iterative steps in the Newton solution of Eqn 9.13. This slows the rate of convergence, but it also protects the procedures from poorly chosen starting values. The effect of the correction term and ridge can be controlled by the value assigned to σ_u^2. Large values increase the influence of the prior and small values can decrease its influence.

By incorporating information from the prior distribution as well as from the data, the MAP standard error (or posterior standard deviation) is generally smaller than that of the ML estimator. In respect to sampling the data points of a given case, the error covariance matrix of the MAP estimator is the inverse of the information matrix evaluated at the solution point. After the gradient and information matrix are modified as in Eqns 9.13 and 9.15, the iterative MAP computations are the same as those for ML estimation.

The use of MAP estimation is essential if the growth model is a symmetric function in some of its parameters. Examples are the JPA-2 model (Jolicoeur *et al.*, 1992) and the BTT model (Bock *et al.*, 1994). These multiple-component models require constraints on their parameters to prevent the components from exchanging places when the data are sparse or irregular. Without constraint, the components attributed to early growth and mid-childhood growth could exchange places. Goodness of fit would not be affected, but interpretation of the parameter estimates would be confounded. Moreover, comparisons of the distributions of the estimates among samples of children would be meaningless.

Example 2: Fitting the JPA-2 model

The eight-parameter JPA-2 model is

$$f(x) = a(1 - D^{-1})$$

where

$$D = 1 + \Sigma_k^3 [b_k(x + e)]^{c_k}$$

As in Bock (1995), we have improved the conditioning of the model by expressing the coefficient of $(x + e)$ as the reciprocal of the coefficient employed by Jolicoeur *et al.* (1992). This model gives a good account of growth from near birth to maturity, with smooth asymptotic growth following the adolescent growth spurt. It shows only slight acceleration of growth in mid-childhood. Let

$$B_k = \partial D/\partial b_k = c_k(x + e)v_k^{c_k - 1}$$
$$C_k = \partial D/\partial c_k = v_k^{c_k} \log v_k$$
$$E = \partial D/\partial e = \Sigma_k^3 b_k c_k v_k^{c_k - 1}$$

where $v_k = b_k(x + e)$.

The first derivatives of the function with respect to the parameters are

$$\frac{\partial f(x)}{\partial a} = 1 - D^{-1}$$
$$\frac{\partial f(x)}{\partial b_k} = aB_k D^{-2}$$
$$\frac{\partial f(x)}{\partial c_k} = aC_k D^{-2}$$
$$\frac{\partial f(x)}{\partial e} = aED^{-2}$$

The penalized gradients for purpose of MAP estimation of the parameters matrix are

$$G_a = \frac{1}{\sigma_u^2}\sum_{j=1}^{n}\frac{\partial f(x_j)}{\partial a} - \sum_{\ell}^{8}\Sigma^{(1,\ell)}(a - \mu_a)$$

$$G_{b_k} = \frac{1}{\sigma_u^2}\sum_{j=1}^{n}\frac{\partial f(x_j)}{\partial b_k} - \sum_{\ell}^{8}\Sigma^{(1+k,\ell)}(b_k - \mu_{b_k})$$

$$G_{c_k} = \frac{1}{\sigma_u^2}\sum_{j=1}^{n}\frac{\partial f(x_j)}{\partial c_k} - \sum_{\ell}^{8}\Sigma^{(4+k,\ell)}(c_k - \mu_{c_k})$$

$$G_e = \frac{1}{\sigma_u^2}\sum_{j=1}^{n}\frac{\partial f(x_j)}{\partial e} - \sum_{\ell}^{8}\Sigma^{(8,\ell)}(e - \mu_e),$$

and the k, ℓ element of the posterior information matrix is

$$I(\nu_k, \nu_\ell) = \frac{1}{\sigma_u^2}\sum_{j=1}^{n}\frac{\partial f(x_j)}{\partial \nu_k}\cdot\frac{\partial f(x_j)}{\partial \nu_\ell} + \Sigma^{(k,\ell)} \qquad (9.16)$$

Estimated by the method of the following section, the population means and covariance matrices for male and female cases in the Fels study are shown in Table 9.3.

In Table 9.4, the small number of iterations required to make all corrections to zero to four decimal places, and the presence of only one correlation greater than 0.98 show the relative good conditioning of MAP estimation for this heavily parameterized model. Table 9.5 and especially Table 9.6 show the residuals to be near the minimum measurement error level. The exception at age 5.0 in the result for the male case is probably a measurement procedure error. The string of somewhat larger underestimates suggests a mid-growth spurt in this case that JPA-2 is not able to model. An obvious mid-growth spurt is less common in girls and is not evident here in the female case.

Consistent estimation of the population mean and covariance matrix of the model parameters

If measurements of a large sample of N cases from a suitable longitudinal growth study are available, the population mean covariance matrix of the model parameters can be estimated from the MAP estimate and posterior covariance matrix for each case. A consistent estimator of the multivariate mean of the

Table 9.3 *Population means, standard deviations and correlation matrices of the JPA-2 parameters estimated from data of the Fels Longitudinal Growth Study*

	Parameters							
	a	b_1	b_2	b_3	c_1	c_2	c_3	e
Males n = 177								
Mean								
	181.1053	0.2635	0.0962	0.0705	0.6822	3.62476	19.91257	1.3930
Standard deviation								
	6.46311	0.07433	0.01233	0.00575	0.16730	0.80727	2.394506	0.868865
Correlation								
a	1.0000							
b_1	−0.1072	1.0000						
b_2	−0.0189	0.3241	1.0000					
b_3	−0.1738	0.6892	0.5053	1.0000				
c_1	0.0268	−0.1972	−0.7590	−0.1723	1.0000			
c_2	−0.0225	0.3566	−0.5186	0.0872	0.5861	1.0000		
c_3	0.0085	−0.0039	−0.0420	−0.2637	0.0438	0.1806	1.0000	
e	−0.0063	−0.9180	−0.3641	−0.6541	0.2534	−0.2349	0.0363	1.0000
Females n = 161								
Mean								
	166.5005	0.3109	0.1090	0.0816	0.7194	3.5838	16.0614	1.5101
Standard deviation								
	5.50440	0.11078	0.01627	0.00787	0.19643	0.86406	2.317119	0.92065
Correlation								
a	1.0000							
b_1	−0.0510	1.0000						
b_2	−0.0753	0.2613	1.0000					
b_3	−0.1229	0.6178	0.4444	1.0000				
c_1	0.0456	−0.3014	−0.7019	−0.0961	1.0000			
c_2	0.0316	0.3039	−0.4680	0.0785	0.4713	1.0000		
c_3	−0.0758	−0.0401	0.0564	−0.3258	−0.0117	0.1888	1.0000	
e	−0.0317	−0.7933	−0.3643	−0.6768	0.2039	−0.2543	0.0606	1.0000

parameters is simply the sample mean of the MAP estimate of each parameter:

$$\hat{\mu}_k = \frac{1}{N} \sum_{i=1}^{N} \bar{\nu}_{ik} \tag{9.17}$$

The corresponding estimator of the k, ℓ element of the covariance matrix is

$$\hat{\Sigma}(\nu_k, \nu_\ell) = \frac{1}{N} \sum_{i=1}^{N} \left[(\bar{\nu}_{ik} - \hat{\mu}_k)(\bar{\nu}_{i\ell} - \hat{\mu}_\ell) + I_i^{-1}(\bar{\nu}_{ik}, \bar{\nu}_{i\ell}) \right] \tag{9.18}$$

Table 9.4 *Maximum a posteriori estimates, standard errors and solution characteristics; no ridge conditioning required, convergence criterion 0.000001, residuals assumed uncorrelated*

	Parameters							
	a	b_1	b_2	b_3	c_1	c_2	c_3	e
		Case no. 599 male				*Number of iterations = 16*		
Starting values	181.11	0.26	0.10	0.07	0.68	3.62	19.91	1.39
Estimates	178.0912	0.4098	0.1317	0.077	80.3347	2.5742	23.7602	0.0524
Standard errors	0.2379	0.0630	0.0095	0.003	40.1033	0.4489	1.2864	0.4185
Last corrections	0.0000	0.0000	0.0000	0.000	00.0000	0.0000	0.0000	0.0000
Error correlations								
a	1.0000							
b_1	−0.0193	1.0000						
b_2	0.0741	0.6944	1.0000					
b_3	0.1359	0.8213	0.9206	1.0000				
c_1	−0.1468	−0.7100	−0.9892	−0.9225	1.0000			
c_2	−0.1973	−0.5117	−0.9514	−0.8667	0.9391	1.0000		
c_3	−0.4958	−0.2236	−0.4360	−0.5986	0.4640	0.5590	1.0000	
e	−0.1350	−0.8985	−0.9246	−0.9620	0.9294	0.8207	0.4229	1.0000
		Case no. 616 female				*Number of iterations = 9*		
Starting values	166.50	0.31	0.11	0.08	0.72	3.58	16.06	1.51
Estimates	165.5422	0.3829	0.1187	0.0872	0.5994	3.7107	16.8881	0.8784
Standard errors	1 0.2379	0.0630	0.0095	0.0034	0.1033	0.4489	1.2864	0.4185
Last corrections	0.0000	0.0000	0.0000	0.0000	0.0000	0.0000	0.0000	0.0000
Error correlations								
a	1.0000							
b_1	−0.0300	1.0000						
b_2	−0.0018	0.9337	1.0000					
b_3	0.0480	0.9422	0.9540	1.0000				
c_1	−0.0451	−0.9368	−0.9949	−0.9590	1.0000			
c_2	−0.1249	−0.6989	−0.8880	−0.8568	0.8866	1.0000		
c_3	−0.3919	−0.4318	−0.5369	−0.6758	0.5657	0.7245	1.0000	
e	−0.0097	−0.9950	−0.9542	−0.9617	0.9544	0.7497	0.4812	1.0000

Table 9.5 *Maximum a posteriori fit of the JPA-2 model for a case from the Fels Longitudinal Growth Study, case no. 599 (male)*

Age	Height (cm)		Residual	Velocity (cm/yr)	
(yrs)	Observed	Fitted	(cm)	Fitted	Observed
1.00	76.0	76.9	−0.91	14.65	
1.51	83.2	83.3	−0.09		
2.00	89.7	88.0	1.68	8.97	10.31
2.50	93.4	92.3	1.11		
2.99	95.9	96.1	−0.20	7.50	7.16
3.49	100.5	99.8	0.73		
4.02	103.0	103.5	−0.51	7.20	6.47
4.48	106.9	106.7	0.17		
5.00	108.1	110.2	−2.15	6.79	5.20
5.46	112.0	113.4	−1.40		
6.01	117.1	117.0	0.07	6.55	3.17
6.50	120.5	120.2	0.30		
7.01	123.3	123.4	−0.12	6.21	6.50
7.50	127.0	126.4	0.60		
8.00	129.7	129.3	0.35	5.79	5.74
8.51	132.8	132.3	0.54		
8.99	136.2	134.8	1.35	5.29	5.78
9.49	138.7	137.4	1.27		
10.00	139.7	140.0	−0.26	4.80	3.94
10.48	142.6	142.2	0.37		
11.00	143.5	144.6	−1.07	4.55	4.13
11.50	146.8	146.9	−0.09		
12.01	148.3	149.5	−1.16	5.42	4.58
12.47	151.9	152.3	−0.35		
13.00	156.3	156.4	−0.06	9.07	9.32
13.50	161.5	161.4	0.08		
14.00	167.1	166.9	0.15	10.45	10.81
14.49	172.2	171.4	0.75		
14.96	173.9	174.4	−0.51	4.66	3.80
15.49	176.0	176.3	−0.34		
16.01	176.4	177.2	−0.82	1.23	1.76
16.51	177.8	177.7	0.13		
16.95	177.7	177.9	−0.16	0.30	
17.98	178.9	178.0	0.82	0.07	

where $I_i^{-1}(\nu_{ik}, \nu_{i\ell})$ is the k, ℓ element of the inverse information matrix of the model parameters at the solution point for case i.

Similarly, the variance of the residuals is estimated by

$$\hat{\sigma}_u^2 = \frac{1}{\Sigma_i^N n_i} \sum_{i=1}^{N} \sum_{j}^{n} \left[u_{ij}^2 + \Sigma_k^m \Sigma_\ell^m \frac{\partial f_i(x_j)}{\partial \nu_k} \cdot \frac{\partial f_i(x_j)}{\partial \nu_\ell} I_i^{-1}(\bar{\nu}_{ik}, \bar{\nu}_{i\ell}) \right] \tag{9.19}$$

Table 9.6 *Maximum a posteriori fit of the JPA-2 model for a case from the Fels Longitudinal Growth Study, case no. 616 (female)*

Age	Height (cm)		Residual	Velocity (cm/yr)	
(yrs)	Observed	Fitted	(cm)	Fitted	Observed
1.99	84.7	85.8	−1.10	9.38	
2.50	91.2	90.3	0.86		
3.00	94.7	94.3	0.37	7.65	6.97
3.49	98.1	97.9	0.17		
4.02	102.5	101.6	0.86	6.92	6.20
4.49	104.3	104.8	−0.53		4.91
4.97	107.2	108.1	−0.87	6.58	6.28
5.47	110.5	111.3	−0.84		
6.09	115.5	115.3	0.20	6.40	7.43
6.48	118.0	117.8	0.23		
6.96	120.8	120.8	0.00	6.20	5.98
7.48	124.0	124.0	0.00		
8.00	128.3	127.2	1.15	6.00	6.30
8.48	130.3	130.0	0.33		
8.97	132.8	132.9	−0.13	6.05	4.50
9.48	134.8	136.0	−1.23		
9.97	139.6	139.2	0.36	6.78	7.62
10.49	142.5	142.9	−0.44		
10.99	147.4	146.8	0.63	8.04	7.84
11.60	151.2	151.7	−0.55		
11.98	155.2	154.8	0.43	7.43	6.49
12.49	158.4	158.2	0.22		
13.00	160.5	160.9	−0.36	4.32	4.41
13.51	162.9	162.7	0.19		
14.03	163.9	163.8	0.02	1.79	
14.99	164.6	164.9	−0.35	0.65	
16.00	164.8	165.3	−0.56	0.23	0.55
17.01	165.7	165.5	0.20	0.09	0.62
18.11	166.1	165.5	0.54	0.03	

(See Bock (1989) for derivations of these results.) Because the population mean and covariance matrix are required in the prior distribution for estimating ν_{ik}, a 'bootstrap' procedure is required in their use. Initially, one may set all of the off-diagonal ($k \neq \ell$) elements of the covariance matrix equal to zero; then set the diagonal elements and the means equal to some plausible values. Provided the number of well-spaced data points per case exceeds, say, 20, a pass through the cases with this provisional prior will give a good approximation to the population quantities. A second or third pass, each time substituting the resulting provisional prior will yield a sufficiently accurate estimate of the population mean and covariance matrix for practical use in MAP estimation of the model parameters.

Maximum marginal likelihood estimation of the population means and covariances of the model parameters

The estimation of population quantities described in the previous section substitutes MAP estimates of the individual case posterior means and covariances in place of Bayes estimates. Although the MAP estimates are much less computationally demanding, the Bayes estimates have the advantage of being non-iterative and more stable than MAP estimates. Moreover, they are sufficient statistics for ML estimation for moments of the marginal distribution of the model parameters. The burden of Bayes estimation in this situation is tolerable, however, because they are computed once and for all in large sample data from longitudinal growth studies and subsequently used in fitting the growth model to individual cases from the population as needed. A full treatment of maximum marginal likelihood (MML) estimation is beyond the scope of this volume, but we sketch the general approach in this section.

Assuming the marginal distribution to be multivariate normal, one can derive the likelihood equations and information matrices required for Newton–Gauss iterative computation of MML estimates at the population level. The matrix expressions given in Bock (1989) for this purpose are for linear models, but they can be generalized to non-linear models merely by replacing the model matrix with the derivatives of the non-linear model with respect to its parameters. The sufficient statistics for MML estimation are the posterior mean vector and the posterior covariance matrix to each case in the sample. Their elements, given the n-element vector of measurements for case i are:

$$\nu_k|y_i = \frac{1}{P_i}\int_{-\infty}^{\infty} \nu_k L_i(\nu)g(\nu)d\nu \tag{9.20}$$

and

$$\Sigma_{k,\ell}|y_i = \frac{1}{P_i}\int_{-\infty}^{\infty} (\nu_k - \nu_k|y_i)(\nu_\ell - \nu_\ell|y_i)L_i(\nu)g(\nu)d\nu \tag{9.21}$$

where

$$P_i = \int_{-\infty}^{\infty} L_i(\nu)g(\nu)d\nu$$

is the marginal probability of the observation. The integration is over the m-dimensional parameter space.

The computational difficulty is that when $g(\nu)$ is multivariate normal, there is no closed form for these integrals. They must be evaluated numerically, preferably by multidimensional Gauss–Hermite quadrature in which the number of points increases exponentially with the number of parameters in the model. As experience seems to indicate that the human growth curve requires as many as eight parameters for good representation, this number is potentially very large. The problem is that the posterior distributions of the cases are scattered randomly about the quadrature space, and these distributions are also relative compact when the number of measurements per case is perhaps 30 or more. To evaluate their volumes and first two moments with reasonable accuracy requires many points per dimension in order that at least a few of them will have an appreciable density in the posterior distribution. The total number of such points in the quadrature space of eight dimensions would then be prohibitive even with modern computing power.

Naylor and Smith (1982) have shown, however, that in this situation the location and spacing of the points can be adapted to any specific posterior distribution by corresponding linear transformation of the quadrature space. With this device, as few as three points per dimension can provide sufficiently accurate evaluation of the first and second moments to allow dependable estimation of population characteristics when accumulated over a large sample of cases. Although $3^8 = 6561$ points may still seem large, it is well within the range of a computer operating at, say, 1 GHz. In addition, S. H. C. du Toit and R. Cudeck (unpublished data) have shown that if some of the parameters appear linearly in the model, the dimensionality of the integration can be reduced to that of the non-linear parameters. In the case of the JPA-2 model, with one linear parameter, the number of points is reduced to 2187; for the BTT model, with three linear parameters, the number is 729 points. These numbers present no difficulties for computation, even with sample sizes in excess of 100 and perhaps 10 iterations of the Newton–Gauss solution of the MML equations.

It is also possible to employ Monte Carlo integration for this purpose, but that introduces random variation between steps of the iterative solution and makes more difficult the ascertainment of convergence. When performed by the usual Markov chain–Monte Carlo (MCMC) method, the computations are also considerably heavier than adaptive quadrature (ADQ). Segawa (1998) compared MCMC estimates of the population means and covariances in a sample of 100 cases from the Fels Longitudinal Growth Study with ADQ estimates computed by Stephen du Toit and found the results almost indistinguishable. For further details of adaptive quadrature in this type of application see Rabe-Hesketh *et al.* (2002) and Bock and Schilling (1997).

Allowance for non-homogeneous residuals at different ages

The growth record of an individual child would not ordinarily contain enough measurements to allow useful estimation of residual error variance at different ages. When data for a large sample of cases are available, however, one can allow for non-homogeneous variance in an average sense. The approach is first to fit the model to each case in the sample, assuming homogeneous error, then to group the residuals in successive age intervals and estimate the within-interval variance. Weights proportional to the reciprocals of these variance estimates can then be incorporated in the likelihood for purposes of ML or MAP fitting of the models for other cases from the sampled population. The weights could be improved by one or two iterations of this procedure in the sample cases before applying them to new cases.

Suppose the weight in interval j is $w_j = c/s_j^2$, where s_j is the standard deviation corresponding to unbiased variance estimates for the interval, and c is a normalizing constant such that $\Sigma_j^n w_j = 1$. With this normalization the overall residual error variance, when fitting the model using weights, is a consistent estimator of σ_u^2. In the calculation of the likelihood, gradient, and Hessian matrix in Eqns 9.2, 9.3 and 9.4, these weights are treated as known constants multiplying the residuals, u_j. Apart from the presence of the weights, all other aspects of ML and MAP estimation remain the same. ML estimation in this situation is identical to weighted least squares.

Allowance for autocorrelated residuals

The extent to which autocorrelated residuals affect the fitting of growth models depends upon goodness of fit and spacing of the measurements. If the growth curve follows the main trend of the data points but does not capture shorter episodes of accelerated or decelerated growth, the residuals will show successive runs of positive and negative residuals that give rise to autocorrelation. In children that exhibit these effects the periods of altered growth velocity may extend between one and two years. More frequent measurements will therefore show appreciable autocorrelation of residuals in these circumstances. Assuming the residuals to be independent when they are in fact autocorrelated has several adverse effects. It leads to overstating the goodness of fit of the model and underestimating the size of confidence intervals for the model parameters. It also allows closely spaced measurements to have more influence on the shape of the growth curve than their information content merits.

If the growth model accounts for all of the main trend in the data, the residuals might reasonably be regarded as a weakly stationary time series, that is, one having constant mean, variance and autocorrelation of the same lag. If the measurements are also equally spaced and the autoregressive process is of order one or two, it might be possible to apply standard statistical methods to estimate autocorrelations for individual cases. But when the measurements are made irregularly or at less than one-year intervals, as would frequently be true of clinical applications, estimating autocorrelation among the residuals jointly with the estimation of the model parameters, is not a practical proposition.

As in the case of heterogeneous residual error variance discussed above, however, we can allow for autocorrelation of residuals in an average sense by considering it to have the same value for each member in some population of cases. The approach would be to estimate, say for same-sex cases, a common autocorrelation function in real time that could be incorporated routinely in the fitting of individual growth models for measurements taken at arbitrary age levels.

A way of estimating a common autocorrelation function exists in the reciprocal relationship between the spectrum of a stationary time series and the corresponding autocovariance function. Suppose each case in the sample has been measured at the same n equally spaced ages. Let the residual for case i at time j be $u_{ij} = y_{ij} - f(t_j)$. Then the succession of residuals may be approximated by a finite Fourier series of order r. Least-squares estimation of the coefficients of the terms in the series is extremely simple because the cross-product terms in the calculations are all zero. Thus the estimators of coefficients A_h and B_h $h = 1, 2, \ldots, r$, are

$$A_{ih} = \sum_{j=1}^{n} u_{ij} W_h \cos(Cht_j^*) \tag{9.22}$$

$$B_{ih} = \sum_{j=1}^{n} u_{ij} W_h \sin(Cht_j^*) \tag{9.23}$$

where $W_h = (1 + \cos \frac{Cht^*}{2})$ is the Tukey 'window' diminishing the influence of extreme lags: $C = 2\pi/(t_U - t_L + 1)$, time $t_j^* = t - t_0$ is age measured from the origin $t_0 = (t_U + t_L)/2$, and t_U and t_L are the youngest and oldest ages. Coefficient A_0 is the mean of the residuals. The spectrum for case i is the additive partition of the sum of squares of the residuals into $r + 1$ terms attributable to successive harmonics of the Fourier series. Because the sine and cosine functions are orthogonal, the ordinates of the point spectrum at the successive harmonics are the squares of the corresponding coefficients:

Harmonics (h)	Spectrum h
0	A_{i0}^2
1	$A_{i1}^2 + B_{i1}^2$
2	$A_{i2}^2 + B_{i2}^2$
$\vdots$	$\vdots$
r	$A_{ir}^2 + B_{ir}^2$

The mean square of the coefficients then estimates the ordinates of the assumed common spectrum:

$$\frac{1}{N}\sum_{i=1}^{N}\left(A_{ih}^2 + B_{ih}^2\right)$$

If the number of terms in the series equals $n/2 - 1$, the sum of the ordinates is exactly equal to the sum of squares of the residuals. Since part of the residual variation is measurement error (which can safely be assumed independent from one age level to another), only part of the variance can be a source of autocovariance. For that reason r should be set so that the variance attributable to the $n - r/2 + 1$ higher harmonics is not smaller than the measurement error variance. In general, r will be much smaller than $n/2$ in growth studies spanning ages near birth to maturity; $r = 5$ or 6 is typical.

From the common spectrum we may obtain the autocovariance function (ACVF), $\gamma(t)$, and autocorrelation function, $\lambda(t)$, for a continuously measured lag t. To that end, let γ_h and λ_h be the values of these functions at lag h of the n discrete, equally spaced measurements.

Then

$$\gamma_h = \frac{1}{n}\sum_{j=1}^{n} u_j u_{j+h}, \qquad h = 0, 1, 2, \ldots, r$$

and

$$\lambda_h = \gamma_h/\sigma_0^2$$

where $\sigma_0^2 > \gamma_h$ is the component of the residual variance not attributed to measurement error.

The essential result relating ACVF and the spectrum in discrete time is

$$s_h = \sum_{k=a}^{b} \gamma_k \cos\frac{2\pi kh}{n}$$

where $a = 1 + n/2$, $b = n/2$, and $h = 0, 1, \ldots, n/2$, for n even, and $a = b = (n-1)/2$ and $h = 0, 1, \ldots, (n-1)/2$, for n odd. In this formula, s_h are the ordinates of the mean-squares spectrum.

Since $\gamma(t)$ is an even function, these results simplify to

$$s_h = \gamma_0 + 2\sum_{k=1}^{b} \gamma_k \cos \frac{2\pi kh}{n}$$

where $b = n/2$ for n even, and $b = (n-1)/2$ for n odd.

The Fourier transform of the spectrum then gives back the ACVF:

$$\gamma_k = s_0 + \sum_{h=1}^{b} s_h \cos \frac{2\pi kh}{n}$$

$$\tilde{\gamma}(t_{jk}) = s_0 + \sum_{h=1}^{r} s_h \cos Cht_{jk}$$

and the corresponding autocorrelation is $\tilde{\lambda}(t_{jk}) = \tilde{\gamma}(t_{jk})/\sigma_0^2$. Correlations for lags greater than five years are set equal to zero. The resulting autocorrelation matrix is positive-definite for any lags in this range, including $t_{jk} = 0$.

To allow for autocorrelation of the residuals when fitting the growth model for a particular child, we compute from the correlation between residuals j and k and insert it in the jh and hj rows and columns of an $n \times n$ symmetric matrix Λ. Elements corresponding to lags greater than r are set to zero.

On the assumption that the correlation of the residuals for case i are multivariate normal with correlations given by Λ, the likelihood of model parameter ν_k is

$$L(\nu_k) = \frac{1}{\sigma_u\sqrt{2\pi}} \exp\left(-\frac{1}{2\sigma_u^2}\sum_{h}^{n}\sum_{j}^{n} \lambda^{(jh)} u_j u_h\right)$$

where $\lambda^{(jh)}$ is the jh element of the inverse matrix, Λ^{-1}. The corresponding term in the gradient matrix for ML and MAP estimation of ν_k in Eqns 9.3 and 9.13 is then

$$\frac{1}{\sigma_u^2}\sum_{h}^{n}\sum_{j}^{n} \lambda^{(jh)} u_j \frac{\partial f(x_h)}{\partial \nu_k}$$

Similarly, the corresponding term for parameters k and l in the Hessian matrix is

$$-\frac{1}{\sigma_u^2}\sum_{h}^{n}\sum_{j}^{n}\left[\lambda^{(jh)}\frac{\partial f(x_j)}{\partial \nu_k} \cdot \frac{\partial f(x_h)}{\partial \nu_\ell} + \lambda^{(jh)} u_j \frac{\partial^2 f(x_h)}{\partial \nu_k \partial \nu_\ell}\right]$$

Table 9.7 *Average spectral ordinates*

		Harmonics					
	N	0	1	2	3	4	5
Males	67	0.0117	0.3162	1.1017	1.9525	0.9768	0.8540
Females	64	0.0097	0.3551	0.6996	1.0980	1.1868	0.6948

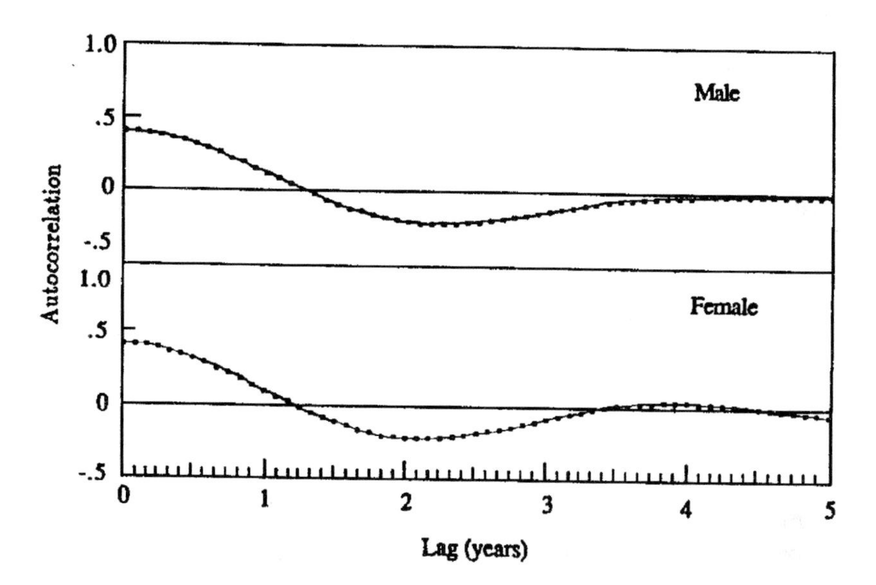

Figure 9.1 Estimated autocorrelation functions.

All other aspects of ML and MAP estimation remain the same. ML estimation in this situation is identical to generalized least squares.

Example 3: Estimation of a common autocorrelation function

The calculations for this example are based on 66 male and 64 female cases in the Fels archive for which observed heights were available for each year and half year (± 0.1 year) between the ages of 2 and 18 years. To meet the requirement of equal spacing, all age points were set exactly to their yearly and half-yearly values. From these data spectral ordinates were computed for the first five harmonics and averaged within each sex group. The resulting values are shown in Table 9.7. Inversion of the spectra yielded the autocorrelations in monthly lags up to 5 years. Smooth curves connecting these values are shown in Figure 9.1. The value for lag zero represents the average coefficient of determination of the residuals by the fitted finite series, i.e. the covariance at that point is the sum of the six average ordinates in Table 9.7. The variation not accounted for is assumed to be independent measurement error and quotidian variability.

Readings

A review of many aspects of non-linear parameter estimation may be found in Bard (1974). Bock (1975) contains an introduction to matrix algebra and matrix differentiation, and gives worked examples of matrix inversion. Magnus and Neudecker (1988) present a comprehensive system of matrix differentiation. Gelman *et al.* (1995) contains a modern treatment of empirical Bayes methods including Monte Carlo integration. Jenkins and Watts (1968) and Fuller (1976) discuss spectral analysis and its applications. Stroud and Sechrest (1966) treat Gaussian quadrature formulas, including extensive tables of points and weights for Gauss–Hermite quadrature.

REFERENCES

Bard, Y. (1974). *Nonlinear Parameter Estimation*. New York: Academic Press.

Bock, R. D. (1975). *Multivariate Statistical Methods in Behavioral Research*. New York: McGraw-Hill.

(1989). Measurement of human variation: a two-stage model. In *Multilevel Analysis of Educational Data*, ed. R. D. Bock, pp. 319–42. New York: Academic Press.

(1995). Growth models. In *Essays on Auxology*, eds. R. Hauspie, G. Lindgren and F. Falkner, pp. 28–38. Welwyn Garden City, UK: Castlemead.

Bock, R. D. and Schilling, S. G. (1997). High-dimensional full-information item factor analysis. In *Latent Variable Modeling and Applications to Causality*, ed. M. Berkane, pp. 164–76. New York: Springer-Verlag.

Bock, R. D., du Toit, S. H. C. and Thissen, D. (1994). *AUXAL: Auxological Analysis of Longitudinal Measurements of Human Stature*. Chicago, IL: Scientific Software International.

Eveleth, P. B. and Tanner, J. M. (1976). *World-Wide Variation in Human Growth*. Cambridge, UK: Cambridge University Press.

Fuller, W. A. (1976). *Introduction to Statistical Time Series*. New York: John Wiley.

Gelman, A., John, B. C., Hal, S. S. and Rubin, D. B. (1995). *Bayesian Data Analysis*. New York: Chapman and Hall.

Jenkins, G. M. and Watts, D. G. (1968). *Spectrum Analysis and its Applications*. San Francisco, CA: Holden-Day.

Jenss, R. M. and Bayley, N. (1937). A mathematical method for studying the growth of a child. *Human Biology*, **9**, 556–63.

Jolicoeur, P., Pointier, J. and Abidi, H. (1992). Asymptotic models for longitudinal growth of human stature. *American Journal of Human Biology*, **11**, 461–8.

Magnus, J. R. and Neudecker, H. (1988). *Matrix Differential Calculus*. New York: John Wiley.

Naylor, J. C. and Smith, A. F. M. (1982). Applications of a method for the efficient computation of posterior distributions. *Applied Statistics*, **31**, 214–25.

Rabe-Hesketh, S., Skrondal, A. and Pickles, A. (2002). Reliable estimation of generalized linear mixed model models using adaptive quadrature. *Stata Journal*, **2**, 1–21.

Segawa, E. (1998). Applications of hierarchical nonlinear models to children's growth in vocabulary and height. Ph.D. dissertation, University of Chicago.

Stroud, A. H. and Sechrest, D. (1966). *Gaussian Quadrature Formulas*. Englewood Cliffs, NJ: Prentice Hall.

Part III

Methods for population growth

10 *Univariate and bivariate growth references*

Edward A. Frongillo
Cornell University

This chapter reviews methods for the construction of human growth references. Specifically, the chapter aims to address the following questions:

1. What are the purposes and uses of growth references?
2. How are growth references presented?
3. What design features are needed for studies to construct growth references?
4. What analytical methods are used to construct growth references?
5. What are some recent examples of how growth references were constructed?
6. What are future challenges for the construction of growth references?

Purposes and uses of growth references

Rationale for references

How do we know if our children are healthy, well nourished and developing? Parents and others who interact closely with children have many and diverse ways of making such assessments. One common way is assessing how children are growing. Assessing the growth of children is important for judging their overall progress at both population and individual levels. Growth information is useful because it relates to multiple dimensions of health and development, underlying socio-economic and environmental determinants of poor health and development, and basic causes of poor socio-economic conditions and poor environments (Frongillo and Hanson, 1995). Growth measurements are relatively simple and inexpensive to make, although important training is required to obtain good-quality measurements. Furthermore, growth information provides a universal language for understanding and comparing child well-being.

Methods in Human Growth Research, eds. R. C. Hauspie, N. Cameron and L. Molinari. Published by Cambridge University Press.

In assessing the growth of a child (or a group of children), interest is not so much the size of the child at any one time, but how the child's size is changing over time.

Assessment of child growth generally involves comparison to a reference population, making use of an appropriate reference growth chart. A reference growth chart is a tool providing a common basis for purposes of comparison. The concept of a reference is different from that of a standard, which 'embraces the notion of a norm or desirable target, and thus involves a value judgement' (World Health Organization, 1995).

Descriptive versus prescriptive references

The reference population should reflect the growth expected for children. There are two approaches for thinking about the expected growth. One approach defines the expected growth descriptively, meaning that the reference portrays the growth that is expected of children typically or on average, most often according to sex and age. The other approach defines the expected growth prescriptively, meaning that the reference portrays the growth that is expected of children who are healthy and well nourished, and who have received proper care. A combination of these two approaches might also be used to construct a reference growth chart.

It has been commonly assumed that larger size or faster growth is generally indicative of better health, but this notion is changing for two reasons. First, the idea of constructing a reference prescriptively has raised the issue that maximal growth may not be the growth that corresponds to optimal care. That is, optimal care may not result in maximal growth, especially in environmental conditions that do not constrain growth. For example, breast-fed infants are smaller than formula-fed infants in the second half of the first year and into the second year of life (World Health Organization Working Group on Infant Growth, 1994, 1995; Dewey, 1998; Frongillo, 2001). Second, the rapid development of obesity in many countries underscores that maximal size, at least in weight, is not consistent with optimal health.

Purposes of anthropometry

The purposes of using and interpreting anthropometric information can be at the population or individual level (World Health Organization, 1995). Purposes for populations are estimating prevalence, determining causes, targeting, monitoring and evaluating (Table 10.1). Purposes for individuals are screening,

Table 10.1 *Population and individual uses for anthropometry*

Population uses	
Prevalence	How many people have the problem?
Determine causes	Why do people have the problem?
Targeting	Who has the problem? Who will benefit from a solution?
Monitoring	How is the situation changing?
Evaluation	Who has benefited and how?
Individual uses	
Screening	Is the person at risk of the problem?
Diagnosis	Does the person have the problem? Will the person benefit from a solution?
Monitoring	Is the person's situation improving?

diagnosis, and monitoring. For example, the growth of individual infants is assessed for a variety of purposes:

- Detection of growth faltering, failure to thrive, and excessive growth
- Assessment of the adequacy of intake of human milk or human milk substitutes
- Assessment of the appropriate age for introduction of complementary foods
- Evaluation of the adequacy of the weaning diet
- Assessment of the impact of illness and response to treatment
- Screening for high-risk individuals needing special services
- Assessment of response to counselling on improvement to feeding and other health-related practices (World Health Organization, 1995).

The World Health Organization (1995) structured its review of the uses and interpretations of anthropometry according to several age groups and conditions: pregnant and lactating women, newborn infants, infants, children, adolescents, overweight adults, thin adults, and adults 60 years of age and older. For each of these groups, anthropometric information has both general uses as well as uses specific to that age group.

To achieve these purposes, growth references are required. Since the use of a growth reference implies that a comparison is to be made, on what basis should they be made? At one extreme is having a single, international reference. Having a single, international reference allows comparisons across the world's populations to be made. If relatively small variability in unconstrained growth occurs across racial and ethnic groups, as has been shown, then it is possible to have this uniform global application without loss of its usefulness for local applications. At the other extreme is having references developed for specific locations. Location-specific references have three major challenges. First, a reference developed from a population with growth deficits has less screening value. Second, there are potentially significant secular changes in growth over a short time if a population has not neared the end of its secular change already. Third, developing and implementing a new reference is difficult and costly.

For the purposes of global monitoring of child growth status and comparing regions and countries, the adoption of a single, international reference has brought great dividends over the past 25 years. For example, many studies have now been completed comparing growth status and improvements in growth (de Onis *et al.*, 2000), examining factors that influence and relate to growth (Frongillo and Hanson, 1995; Frongillo *et al.*, 1997; Smith and Haddad, 2001; Pelletier and Frongillo, 2003), and forecasting future growth status (de Onis *et al.*, 2000). These studies would have been impossible without a uniform way of reporting anthropometric information based on a common reference. Nevertheless, there continues to be considerable interest in the development of national and regional references (Whitehead and Paul, 2000).

Presentation of growth references

Indices of growth

Growth information can be used in its raw form or as indices. Indices are combinations of measurements, and are useful for the interpretation of measurements. Examples are weight-for-age, height-for-age, weight-for-height and body mass index (weight-for-height-squared). Measurements or indices transformed using reference data create normalized indices. Typical normalized indices are generated by matching raw data measurements with reference values on sex and age (or on height if the measurement is weight). Normalized indices can take the form of percentage of the median, centiles, or *Z*-scores. Percentage of median, taking the ratio (times 100) of the observed measurement to the median reference value, is an older form of a normalized index that is now not favoured. Instead, centiles and *Z*-scores are preferred both because of their mathematical properties and ease of interpretation.

Centiles refer to numerical values from 0 to 100 that represent the rank of an individual on a given reference distribution. Centiles have a straightforward interpretation for clinical applications, but have some disadvantages (World Health Organization, 1995). First, the same intervals of centile values correspond to different intervals in the measurements, depending on which part of the distribution is involved. Therefore, summary statistics (i.e. means and standard deviations) for centiles are not useful. Second, at the extremes of the distribution, centiles change little even though the measurements change substantially.

The most commonly used indices are Z-scores. Z-scores are obtained by taking the value of a growth measurement, subtracting the age- and sex-specific (or height-specific) median taken from the reference, and dividing by the age- and sex-specific (or height-specific) standard deviation taken from the reference. Z-scores for weight-for-age (and sex), height-for-age (and sex), and weight-for-height (also sex-specific) are commonly used with infants and young children.

The primary purpose of Z-scores is descriptive. Z-scores allow a set of measurements from children in a sample who may vary in age and sex to be combined together. The use of Z-scores has greatly facilitated comparing the growth of groups of children from one place to another and across ages, and is also helpful for assessing the growth of individual children (World Health Organization, 1995).

The use of Z-scores for analytic purposes, however, can be problematic, especially when the pattern of growth in the reference, both in terms of central tendency (i.e. median) and variability (i.e. standard deviation), differs from the sample at hand over time because of differential feeding mode and other factors (World Health Organization Working Group on Infant Growth, 1994, 1995; Frongillo, 2001). When sex and age adjustment is needed in analyses, this adjustment might better be made by the inclusion of covariates for sex and age (and possibly powers of age) in the analysis (World Health Organization, 1995). Furthermore, the growth of male and female infants should often be assessed separately because each sex may respond differently to different conditions, such as different infant formulas (Nelson *et al.*, 1989; Roche *et al.*, 1993). When sex is an effect modifier, growth should be assessed separately for male and female infants and there is no need to adjust for sex as a covariate or through use of Z-scores.

Distance, velocity and conditional references

Most growth references document the distribution of attained size rather than growth per se. When used with a particular child's data, they indicate whether the child is tall or short, heavy or light, fat or thin, expressing the result either as a centile or a Z-score. A chart of attained size is also known as a distance chart because it measures the distance the child has travelled on its journey forward in time from conception.

To assess growth directly, a chart should indicate the expected average and variability in centile or Z-score over time. Most growth references do not do this. Such a chart is called a velocity chart; it measures the rate of change of size or distance. Conventionally, velocity is defined in measurement units per unit time, e.g. height velocity in cm/year or weight velocity in kg/month, obtained as the difference in two growth measurements over an interval divided by the corresponding difference in time. But, a simpler alternative is to view velocity as centile crossing, when it is measured naturally in terms of the rate of change of a standard deviation score (SDS) per time (Cole, 1998). This rate of change can itself be expressed as an SDS for velocity.

Traditional velocity charts are unconditional, meaning that the growth over an interval is not adjusted for the starting point, the size at the beginning of an interval. But, in general, velocity is inversely related to the starting point, a phenomenon known as regression to the mean (Royston, 1995; Cole, 1998). The consequence is that the unconditional chart will display an expected value of growth for a given child that is either too large or too small, depending on the attained size of the child at the start of the interval. To account for this problem, conditional velocity charts can be constructed that estimate the expected velocity as a function of the starting size (Royston, 1995; Cole, 1998). Conditional velocity and conditional size are essentially the same concept (Royston, 1995).

The idea of conditioning to estimate expected size, taking into account prior size, has been extended to include other covariates. Thompson and Fatti (1997) developed a multivariate normal approach which allows conditioning on multiple variables, including prior size. For example, a weight chart for women was developed that adjusts the overall location of the chart according to the height, age and parity of each individual. This idea may allow the development of better screening tools that are adapted to each individual's characteristics, with potentially greater sensitivity and specificity.

Modes of display

Growth reference information about distance and velocity can be displayed in two basic ways, visually and numerically. For many users (particularly health workers and the public) and uses (particularly individual), visual display in the form of graphs has been most common. Graphs are typically bivariate, with growth information on the vertical axis and age (or height if weight-for-height) information on the horizontal axis (for example, see Kuczmarski *et al.*, 2000). Well-designed growth charts include a graph showing the reference, but also have other features. Growth charts that function well in practice with health workers and the public are designed to enhance their educational value. For example, they are large enough to be easily read, have clear grids for recording and interpreting information more easily, provide culturally relevant details, accentuate the vertical axis so that changes in position can be easily detected, indicate categories of nutritional status with shading or colours, and provide reminders of key behaviours for certain ages (Griffiths *et al.*, 1996). The development of electronic hand-held devices, new computer storage media, and the World Wide Web has raised the potential of providing new tools for the display and use of growth references.

For other users (particularly agency officials and scientists) and uses (particularly population), numerical displays are common. Growth information about a population that incorporates growth references may be displayed as averages of measurements or Z-scores, or as the prevalence of categories of nutritional status (World Health Organization, 1995). For scientists, the references themselves may be displayed in tables (e.g. World Health Organization, 1983), provided as equations (e.g. Dibley *et al.*, 1987), or be incorporated into computer software (e.g. Mei *et al.*, 1998; Kuczmarski *et al.*, 2000).

Conditional velocity charts are appealing in principle because of the potential to assess growth over time directly. Designing the effective display of conditional velocity charts, however, is a major challenge. Cole (1997a, 1998) has demonstrated a design that is biologically and statistically sound, and displays the necessary information. Further research and development will be required to determine whether such a design will be understood and used effectively in practice and how best to refine the design.

Designing studies for the construction of growth references

Reference populations

A key consideration for the construction of a growth reference is the reference population from which the data will be obtained. The reference population chosen should exhibit the expected growth, given the purposes for which the reference will be used. For example, several criteria were defined for the selection of the reference population for the World Health Organization Multicentre Growth Reference Study (de Onis *et al.*, 2001):

- Socioeconomic status that does not constrain growth
- Low altitude
- Low mobility of the target population
- Minimum of 20% of mothers willing and able to follow specified infant feeding recommendations
- Existence of breast-feeding support system

- Local presence of qualified collaborative institutions
- Rate of hospital deliveries
- Sufficient number of eligible births
- Feasibility.

These criteria were used to select sites that had sufficient numbers of individuals in the reference population of interest. The Steering Committee also considered other site-level factors that were not selection criteria: mean birth weight, maternal height, complementary feeding practices, health-related behaviours, environmental hazards, geographic distribution, and fundability of the study to collect the data. Other criteria applied to the selection of mothers and infants into the study were used to further define the reference population:

- No health, environmental or economic constraints on growth
- Mother willing to follow feeding recommendations
- Term birth
- Single birth
- No maternal smoking.

These criteria and factors were identified either as being potentially important to shaping growth patterns or to the ability to collect data effectively.

In the development of the 2000 CDC Growth Charts (Kuczmarksi *et al.*, 2000), the reference population was the civilian, non-institutionalized population of the United States. The reference population excluded infants with very low birth weight (<1500 g).

Anthropometric status varies substantially across regions and countries (Frongillo and Hanson, 1995; Frongillo *et al.*, 1997). This variation could either be due to differential socio-economic conditions or to genetic differences. Genetic differences are unlikely to play an important role, however, since young children across most ethnic groups show very similar linear growth potential given adequate living conditions. This conclusion is based on earlier studies going back to Habicht *et al.* (1974), many of which did observe some ethnic variability in growth. In these studies, this ethnic variability was downplayed relative to that due to other causes or in the context of justifying the appropriateness of a single international reference (Frongillo and Hanson, 1995; Frongillo *et al.*, 1997). Beaton *et al.* (1990) summarized this perspective: 'Across populations of different socio-economic status, the differences in growth velocity and size are above all caused by environmental circumstances and not by ethnic differences in growth potential, at least up to five years of age.' Recent studies have confirmed that growth patterns of well-fed, healthy children from diverse ethnic backgrounds are similar (World Health Organization Working Group on the Growth Reference Protocol and World Health Organization Task Force on

Methods for the Natural Regulation of Fertility, 2000; Bhandari *et al.*, 2002). This similarity means that it is feasible to construct growth references by combining together data from multiple sources or locations. Two main possible reasons for doing this are to increase sample size and to ensure geographic representation.

Study designs

The distinction made earlier between size and growth, i.e. distance and velocity, is important for understanding the advantages and disadvantages of different survey designs. Most growth references are based on cross-sectional studies in which each child is measured only once. Consequently, these growth references are distance charts that do not have direct information about the growth of individuals. Since growth is the change (or rate of change) in anthropometry over time, individuals contribute growth information directly to a reference only if they are measured more than once. With a cross-sectional study, the rate of growth is inferred from trends in average size over time, and no estimate of the variability in growth rate in the population can be made. A longitudinal study, in contrast, directly measures growth over time in individuals. Each individual is measured more than once. This allows the direct estimation of growth rate and variability in growth rate, and the construction of velocity and conditional references. As documented in the next section on sample size, for a given obtained sample size, a cross-sectional study, relative to a longitudinal study, measures distance more precisely and growth rate less precisely.

Precision refers to the statistical efficiency once a sample is obtained. But study efficiency also depends upon the costs of obtaining a sample and collecting the data. These costs include subject recruitment, measurement time, staff retention and capital costs. These costs may be quite different for cross-sectional and longitudinal studies.

Cross-sectional studies typically can be carried out more quickly and are easier to manage. But obtaining the sample may be costly if it is difficult to identify and locate individuals in the reference population. If it is difficult to find individuals, it may be inefficient to measure each individual only once.

Longitudinal studies are challenging logistically because of the need to follow individuals over time and to maintain a management system to track individuals. The occurrence of drop-outs, missed measurements and late measurements complicates data management and analysis. If it is difficult to find individuals, however, a longitudinal study may be more efficient logistically because the investment made to identify and locate individuals pays off in being able to obtain multiple measurements.

A mixed longitudinal design is a compromise between a cross-sectional and longitudinal design that in principle should allow estimation of both distance and velocity reasonably well. In a mixed longitudinal design, individuals are measured more than once, but not throughout the entire age range of interest for the reference. Theoretically, it should be possible to find an optimal mixed longitudinal design that balances the precision of its estimates of distance and velocity with the costs, because cross-sectional and longitudinal designs are special cases of the mixed longitudinal design. A cross-sectional design measures each individual once, whereas a longitudinal design measures each individual on every measurement occasion. The optimal design may be cross-sectional, longitudinal, or mixed between them.

A mixed longitudinal design might take measurements at regular fixed time intervals. If the interval is short, e.g. one week, then adjacent measurements are highly correlated and have much redundant information. But, if the interval is long, e.g. one year, then the correlation and redundancy are less.

Despite the potential efficiency of mixed longitudinal designs, these designs are not used very often compared to cross-sectional and longitudinal designs, and their statistical properties have been little studied (Helms, 1992). From an analysis done for the World Health Organization Multicentre Growth Reference Study, it appears that, compared to a cross-sectional design, a mixed longitudinal design can save recruitment costs, and trades off increased precision in estimating velocity for decreased precision in estimating distance.

Sample size

Sample size is an important consideration when constructing growth references because it is a major determinant of the precision of the reference curves. Other determinants are study design, timing of measurements and analytic method. For about 25 years, it has been understood that a sample size of about 200 individuals for each age and sex gives adequate precision (Waterlow *et al.*, 1977; World Health Organization, 1995). For example, a sample size of 200 gives a precision of about 1.54 centile units for the 5th centile. This sample size is also adequate to separate environmental from genetic effects on growth (World Health Organization, 1995).

A more recent examination of this issue done by World Health Organization in preparation for the World Health Organization Multicentre Growth Reference Study (de Onis *et al.*, 2001) confirmed the adequacy of 200 for each age and sex. Four criteria were considered for estimating the needed sample size: (1) precision of a given centile at a particular age, (2) precision of the slope of the median curve over a given range of ages, (3) precision of the median curve overall and the influence (i.e. leverage) of data at particular ages, and (4) precision of the correlation between measurements in the same individuals at different ages. For each criterion, a sample size estimate was made. From these calculations, it was determined that an adequate sample size would be 200 individuals for the longitudinal portion of the study, and 200 individuals every three months for the cross-sectional portion of the study (de Onis *et al.*, 2001).

For the first criterion (centile at a given age), a sample size of 200 was estimated to provide precision values (expressed as the standard error of the mean, divided by the mean, times 100) of 0.28 and 0.64 for the median and 2nd centile for length, respectively (World Health Organization Working Group on the Growth Reference Protocol, 1997). The corresponding precision values for weight were 0.85 and 1.77.

For the second criterion (slope of median curve), with a sample size of 200, the standard error of the slope of the median curve (in SDS units) was estimated to be 0.045, 0.14 and 0.24 for a longitudinal design (with correlation between measurements of 0.8), cross-sectional design with measurements at two ages, and cross-sectional design with uniform ages throughout an interval, respectively (World Health Organization Working Group on the Growth Reference Protocol, 1997). These results illustrate that, for estimation of the slope of median curve, a longitudinal design is much more precise than a cross-sectional design. These results also illustrate that, of the two cross-sectional designs, measuring children at set nominal ages is more precise than measuring ages uniformly throughout an age range. For cross-sectional designs, however, uniform ages are more practical because it is easier to select and measure children at any age in the specified range. Also, uniform ages have the advantage of better estimating the shape of the curve when the curve is not a straight line and the age range is wide.

For the third criterion (precision of median curve overall and influence), analysis demonstrated that, for estimating the median curve over the age range of 0 to 5 years with a sample size of 200, the imprecision was uniformly low throughout the range with the exception, not surprisingly, of the extremes of the range (World Health Organization Working Group on the Growth Reference Protocol, 1997). This problem is easily resolved by increasing the sample size at birth and extending the age range for measurements past that which is needed for construction of the reference curves.

For the fourth criterion (correlation between measurements), a sample size of 200 gave imprecision values (expressed as the width of a 95% confidence interval) of 0.028, 0.055, 0.103 and 0.146 for correlation coefficients of 0.95,

0.90, 0.80 and 0.7, respectively (World Health Organization Working Group on the Growth Reference Protocol, 1997). These results are relevant only for longitudinal designs.

Timing of measurements

In general, more measurements are needed during periods of most rapid acceleration or deceleration (i.e. changes in velocity) to capture accurately the patterns of growth. For example, in the longitudinal portion of the World Health Organization Multicentre Growth Reference Study, measurements were taken biweekly during the first two months, monthly thereafter during the first year, and bimonthly in the second year (de Onis *et al.*, 2001). In the cross-sectional portion of the study, measurements were planned to have an adequate number every three months.

Usual practice in longitudinal studies has been to attempt to take measurements as closely as possible to set nominal times. This approach has two principal advantages. First, it is relatively easy to organize and examine the data at each age. Second, having set measurement times makes it easier to plan and implement data collection. A potential disadvantage is that growth is monitored in discrete rather than continuous time. No or little information about growth status will be available between the nominal times. If the nominal times are close enough together, then this disadvantage disappears.

Issues for the timing of measurements for cross-sectional studies were discussed in the section on sample size above. Note that, in some ways, the advantages and disadvantages for longitudinal and cross-sectional studies are reversed.

Analytic methods for construction of growth references

Aims of analytic methods

The development of analytic methods for construction of growth references has drawn considerable attention in the statistics and human biology literature, especially recently corresponding with advances in statistical and computing technology. Since most growth references relate distance or velocity to age (an exception being weight-for-height references), this discussion of methods will focus on the construction of age-related references. It will build from the review of methods by Wright and Royston (1997a), and discuss developments that have occurred since then. Since most references are distance references, the construction of these references will be considered first, with discussion of velocity references from longitudinal data to follow.

The aim of analytic methods for age-related distance references is to construct a summary of the size expected in a reference population as a function of age. (As discussed above, this could be generalized to include other variables besides age, including prior size.) That is, the aim is to estimate the distribution of size at each age. Since age-related distance references involve two variables, size and age, it is helpful to examine methods from the perspective of each variable.

Empirical versus model-based estimation of size

The simplest way to estimate the distribution of size is the empirical distribution function. The empirical distribution function uses the ordered values of size (i.e. the order statistics) at each age to estimate centiles directly from the data (Wright and Royston, 1997a). The estimator is biased for small samples, but estimates accurately for moderately large samples. Besides simplicity, the major advantage of this approach is that the empirical distribution function, if the sample size is large enough to allow reasonably precise estimation, will produce an accurate estimate of the shape of the distribution of size throughout the range of size. The major disadvantage is statistical inefficiency, because of the variability of the individual order statistics (Gasser *et al.*, 1996; Wright and Royston, 1997a). This disadvantage can be partly overcome by averaging over the order statistics and weighting according to their proximity to the sample centiles using smoothing techniques such as kernel estimation (Wright and Royston, 1997a).

The alternative to the empirical approach is to assume that the distribution of size can be approximated by a member of a family of suitable distributions. That is, the distribution of size is modelled by a suitable distribution. If the distribution of size is symmetric and bell-shaped, then the normal distribution with two parameters (mean and variance) may be used to approximate the distribution. If the distribution of size is positively skew, then the log-normal distribution with two parameters may be used. More flexible distributions have more parameters. The most common three-parameter distribution is the Box–Cox or power family that has an extra parameter for the skew. Four-parameter distributions, particularly the shifted-power Box–Cox family and the Johnson family, have been used, as have some other approaches (Wright and Royston, 1997a). With each of these families of distributions, centiles are estimated by fitting the data to the distribution, transforming to the normal scale to estimate the centiles, and then back-transforming to the original scale. The advantage of the model-based, in contrast to the empirical, approach is that the estimator of the distribution of size is more efficient because all of the data are used together to estimate the distribution. The disadvantage of the model-based approach is that, if the data do not fit the assumed distribution in all respects, then some portions of the estimated distribution may be biased (Gasser *et al.*, 1996). For example, if the distribution of size (perhaps after transformation to account for positive skewness) has positive kurtosis, then the estimated centiles at each extreme of the distribution will be biased, i.e. the width of the estimated reference will be too narrow.

Incorporating variation across age for distance references

The simplest way to account for variation in the distribution of size at each age is to calculate the empirical estimates for each centile of interest at each age, and then combine these to make the reference. But, even for large sample sizes, the resulting curves are too rough to be acceptable. For example, curves for centiles in a reference should not cross, but this can easily occur with the empirical approach. This approach can be improved by smoothing locally within windows of age using a variety of techniques (Wright and Royston, 1997a).

Another approach is to normalize the data at each age, and then use polynomial or other mathematical forms to fit the mean and standard deviation as a function of age. The standard deviation is estimated from the residuals (Altman, 1993). Centiles are then estimated under the assumption of normality, and then transformed back to the original scale (Wright and Royston, 1997a).

Cole (1988) developed the LMS method that brought together the three-parameter Box–Cox family with smoothing across age. In the LMS method, three parameters are defined for the skewness, median and coefficient of variation. In the original form of the LMS method, these parameters are estimated separately within each age group by maximum likelihood, and then they are smoothed across age using any regression smoothing technique such as polynomial regression or kernel estimation. Centiles are calculated on the transformed scale, and then are back-transformed to the original scale. Categorizing the age variable must be done subjectively, with the choice of categorization affecting the estimated reference curves. Cole and Green (1992) enhanced the original LMS method to use maximum penalized likelihood to estimate the age-related curves for each parameter by natural cubic splines. The advantage of this enhancement is that age categorization is not required, although some subjectivity remains in that values of the smoothing parameters must be chosen (Wright and Royston, 1997a). An important advantage of the LMS method is that centiles and *Z*-scores can be constructed directly.

Healy *et al.* (1988) developed a two-stage method. First, a set of centile positions is chosen, for example the set of seven centiles: 3, 10, 25, 50, 75, 90 and 97.

Empirical centile curves for each are estimated in windows as discussed earlier. Then, each of the resulting empirical centile curves is smoothed using polynomial regression. The polynomial coefficients are also smoothed, restricting the distance between centiles and preventing the resulting curves from crossing.

Recently, a method has been developed that is similar to the LMS method in some ways (Wright and Royston, 1997b; Royston and Wright, 1998). The distribution of size is described by either an exponential–normal (three parameters) or modulus–exponential–normal (four parameters) distribution; the latter distribution can account for kurtosis. Smoothing across ages is accomplished using fractional polynomials. This method provides explicit formulae for centiles and *Z*-scores. Unlike the LMS method, this method can be implemented in standard software that has a maximum likelihood routine.

Heagerty and Pepe (1999) proposed a semi-parametric method that models the mean and variance as flexible regression spline functions, and allows the unspecified distribution to vary smoothly as a function of covariates. This approach separately estimates three parameters similar to the LMS method, but no parametric form is assumed for the distribution. Estimation with either cross-sectional or longitudinal samples uses estimating equations for the location and scale functions and kernel smoothing for standardized residuals.

Sorribas *et al.* (2000) applied to growth references a flexible parametric family of distributions called S-distributions. S-distributions approximate many other distributions closely. Their approach differs from the LMS method, for example, by not transforming the original data and not requiring the assumption of normality for the transformed data. The result is a smooth family of parametric distributions in the original scale of the data.

Rigby and Stasinopoulos (2000) developed mean and dispersion additive models that can be used with skew or kurtotic distributions, and applied these to the construction of age-related centiles. In this application, the mean and variance (or the logarithm of the variance) were modelled as parametric or non-parametric functions of age. Normality of the residuals was assumed, with the Box–Cox transformation employed if residuals were not normal.

Comparison of methods for constructing distance references

Royston and Wright (1997a) compared three methods (empirical, LMS, Healy) on three data sets. For two of the data sets, each of the three methods gave satisfactory and similar results. The third data set had a pronounced dip in the age trend just before puberty, and only the LMS method produced satisfactory results. Moussa (2002) compared the LMS and exponential–normal methods and found that they gave similar results when the same age-smoothing procedure was used. This is not surprising since both distributions have three parameters.

Assessing fit of distance references

Relatively little research has been done on the topic of assessing whether a constructed reference provides a good fit to the sample of data, and whether the fit can be improved. Along with visual inspection of the reference against the data, a variety of statistical approaches have been used to assess fit. These include comparison of observed and expected frequencies in age and centile groups, quantile–quantile residual plots and the Shapiro–Wilk W-statistic to assess normality, grid tests, permutation bands and B-tests, and comparison of the fitted model with one that uses indicator variables for age groups (Wright and Royston, 1997a; Royston and Wright, 1998, 2000). Van Buuren and Fredriks (2001) presented application of a worm plot to assess the age-conditional normality of the transformed data under a variety of LMS models.

Besides assessing the accuracy of the fit, it is also useful to estimate the precision of reference centiles. With some methods, calculation of standard errors of centile estimates is straightforward. With other methods, bootstrapping can be used.

Methods for constructing velocity and conditional references

Whereas distance references can be constructed from either cross-sectional or longitudinal data, velocity and conditional references require longitudinal data. The earlier methods for constructing unconditional velocity references are basically the same as those used for distance references. This is because the increment or difference between two measurements is a single random variable that can be analysed in much the same way as a random variable for size. Construction of such a reference involves additional decisions about the length and starting points of the intervals. Intervals that are too short will not be useful because of the relative imprecision of both the increments in the reference and those for an individual. Unconditional velocity references have also been estimated using multilevel models (Royston, 1995).

As discussed earlier, conditional velocity references are much preferred to unconditional ones. A number of methods have been recently used to construct conditional references. Royston (1995) and Reinhard and Wellek (2001) used multilevel models to account for the multiple measurements contributed by each individual. Wright *et al.* (1994), Bailey (1994) and Cole (1994) regressed SDS at a later time on SDS at an earlier time; these regressions take a simple form since the regression coefficient is the correlation coefficient between the SDS. Thompson and Fatti (1997) applied multivariate normal theory to construct references that are conditional on prior measurements and other variables. Wade and Ades (1998) extended an LMS-based maximum likelihood method by incorporating into the likelihood the matrix of the correlations between repeated measurements expressed as SDS.

Recent examples of growth references

About 12 years ago, both National Center for Health Statistics/Centers for Disease Control and Prevention (NCHS/CDC) and the World Health Organization independently but co-operatively began processes to extensively review the uses and interpretations of growth information. Included in these reviews was consideration of the limitations of the then-current references (World Health Organization Working Group on Infant Growth, 1994, 1995; Mei *et al.*, 1998; Kuczmarski *et al.*, 2000). Although these reviews differed in process and outcomes, both reviews reached a consensus that there were a number of important limitations of the 1978 NCHS/CDC (Hamill *et al.*, 1977) and 1983 World Health Organization references. As a result of the recommendations from these reviews, both NCHS/CDC and the World Health Organization set out to construct new reference growth charts.

2000 CDC Growth Charts

In May 2000, NCHS/CDC released revised reference growth charts for the United States (Kuczmarski *et al.*, 2000). These charts were created with improved data and statistical curve smoothing techniques. Data were taken from five national health examination surveys collected from 1963 to 1994 and five supplementary sources. These were combined into one analytic data set to produce the reference growth charts. These data better represent racial and ethnic diversity in the United States than the previous reference, and contain a mixture of growth data from infants who were breast-fed and formula-fed. The new reference growth charts were largely constructed using a descriptive approach, meaning that the reference portrays the growth that is expected of children in the population typically or on average, according to sex and age. Some aspects of a prescriptive approach were also taken, meaning that the reference portrays the growth that is expected of children who are healthy and well nourished, and who have received proper care. Specifically, some data were excluded to avoid the influence of an increase in body weight that was observed in the more recent data.

A two-step method was used to construct the references. In the first step, empirical centiles were estimated without using any distributional assumptions.

Selected empirical centiles were smoothed using one of several parametric or non-parametric models. In the second step, a modified LMS procedure was used with the smoothed curves to provide Z-scores closely matched to the empirically smoothed centile curves.

The World Health Organization Multicentre Growth Reference Study

The international infant growth reference currently used (World Health Organization, 1983) was constructed primarily from the growth of formula-fed infants. A World Health Organization review found that there was considerable evidence that the growth patterns of breast-fed and formula-fed infants differ, especially in the first year of life (World Health Organization Working Group on Infant Growth, 1994, 1995). These differences mean that the current international reference probably underestimates obesity in well-off populations and underestimates underweight and stunting in malnourished populations for ages 0 to 6 months. Furthermore, diagnosis of poor weight gain often occurs at 3 months with the current reference (leading to undesirable too-early introduction of complementary foods), whereas this diagnosis would probably occur at 5 months with a reference made from breast-fed infants. Consequently, motivated in part by this evidence, the World Health Organization and its member states strongly endorsed the construction of a new reference, and recommended that it should reflect the growth expected of a population defined on the basis of having followed widely endorsed health and nutritional recommendations.

To achieve this, the World Health Organization has been conducting an intensive six-country study of children living in healthy environments to collect the necessary data (Garza and de Onis, 1999; de Onis *et al.*, 2001). The new reference growth charts from this study will probably be completed in 2005. An important part of the rationale for constructing this new international growth reference is that it will improve the nutritional management of infants and lead to better support for breast-feeding and other accepted health practices.

Cole et al. *(2000) reference for body mass index*

Besides reference growth charts that have been or are being developed by technical agencies such as NCHS/CDC and the World Health Organization, other scientists have been working to improve reference information. For example, Cole *et al.* (2000) aimed to develop an internationally acceptable definition of child overweight and obesity. The study used data from six large nationally representative cross-sectional growth studies from Brazil, Great Britain, Hong Kong, the Netherlands, Singapore and the United States. The main outcome measure was body mass index. Although cut-off points for overweight and obesity for adults have been developed based on relations of body mass index with morbidity and mortality data, this is not possible to do for children. The idea implemented in this paper was to instead extrapolate from cut-off points for adults to define statistically analogous cut-off points for children.

Other recent growth references

Many other growth references have recently been developed. The largest effort of these is the Euro-Growth Study which collected data on 2245 infants at 22 study sites in 11 countries (van't Hof *et al.*, 2000a). From these data, references have been developed for length, weight and circumferences (Haschke *et al.*, 2000), increments of these measures (van't Hof *et al.*, 2000b), and body mass index and weight-for-length (van't Hof *et al.*, 2000c). Recent examples of national growth references are those from the United Kingdom (Cole *et al.*, 1998; Wright *et al.*, 2002), Sweden (Lindgren *et al.*, 1995), Italy (Cacciari *et al.*, 2002), India (Khadgawat *et al.*, 1998) and Hong Kong (Leung *et al.*, 1996). Recent examples of specialized references are ones for mid-upper arm circumference (de Onis *et al.*, 1997), birth weight for twins (Min *et al.*, 2000), children with Turner syndrome (Rongen-Westerlaken *et al.*, 1997), birth weight-for-gestational age (Kramer *et al.*, 2001), body mass index for women in developing countries (Nestel and Rutstein, 2002), body composition for children (Ruxton *et al.*, 1999) and the age at childhood (i.e. post-infancy) onset of growth (Liu *et al.*, 2000).

Future challenges for growth references

Construction of growth references

An important challenge for the future is the basic approach to constructing growth references and definition of the reference population. As more is learned about the functional consequences of growth status and as there continue to be secular trends in growth status (e.g. greater prevalence of overweight and obesity), it is likely that the decision as to whether the approach should be descriptive or prescriptive and how the reference population should be defined will be increasingly complex. For example, the 2000 CDC Growth Charts are primarily descriptive, but a decision was made to not include some recent data that would have shifted the reference towards greater weights. From a descriptive approach, a reference population would have to change as the overall population changes. From a prescriptive approach, a reference population would not change, but the reference population would increasingly become a selected subset of the overall population.

Given cross-sectional and longitudinal data that are now or will be soon available, it will likely be more difficult to justify the collection of new growth data for constructing growth references. As more is learned about human growth, methods for making best use of existing data will be increasingly important. For example, Hermanussen and Burmeister (1999) have explored methods for constructing synthetic growth charts from existing data.

Methods for constructing distance references have now reached some maturity, but methods for constructing velocity and especially conditional references are in relative infancy. Given the recent availability of extensive longitudinal data, more research and development work will be needed to realize the potential of creating true growth references that allow the assessment of changes in size over time. Similarly, methods for constructing multivariate conditional references are also in relative infancy. The creation of conditional growth references will require considerable conceptual thinking and empirical work to guide decisions about lengths of intervals, how to accommodate measurement error to ensure adequate screening value (Cole, 1997b), what individual characteristics are most useful as conditioning variables, and how best to display these references for users.

Use of growth references

The construction and release of new reference growth charts has three important implications for health-care workers, child caregivers, agency officials, scientists, and others who use growth information. First, after more than two decades of having a single growth reference widely endorsed and used internationally, there are now multiple reference growth charts available. Consequently, there is the potential for confusion in the choice and use of these reference growth charts. The 1983 World Health Organization Reference Growth Charts continue to be preferred for international use. When the new World Health Organization reference is released, that new reference will be adopted for international use. Even in counties using national references, the new World Health Organization reference may sometimes be preferred for assessing the growth of breast-fed infants.

Second, the adoption of any new reference growth charts means that many decisions must be made on how to transition from the old to the new. For example, existing copies of the previous growth charts need to be replaced, growth information for some children may need to be transferred to the new charts, and computer software may need modification. Guidance and training is required to help people who use growth information to make the transition to the new charts.

Third, any new reference growth charts will probably portray the expected growth differently than did the previous charts that they are replacing. That is, the growth of the reference populations will be somewhat different. For example, below the age of 24 months, the weight-for-age curves in the 2000 Revised CDC Reference Growth Charts for the United States are generally higher than those for the 1978 NCHS/CDC charts. This will result in classifying infants more often as underweight and less often as overweight than before. Other examples are given in Kuczmarksi *et al.* (2000). Similarly, when the new World Health Organization reference is released, breast-fed infants will be less likely to appear to be faltering early in the first year, while formula-fed infants may be more likely to be identified as being overweight, with the risk that parents may react by placing the child on a low-energy diet, which is not recommended during infancy. Nevertheless, the potential risks to formula-fed infants of changing the growth reference are judged to be much lower than those faced by breast-fed infants whose management is based on the current international reference.

Availability and use of anthropometric data world-wide has grown, due to major international initiatives aimed at increasing access to growth charts and software for a variety of users (World Health Organization, 1995). Maximizing the appropriate uses and interpretation of anthropometry requires understanding and interaction among different users: decision-makers, health workers, caregivers and scientists.

In most developing countries, growth monitoring and promotion programmes have been implemented at the community level, generating both individual and group-level data (Peterson *et al.*, 1994). At the core of these programmes is the use and interpretation of growth charts by health workers and caregivers. Growth charts are used as a dynamic, visible tool, to educate and motivate health workers and parents to take actions to improve or maintain the child's growth, and to screen for growth faltering and targeting of appropriate interventions (Ruel *et al.*, 1991; Owusu and Lartey, 1992; McAuliffe *et al.*, 1993). Previous studies have shown that several conditions are necessary for growth charts to be successful (Grant and Stone, 1986; Aden *et al.*, 1990; Fagbule *et al.*, 1990; Ruel *et al.*, 1991; Owusu and Lartey, 1992; McAuliffe *et al.*, 1993). First, health workers must be able to assess child's age, weight and stature accurately. Second, health workers must be able to record data and interpret growth charts properly. Finally, once growth faltering is detected, workers must be able to identify clinically and culturally appropriate and affordable actions. The quality of growth information collected for individual-level purposes directly affects the validity of aggregate figures used by decision-makers and programme supervisors for population purposes. Given that growth charts play a central role in this monitoring process, it is essential that explicit research and development effort is devoted to understanding how best, from a user perspective, to construct, design, display, communicate about and disseminate the new growth reference charts, and train health workers on their use.

ACKNOWLEDGEMENTS

Anna Milman assisted with the preparation of materials for this chapter. Elaine Borghi provided some references. Some information about study designs was adapted from an unpublished report on the use of mixed longitudinal designs prepared for the World Health Organization Multicentre Growth Reference Study by Tim J. Cole and Edward A. Frongillo. Some information about implications for uses was adapted from a proposal written by Karen Peterson, Ana Terra de Souza, and Edward A. Frongillo. Dominic Frongillo provided comments on an earlier draft.

REFERENCES

Aden, A. S., Brännström, I., Mohamud, K. A., Persson, L. A. and Wall, S. (1990). The growth chart: a road to health chart? Maternal comprehension of the growth chart in two Somali villages. *Paediatrica and Perinatal Epidemiology*, **4**(3), 340–50.

Altman, D. G. (1993). Construction of age-related reference centiles using absolute residuals. *Statistics in Medicine*, **12**, 917–24.

Bailey, B. J. R. (1994). Monitoring the heights of prepubertal children. *Annals of Human Biology*, **21**, 1–11.

Beaton, G., Kelly, A., Kevany, J., Martorell, R. and Mason, J. (1990). *Appropriate Uses of Anthropometric Indices in Children: A Report Based on an ACC/SCN Workshop*, Nutrition Policy Discussion Paper no. 7. Geneva, Switzerland: United Nations Administrative Committee on Coordination/Subcommittee on Nutrition.

Bhandari, N., Bahl, R., Taneja, S., de Onis, M. and Bhan, M. K. (2002). Growth performance of affluent Indian children is similar to that in developed countries. *Bulletin of the World Health Organization*, **80**, 189–95.

Cacciari, E., Milani, S., Balsamo, A., *et al.* (2002). Italian cross-sectional growth charts for height, weight and BMI (6–20 years). *European Journal of Clinical Nutrition*, **56**, 171–80.

Cole, T. J. (1988). Fitting smoothed centile curves to reference data. *Journal of the Royal Statistical Society A*, **151**, 385–418.

(1994). Growth charts for both cross-sectional and longitudinal data. *Statistics in Medicine*, **13**, 2477–92.

(1997a). Growth monitoring with the British 1990 growth reference. *Archives of Disease in Childhood*, **76**, 47–9.

(1997b) 3-in-1 weight monitoring chart. *Lancet*, **349**, 1020–30.

(1998). Presenting information on growth distance and conditional velocity in one chart: practical issues of chart design. *Statistics in Medicine*, **17**, 2697–707.

Cole, T. J. and Green, P. J. (1992). Smoothing reference centile curves: the LMS method and penalized likelihood. *Statistics in Medicine*, **11**, 1305–19.

Cole, T. J., Freeman, J. V. and Preece, M. A. (1998). British 1990 growth reference centiles for weight, height, body mass index and head circumference fitted by maximum penalized likelihood. *Statistics in Medicine*, **17**, 407–29.

Cole, T. J., Bellizzi, M. C., Flegal, K. M. and Dietz, W. H. (2000). Establishing a standard definition for child overweight and obesity worldwide: international survey. *British Medical Journal*, **320**, 1240–3.

de Onis, M., Yip, R. and Mei, Z. (1997). The development of MUAC-for-age reference data recommended by a World Health Organization Expert Committee. *Bulletin of the World Health Organization*, **75**, 11–18.

de Onis, M., Frongillo, E. A. and Blössner, M. (2000). Is malnutrition declining? An analysis of changes in levels of childhood malnutrition since 1980. *Bulletin of the World Health Organization*, **78**, 1222–33.

de Onis, M., Victora, C. G., Garza, C., Frongillo, E. A. and Cole, T. J. for the World Health Organization Working Group on the Growth Reference Protocol (2001). A new international growth reference for young children. In *Perspectives in Human Growth, Development and Maturation*, eds. P. Dasgupta and R. C. Hauspie, pp. 45–53. Dordrecht, The Netherlands: Kluwer Academic Publishers.

Dewey, K. G. (1998). Growth characteristics of breast-fed compared to formula-fed infants. *Biology of the Neonate*, **74**, 94–105.

Dibley, M. J., Staehling, N., Nieburg, P. and Trowbridge, F. L. (1987). Interpretation of Z-score anthropometric indicators derived from the international growth reference. *American Journal of Clinical Nutrition*, **46**, 749–62.

Fagbule, D. O., Olaosebikan, A. and Parakoyi, D. B. (1990). Community awareness and utilization of growth chart in a semi-urban Nigerian community. *East African Medical Journal*, **67**(2), 69–74.

Frongillo, E. A. (2001). Growth of the breast-fed child. In *Nutrition and Growth*, 47th Nestlé Nutrition Workshop, eds. R. Martorell and F. Haschke, pp. 37–52. Philadelphia, PA: Lippincott.

Frongillo, E. A. and Hanson, K. M. P. (1995). Determinants of variability among nations in child growth. *Annals of Human Biology*, **22**, 395–411.

Frongillo, E. A., de Onis, M. and Hanson, K. M. P. (1997). Socioeconomic and demographic factors are associated with worldwide patterns of stunting and wasting. *Journal of Nutrition*, **127**, 2302–9.

Garza, C. and de Onis, M. (1999). A new international growth reference for young children. *American Journal of Clinical Nutrition*, **70**, 169S–172S.

Gasser, T., Molinari, L. and Roos, M. (1996). Methodology for the establishment of growth standards. *Hormone Research*, **45**(Suppl. 2), 2–7.

Grant, K. and Stone, T. (1986). Maternal comprehension of a home-based growth chart and its effect on growth. *Journal of Tropical Pediatrics*, **32**, 255–7.

Griffiths, M., Dickin, K. and Favin, M. (1996). *Promoting the Growth of Children: What Works – Rationale and Guidance*. Washington, DC: Human Development Department, The World Bank.

Habicht, J. P., Martorell, R., Yarbrough, C., Malina, R. M. and Klein, R. E. (1974). Height and weight standards for preschool children: how relevant are ethnic differences in growth potential? *Lancet*, **i**, 611–15.

Hamill, P. V., Drizd, T. A., Johnson, C. L., Reed, R. B. and Roche, A. F. (1977). NCHS growth curves for children birth–18 years, United States. *Vital Health Statistics*, **11**, 165.

Haschke, F., van't Hoff, M. A. and The Euro-Growth Study Group (2000). Euro-Growth references for length, weight, and body circumferences. *Journal of Pediatric Gastroenterology and Nutrition*, **31**, S14–S38.

Heagerty, P. J. and Pepe, M. S. (1999). Semiparametric estimation of regression quantiles with application to standardizing weight for height and age in US children. *Applied Statistics*, **48**, 533–51.

Healy, M. J. R., Rasbash, J. and Yang, M. (1988). Distribution-free estimation of age-related centiles. *Annals of Human Biology*, **15**, 17–22.

Helms, R. W. (1992). Intentionally incomplete longitudinal designs. I. Methodology and comparison of some full span designs. *Statistics in Medicine*, **11**, 1889–913.

Hermanussen, M. and Burmeister, J. (1999). Synthetic growth reference charts. *Acta Paediatrica*, **88**, 809–14.

Khadgawat, R., Dabadghao, P., Mehrotra, R. N. and Bhatia, V. (1998). Growth charts suitable for evaluation of Indian children. *Indian Pediatrics*, **35**, 859–65.

Kramer, M. S., Platt, R. W., Wen, S. W., *et al.* for the Fetal/Infant Health Study Group of the Canadian Perinatal Surveillance System (2001). A new and improved population-based Canadian reference for birth weight for gestational age. *Pediatrics*, **108**(2), E35.

Kuczmarski, R. J., Ogden, C. K., Gummer-Strawn, L. M., *et al.* (2000). *CDC Growth Charts: United States*, Advance Data from Vital and Health Statistics no. 314. Hyattsville, MD: National Center for Health Statistics. Available: http://www.cdc.gov/growthcharts/.

Leung, S. S. F., Lau, J. T. F., Tse, L. Y. and Oppenheimer, S. J. (1996). Weight-for-age and weight-for-height references for Hong Kong children from birth to 18 years. *Journal of Paediatrics and Child Health*, **32**, 103–9.

Lindgren, G., Strandell, A., Cole, T., Healy, M. and Tanner, J. (1995). Swedish population reference standards for height, weight and body mass index attained at 6 to 16 years (girls) or 19 years (boys). *Acta Paediatrica*, **84**, 1019–28.

Liu, Y. X., Albertsson-Wikland K. and Karlberg, J. (2000). New reference for the age at childhood onset of growth and secular trend in the timing of puberty in Swedish. *Acta Paediatrica*, **89**, 637–43.

McAuliffe, J. F., Falcao, L. and Duncan, B. (1993). Understanding of growth monitoring charts by literate and illiterate mothers in Northeast Brazil. *Journal of Tropical Pediatrics*, **39**, 370–2.

Mei, Z., Yip, R., Grummer-Strawn, L. M. and Trowbridge, F. L. (1998). Development of a research child growth reference and its comparison with the current international growth reference. *Archives of Pediatrics and Adolescent Medicine*, **152**, 471–9.

Min, S.-J., Luke, B., Gillespie, B., *et al.* (2000). Birth weight references for twins. *American Journal of Obstetrics and Gynecology*, **182**(5), 1250–7.

Moussa, M. A. A. (2002). Estimation of age-specific reference intervals from skewed data. *Methods of Information in Medicine*, **2**, 147–53.

Nelson, S. E., Rogers, R. R., Ziegler, E. E. and Fomon, S. J. (1989). Gain in weight and length during early infancy. *Early Human Development*, **19**, 223–39.

Nestel, P. and Rutstein, S. (2002). Defining nutritional status of women in developing countries. *Public Health Nutrition*, **5**, 17–27.

Owusu, B. and Lartey, A. (1992). Growth monitoring: experience from Ghana. *Food and Nutrition Bulletin*, **14**, 97–100.

Pelletier, D. L. and Frongillo, E. A. (2003). Changes in child survival are strongly associated with changes in malnutrition in developing countries. *Journal of Nutrition*, **133**, 107–19.

Peterson, K. E., Ayub, M. and Jalil, F. (1994). *Pakistan Growth Promotion Study*. Washington, DC: The World Bank.

Reinhard, I. and Wellek, S. (2001). Age-related reference regions for longitudinal measurements of growth characteristics. *Methods of Information in Medicine*, **40**, 132–6.

Rigby, R. A. and Stasinopoulos, D. M. (2000). Construction of reference centiles using mean and dispersion additive models. *Statistician*, **49**, 41–50.

Roche, A. F., Guo, S., Siervogel, R. M., Khamis, H. J. and Chandra, R. K. (1993). Growth comparison of breast-fed and formula-fed infants. *Canadian Journal of Public Health*, **84**, 132–5.

Rongen-Westerlaken, C., Corel, L., van den Broeck, J., *et al.* and the Dutch and Swedish Study Groups for Growth Hormone Treatment (1997). Reference values for height, height velocity and weight in Turner's Syndrome. *Acta Paediatrica*, **86**, 937–42.

Royston, P. (1995). Calculation of unconditional and conditional reference intervals for foetal size and growth from longitudinal measurements. *Statistics in Medicine*, **14**, 1417–36.

Royston, P. and Wright, E. M. (1998). A method for estimating age-specific reference intervals ('normal ranges') based on fractional polynomials and exponential transformation. *Journal of the Royal Statistical Society A*, **161**, 79–101.

(2000). Goodness-of-fit statistics for age-specific reference intervals. *Statistics in Medicine*, **19**, 2943–62.

Ruel, M. T., Pelletier, D. L. and Habicht, J. P. (1991). Comparison of two growth charts in Lesotho: health workers' ability to understand and use them for action. *American Journal of Public Health*, **81**, 610–16.

Ruxton, C. H. S., Reilly, J. J. and Kirk, T. R. (1999). Body composition of healthy 7- and 8-year-old children and a comparison with the 'reference child'. *International Journal of Obesity*, **23**, 1276–81.

Smith, L. C. and Haddad, L. (2001). How important is improving food availability for reducing child malnutrition in developing countries. *Agricultural Economics*, **26**, 191–204.

Sorribas, A., March, J. and Voit, E. O. (2000). Estimating age-related trends in cross-sectional studies using S-distributions. *Statistics in Medicine*, **19**, 697–713.

Thompson, M. L. and Fatti, L. P. (1997). Construction of multivariate centile charts for longitudinal measurements. *Statistics in Medicine*, **16**, 333–45.

van Buuren, S. and Fredriks, M. (2001). Worm plot: a simple diagnostic device for modeling growth reference curves. *Statistics in Medicine*, **20**, 1259–77.

van't Hof, M. A., Haschke, F., Darvay, S. and the Euro-Growth Study Group (2000a). Euro-Growth references on increments in length, weight, and head and arm circumferences during the first three years of life. *Journal of Pediatric Gastroenterology and Nutrition*, **31**, S39–S47.

van't Hof, M. A., Haschke, F. and the Euro-Growth Study Group (2000b). Euro-Growth references for body mass index and weight for length. *Journal of Pediatric Gastroenterology and Nutrition*, **31**, S48–S59.

(2000c). The Euro-Growth study: why, who, and how. *Journal of Pediatric Gastroenterology and Nutrition*, **31**, S3–S13.

Wade, A. M. and Ades, A. E. (1998). Incorporating correlations between measurements into the estimation of age-related reference ranges. *Statistics in Medicine*, **17**, 1989–2002.

Waterlow, J. C., Buzina, R., Keller, W., *et al.* (1977). The presentation and use of height and weight data for comparing nutritional status of groups of children under the age of 10 years. *Bulletin of the World Health Organization*, **55**, 489–98.

Whitehead, R. G. and Paul, A. A. (2000). Growth patterns of breastfed infants. *Acta Paediatrica*, **89**, 136–8.

World Health Organization (1983). *Measuring Change in Nutritional Status: Guidelines for Assessing the Nutritional Impact of Supplementary Feeding Programmes for Vulnerable Groups*. Geneva, Switzerland: World Health Organization.

(1995). *Report of WHO Expert Committee: Physical Status – The Use and Interpretation of Anthropometry*, WHO Technical Report Series no. 854. Geneva, Switzerland: World Health Organization.

World Health Organization Working Group on Infant Growth (1994). *An Evaluation of Infant Growth*. Geneva, Switzerland: World Health Organization.

(1995). An evaluation of infant growth: the use and interpretation of anthropometry in infants. *Bulletin of the World Health Organization*, **73**, 165–74.

World Health Organization Working Group on the Growth Reference Protocol (1997). *A Growth Curve for the 21st Century: The WHO Multicentre Growth Reference Study*. Geneva, Switzerland: World Health Organization Department of Nutrition.

World Health Organization Working Group on the Growth Reference Protocol and World Health Organization Task Force on Methods for the Natural Regulation of Fertility (2000). Growth patterns of breastfed infants in seven countries. *Acta Paediatrica*, **89**, 215–22.

Wright, C. M., Matthews, J. N. S., Waterston, A. and Aynsley-Green, A. (1994). What is a normal rate of weight gain in infancy? *Acta Paediatrica*, **83**, 351–6.

Wright, C. M., Booth, I. W., Buckler, J. M. H., *et al.* (2002). Growth reference charts for use in the United Kingdom. *Archives of Disease in Childhood*, **86**, 11–14.

Wright, E. M. and Royston, P. (1997a). A comparison of statistical methods for age-related reference intervals. *Journal of the Royal Statistical Society A*, **160**, 47–69.

(1997b). Simplified estimation of age-specific reference intervals for skewed data. *Statistics in Medicine*, **16**, 2785–803.

11 *Latent variables and structural equation models*

Gino Verleye
Rijksuniversiteit Gent

Marie-José Ireton
National University of Colombia

J. Cesar Carrillo
National University of Colombia

and

Roland C. Hauspie
Free University Brussels

Introduction

A common problem in the analysis of human growth data is to relate biometric variables to genetic and/or environmental or demographic factors. Quite often we refer to techniques such as multiple regression or principal components analysis. Structural equation modelling is a technique that combines the benefits of both approaches. While in principal components analysis all variables score on each factor (component or latent variable), in structural equation modelling, the investigator can decide about the set of variables that will explain a specific latent variable. The investigator also decides on which paths of relationship between observed and latent variables should be investigated by the model and which ones should not. The procedure consists of an explorative phase, essentially based on principal components analysis of the data, allowing identification of the structure of the latent variables or constructs that, at biological level, are best able to explain the various interrelationships. The second phase consists of testing several possible models and gradually coming to an optimal solution that can explain the interrelationships between the explanatory variables and the dependent variables.

Methods in Human Growth Research, eds. R. C. Hauspie, N. Cameron and L. Molinari.
Published by Cambridge University Press.

From the late 1980s on, structural equations with latent variables, or so-called LISREL models, became very popular in social sciences. There are two reasons for this increased attention. The capability to include latent variables (or concepts) in the models is a major step forward compared to models where only manifest variables (observed measures or items) can be used. This latent variable characteristic also allows the inclusion of measurement error structure, as will be seen further on. Furthermore, latent variable models are better suited for bringing theory and data together. The second main reason for the success is the flexibility of the structural part of the structural equations model (SEM). Compared to regression models, we can now include multiple dependent variables, mediator variables and even feedback loops.

Previously, one needed a firm background in covariance algebra and matrix mathematics and especially a lot of patience to program and reprogram the models to be computed. Today, however, interface software translates the graphical representation of the model into program code. The results (parameter estimates, goodness-of-fit, etc.) are reproduced on the graphs for maximum convenience.

The aim of this chapter is to encourage further the use of SEM. From our experience, the nature of growth data and the typical questions regarding causal models in this field may benefit from increased use of SEM. We intend to introduce SEM and related models and procedures in a pragmatic way. For that purpose, we will apply the process to one data set, in the very same way that we initially learned from that data. The focus will not lie in mathematical or computational issues. The interested reader may consult works in the bibliography to yield deeper knowledge of the details of SEM. Our attention will be on added value, interpretation and data requirements.

Latent variables

Traditionally, statistical analysis is performed on variables that are observed directly (e.g. mother's weight, father's height, etc.). Such manifest data analyses have a long and successful tradition. Looking at the theoretical framework that is used to describe and explain human growth, however, we are faced with a higher level of inquiry than manifest variables. We now deal with conceptual variables such as socio-economic status, nutrient quality, family preoccupation, length at childhood (as opposed to length at a certain specific age, . . .). Such latent variables or concepts are brought into the spectrum of statistical methods because of mainly two reasons. The first being the fact that single item measurement means lower reliability. As an example, consider the so-called Likert item 'To what extent are you interested in the health of your child?' (1: totally uninterested; 2: uninterested, . . . , 5: very interested) as it is used in many surveys as sole indicator of the concept 'Mother's preoccupation with child's health status'. If a researcher wants to use this manifest Likert item variable as a predictor in a statistical model, the potential explanatory value of the concept towards growth and developmental measures is limited by the measurement error in the Likert item. By increasing the number of equally useful indicator variables, the reliability of the measured concept will increase so that its predictive power is enhanced. The second reason for the inclusion of latent variables or concepts is the very nature of the theory that is used to explain human growth. The use of SES in a theoretical model implies the use of a latent variable that has been developed and tested to ensure that it represents socio-economic status. In the next paragraphs, we will distinguish two sorts of measurement models: a more explorative procedure known as principal components analysis (PCA) and two confirmatory models that explicitly test the presence and structure of conceptual variables.

Exploratory measurement model: PCA

Principal components analysis is a mathematical technique, based on eigenvalue and eigenvector analysis, which attempts to reduce the number of variables in a data set while preserving as much information as possible. The technique allows data reduction based on grouping of correlated variables. Pragmatically, PCA is a two-step procedure. In the first step, the components (linear combinations of original variables) are derived by computing the eigenvalues and corresponding eigenvectors of the correlation matrix. The total number of components extracted equals the number of original variables. The first component contains the largest amount of information (largest eigenvalue), the last component the smallest amount. In order for the analysis to be substantially meaningful, it is common practice to maintain only those components that contain at least as much information as an average original item. In fact the eigenvalues are the variances of the components and we aim at using components that have a variance at least equal to 1, the variance of the standardized original variables. This criterion implies a cut-off value equal to 1 for the eigenvalues. Regarding the first analysis step, it is important to realize that all components are mutually uncorrelated. Once the solution is interpreted, by labelling the components using those variables that have a high loading (absolute value at least 0.50) on the respective components, one may proceed to step two. In step two, new variables (so called component scores) are computed as linear combinations of all original variables. These uncorrelated and standardized variables can now be used in further analyses. Quite often, one computes unweighted sums of those variables that have major loadings on each meaningful component. These simple sums are not sample-specific as are the component scores, which makes them 'portable' to further research projects. Before computing such simple sums, it is advised to test the homogeneity of each set of variables following Cronbach's (1951) alpha procedure (see below).

Example of a study using principal components analysis

This study was carried out within the more general framework of the project 'Relationship between biometric, socio-economic, geographical and nutritional variables in a population sample of schoolchildren from El Yopal, Casanare–Colombia – BESNE 2000–2002' conducted by two of the authors (M.-J.I and J.C.C). This study's goals are to establish the relationship between socio-economic, ecogeographical and nutritional variables, as well as to determine their impact on growth and development of children in a population sample from El Yopal, Casanare–Colombia.

Between June 2000 and November 2001 we evaluated a population sample composed of schoolchildren of both sexes, between 6 and 18 years of age, apparently healthy, and suffering neither from congenital malformations nor from metabolic or endocrinological alterations.

Socio-demographic characteristics

All the children examined were natives of the department of El Casanare, and resided within their nuclear or extended family. At least one of the parents of each child was also a native of El Casanare.

Biological data

ANTHROPOMETRIC MEASUREMENTS

- First evaluation. Between June 2000 and May 2001, we collected and classified according to sex and age groups the following anthropometric measurements in 1039 individuals (517 females and 522 males): height, sitting height, weight, arm circumference and skinfold thickness (bicipital, tricipital and subscapular skinfolds). We calculated the body mass index (BMI), the total upper arm area (TUA), the upper arm fat area estimated (UFE), the upper arm muscle area estimated (UME) and the arm fat percentage (FP) (Rolland-Cachera *et al.*, 1997).
- Second evaluation. Between June and November 2001 (more specifically, between 350 and 380 days after the first evaluation), a subsample of 496 schoolchildren (252 females and 244 males), recalled from the first evaluation's sample of 1039 individuals, were re-examined from the anthropometric point of view in order to determine the height, sitting height and weight yearly increments.

SEXUAL MATURATION STAGES In the first anthropometric evaluation, we determined the sexual development stage (i.e. breast stages 1 to 5 for girls, genital stages 1 to 5 for boys), using the Tanner–Whitehouse reference (Tanner and Whitehouse, 1976) (see also Chapter 5), and established:

1. Prepuberty stage: B1 for girls and G1 for boys.
2. Puberty stage: B2 to B4 for girls and G2 to G4 for boys.
3. Postpuberty stage: B5 for girls and G5 for boys.

RADIOGRAPHIC STUDIES Hand–wrist X-rays of the 1039 subjects were taken during the first anthropometric evaluation in order to determine the bone age according to the method of Sempé and Pavia (1979).

HAEMOGLOBIN DETERMINATION We determined the haemoglobin value in the blood samples during the first anthropometric evaluation.

Individual nutritional survey

Simultaneously, we conducted a nutritional poll in order to establish the last 24 hours recall.

Family socio-economic survey

A socio-economic poll was conducted synchronously with the first anthropometric evaluation. The children's parents, particularly the mothers, were interviewed with the purpose of determining each family's socio-economic characteristics: income, employment, nutritional expenses, parents' educational level, family structure, housing situation, social security and general living conditions.

Statistical method

We applied the LMS method (Cole, 1988) for estimating centiles, and converting measurements into standard deviation scores (Z-scores). Children were divided in three groups of pubertal stages (see above).

Using data from this large-scale survey we will, in the next paragraphs, try to build a model that relates the socio-economic status (SES) of the family to child growth data. Seven variables are considered as indicators of SES (Table 11.1).

Using 592 available cases, the principal components analysis reveals that the seven indicators cannot be treated as unidimensional. In fact we discover two dimensions in the data. Component 1 has an eigenvalue of 3.37 and contains 48% of the information in the seven variables. The eigenvalue of component 2 equals 1.38 with 20% of information. Both components together cover 68% of information. After Varimax rotation of the loadings matrix (orthogonal

Table 11.1 *Indicators of socio-economic status (SES)*

Variable name	Description	Type
MOFOEX	Monthly food expenses	Ordinal four categories
MOEXELFO	Monthly expenses elementary food	Metric: monetary value
MOEXENFO	Monthly expenses energy food	Metric: monetary value
MOEXFV	Monthly expenses fruits and vegetables	Metric: monetary value
PROFHOF	Professional class head of family	Ordinal four categories
SCHOOLM	Number of years school mother	Ordinal three categories
SCHOOLF	Number of years school father	Ordinal three categories

Table 11.2 *Principal components analysis loadings after Varimax rotation*

Variable[a]	Component 1 loading	Component 2 loading
MOFOEX	0.904	0.219
MOEXELFO	0.799	−0.044
MOEXENFO	0.798	0.231
MOEXFV	0.710	0.351
PROFHOF	0.092	0.849
SCHOOLM	0.107	0.828
SCHOOLF	0.272	0.655

[a] For definitions of variables see Table 11.1.

rotation for easier interpretation), we clearly identify the two dimensions using Table 11.2.

Component 1 can be labelled MONEYFOOD because it represents the amount of money that is spent on food categories per capita in the family. The second component describes the intellectual level of the family and is named INTELFAM. One can now proceed with new analyses since two new columns were automatically added to the data file containing the factor scores for the 592 cases. These are metric standardized variables (mean = 0 and standard deviation = 1).

Confirmatory measurement models

As mentioned above, combining (summing) correlated variables is a way to increase the reliability of the core variables in a study. Before summing correlated variables, it is advisable to compute the reliability of the construct given the data. From common practice in psychometrics, the degree of measurement error in a construct should not be larger than 30%. Cronbach's alpha coefficient (Cronbach, 1951) is the commonly used reliability/homogeneity index of a number of variables, assumed to be equally good indicators of the latent construct. As can be seen from the formula, Cronbach's alpha is a function of the size of the product–moment correlation between the variables:

$$\alpha = \frac{N\bar{r}}{1 + (N - 1)\bar{r}}$$

Where N is equal to the number of variables and $\bar{r}$ is the average intercorrelation among the variables in the set.

Alpha ranges between 0 and 1. The standard cut-off value is 0.70. This means that in such a case, the simple sum of the variables in the set (possibly divided by the number of variables) contains 70% true variance and 30% measurement error. Of course, before computing alpha, all variables in the set must be scaled in the same direction: the scores of negatively formulated items must be reversed. In the case where the variables are measured in different scales, it is advisable to standardize the variables before computing alpha and the sum scales.

Example of calculation of alpha

The alpha values for MONEYFOOD and INTELFAM are 0.73 and 0.86 respectively. This means that instead of using the PCA factor scores (weighted sums), one can also compute unweighted sums. Both simple sum scales have a satisfactory reliability.

Cronbach's alpha tests whether a series of variables cover one single concept. Sometimes however, we need a confirmatory procedure that models multiple concepts and the relationships between them. In that case one has to define a full measurement model and confront it with data. Figure 11.1 contains an example where three, potentially correlated, latent variables (ξ_1 to ξ_3) are measured by two, three and two indicators (x_1 to x_7). The measurement errors are labelled δ_1 to δ_7. The core equation for SEM measurement models is:

$$x = \Lambda_x \xi + \delta$$

where Λ_x is the matrix containing the regression coefficients from the latent variables toward the indicators. The error variance–covariance square matrix is denoted Θ_δ. In the example in Figure 11.1 the off-diagonal elements of Θ_δ are zero since the errors are assumed to be uncorrelated. The graphical symbols for latent variables are circles or ellipses, for manifest variables, square or rectangle boxes are used. Single-headed arrows represent regression effects and double-headed arrows represent covariances. In applying that model to data, we are interested in two things. First, does the model as a whole fit, and second, are all

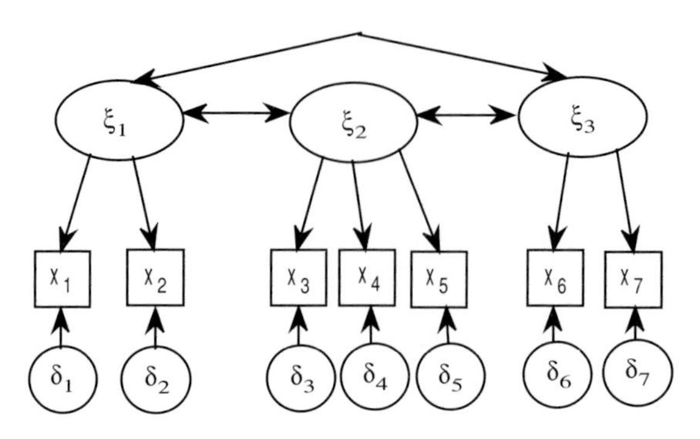

Figure 11.1 Three correlated concepts with multiple indicators.

of the regression paths from the concepts substantially large (at least 0.50) and significantly different from zero.

The overall goodness of fit can be evaluated in different ways. The first measure is the χ^2 value of the model, given the degrees of freedom. One can prove that, if the null hypothesis (stating that the model implied covariance matrix equals the observed covariance matrix) is true, the maximum likelihood based fit function follows a χ^2 distribution (under normality conditions). So if the observed χ^2 value is too large with respect to its degrees of freedom (the difference between the unique number of variances and covariances in the observed covariance matrix and the number of estimated model parameters), the model is rejected. As stated before, this approach tends often to reject well-fitting models because with a lot of cases, the observed χ^2 value is relatively large. Also commonly used with cut-off values of 0.90 are the goodness-of-fit index (GFI) and the adjusted for degrees of freedom goodness-of-fit index (AGFI). The AGFI penalizes models with too many parameters. Both measure the relative amount of the variances and covariances in the observed covariance matrix S that are predicted by $\hat{\Sigma}$, the predicted covariance matrix based on the estimates of the model parameters. AGFI rewards simpler models with fewer parameters. One further popular index of fit is the root mean square residual (RMSR), which is the average residual (not explained) correlation that results from subtracting the model implied correlation matrix from the observed one. Next to these more absolute measures of fit, a number of incremental fit indices were developed. Basically, such incremental fit indices compare the hypothesized model to a no-effect 'baseline model'. The typical baseline model for a measurement model is the situation where no constructs are behind the observed variables and that the covariances between the variables are zero in the population. Four such incremental fit measures have become standards: ρ_1, ρ_2, Δ_1 and Δ_2. It is generally accepted that all four need to be at least 0.85 in order to have an acceptable model. For definitions and characteristics see

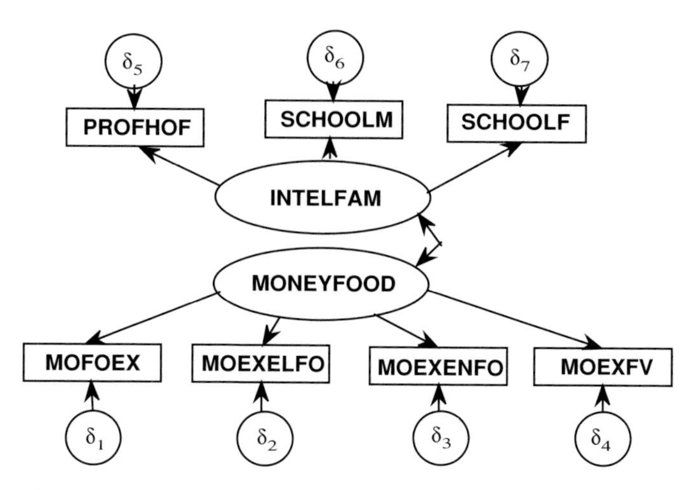

Figure 11.2 The MONEYFOOD and INTELFAM measurement model.

Bollen (1989). The next step in the evaluation process is the inspection of the regression and covariance coefficients together with their respective standard errors. By dividing each parameter by its standard error, the so-called critical ratio (t-value) is computed which should, in absolute value, be greater than 2 for significance at 0.05 Type I error level.

Example of fitting a model

The graphical representation of the measurement model for MONEYFOOD and INTELFAM is shown in Fig. 11.2.

From the results, we can conclude that this model fits the data very well. Except for the χ^2 GFI (due to large N), all fit indices are satisfactory. Every regression weight is significantly different from 0 and the correlation between the two latent variables is substantial (0.42). Table 11.3 contains the estimates. The GFI information is shown in Table 11.4.

The model is computed on a 7×7 covariance matrix containing $7 \times (7 + 1)/2 = 28$ unique parameters. The model contains 15 estimates (5 free regression weights, 9 variances: 7 observed and 2 latent, 1 covariance between INTELFAM and MONEYFOOD), which implies 13 degrees of freedom.

Structural models

With structural equation modelling (SEM), one can estimate so-called structural or path models with or without latent variables. Figures 11.2 and 11.3 are examples of typical models applied on growth data. Figure 11.3 could be

Table 11.3 *Parameter estimates for the measurement model*

Effect	Estimate[a]	Standard error	Critical ratio	Standardized estimate
SCHOOLF ⟵ intelfam	1.000*			0.769
SCHOOLM ⟵ intelfam	0.957	0.077	12.401	0.737
PROFHOF ⟵ intelfam	0.736	0.067	11.028	0.557
MOFOEX ⟵ moneyfood	1.000*			0.981
MOEXELFO ⟵ moneyfood	15941.762	681.419	23.395	0.765
MOEXENFO ⟵ moneyfood	7815.596	441.199	17.714	0.633
MOEXFV ⟵ moneyfood	12506.889	617.905	20.241	0.695
INTELFAM ⟷ moneyfood	0.158	0.020	7.871	0.423

[a] Parameters with an * denote fixed values in order 'to set the metric' and for identification purposes.

Table 11.4 *Goodness-of-fit values for the measurement model*

Goodness-of-fit index	Value	
χ^2, 13 df	82.949	$p = 0.000$
GFI	0.963	
AGFI	0.921	
Δ_1	0.951	
ρ_1	0.921	
Δ_2	0.958	
ρ_2	0.933	

the graphical representation where the impact of socio-economic status of the household (ξ_1) on length at birth (y_1) and length at childhood (η_1) is estimated. Length at childhood is considered a latent variable with two indicators: length at age 3 years and 6 years. A more complex model on this subject containing more variables and concepts can be found in Hauspie *et al.* (1996). Figure 11.4 show a model with a path of five manifest variables with correlated predictors. In fact, this path model represents a multivariate regression model assuming uncorrelated residuals (ςs). This is a typical model that one would use to go one step beyond multiple regression because of three simultaneous dependent variables and correlated predictors. In general, a SEM with latent endogenous and exogenous variables is specified by three equations: two measurement models and the structural equation.

Measurement model for the exogenous variables is

$$x = \Lambda_x \xi + \delta$$

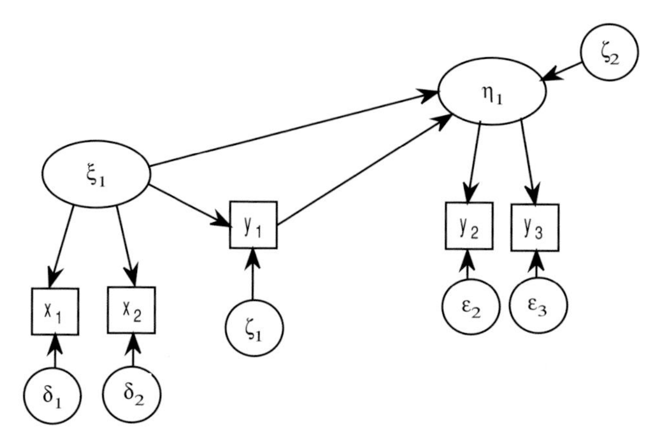

Figure 11.3 Path diagram with latent and manifest variables.

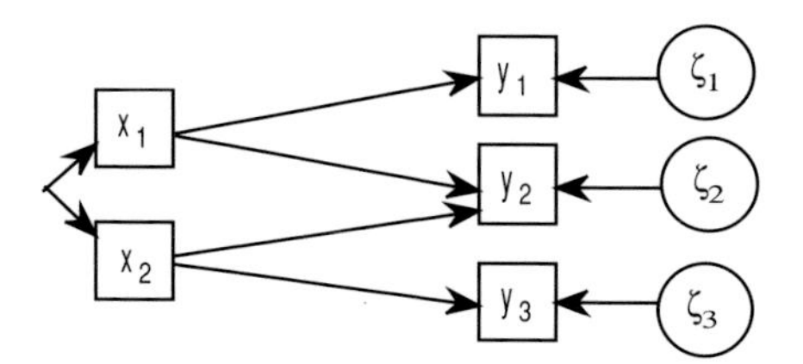

Figure 11.4 Path diagram with correlated predictors.

and the measurement model for the endogenous variables is

$$y = \Lambda_y \eta + \varepsilon$$

In analogy with the measurement model for the exogenous variables, ε contains the measurement errors of the y variables.

The structural model is

$$\eta = \beta\eta + \Gamma\xi + \varsigma$$

In the structural model the ηs are the endogenous latent variables and the ξs are the exogenous latent variables. The regression weights from ξ on η are present in the Γ matrix and β contains the regression coefficients from η on η. In the case of a path model without latent variables, only the structural equation is needed where η is substituted by y and ξ by x. The regression error is contained in ς. So the presence of η on both sides of the equation means that some endogenous variables may explain other endogenous variables (e.g. mediating variables).

Table 11.5 *Exogenous variables*

Variable	Description	Type of variable	Alpha value
DAYFOEX	Daily food expenses for each member of the family	Sum scale	0.8437
SES	Socio-economic status (employment status, income, education)	Sum scale	0.8720
MFAFOE	Monthly family food expenses	Sum scale	0.7573
HEALTH	Father's, mother's and child's health insurance	Sum scale	0.9359
HABITAT	Housing conditions	Sum scale	0.6003
RURURB	Rural or urban tendency (place of birth and residence, duration) (1 = urban, 2 = in between, 3 = rural)	Ordinal three categories	—

Table 11.6 *Expectation maximization correlation matrix for the eight variables*

	DAYFOEX	SES	MFAFOE	HEALTH	HABITAT	RURURB	ZS.HEI	ZS.SITH
DAYFOEX	1.000							
SES	0.601	1.000						
MFAFOE	0.517	0.306	1.000					
HEALTH	0.020	0.045	0.071	1.000				
HABITAT	0.485	0.471	0.215	−0.180	1.000			
RURURB	−0.034	0.116	−0.095	0.127	−0.295	1.000		
ZS.HEI	0.147	0.319	0.268	0.159	0.285	−0.117	1.000	
ZS.SITH	0.034	0.340	0.207	0.182	0.263	−0.153	0.720	1.000

Note: For definition of the first six variables, see Table 11.5. ZS.HEI, *Z*-score height; ZS.SITH, *Z*-score sitting height.

Example of a structural model

On a subset of 82 older boys (postpuberty), a model is estimated that tries to explain differences in height as a latent variable with two indicators (*Z*-scores of height and sitting height) using a number of sum scales regarding environmental and family characteristics. Table 11.5 describes the exogenous variables.

Because some data are missing, the input data for this model is the expectation maximization (EM) correlation matrix in Table 11.6. As can be seen from Tables 11.7 and 11.8, the model fits the data very well. One might rerun the model omitting the non-significant coefficients.

This model uses both height and sitting height as indicators of height as a concept (Fig. 11.5). In the model, se3 is the regression residual while e6 and e7 represent measurement errors. Again, some values are fixed to 1 for identification

Table 11.7 *Estimates for the SEM height model*

Effect	Estimate	Standard error	Critical ratio	Standardized estimate
HEIGHT ⟵ DAYFOEX	−0.095	0.031	−3.079	−0.475
HEIGHT ⟵ SES	0.099	0.031	3.233	0.477
HEIGHT ⟵ MFAFOE	0.081	0.037	2.213	0.263
HEIGHT ⟵ HEALTH	0.271	0.123	2.208	0.232
HEIGHT ⟵ HABITAT	0.086	0.047	1.859	0.244
HEIGHT ⟵ RURURB	−0.153	0.102	−1.495	−0.165
ZS.HEI ⟵ HEIGHT	1.000			0.790
ZS.SITH ⟵ HEIGHT	1.237	0.207	5.967	0.934
DAYFOEX ⟷ SES	6.564	1.439	4.561	0.603
DAYFOEX ⟷ MFAFOE	3.625	0.934	3.880	0.496
DAYFOEX ⟷ HEALTH	0.034	0.218	0.157	0.018
DAYFOEX ⟷ HABITAT	3.004	0.802	3.746	0.471
DAYFOEX ⟷ RURURB	−0.097	0.275	−0.352	−0.040
SES ⟷ MFAFOE	2.100	0.826	2.542	0.299
SES ⟷ HEALTH	0.082	0.207	0.395	0.044
SES ⟷ HABITAT	2.853	0.756	3.774	0.467
SES ⟷ RURURB	0.268	0.262	1.025	0.115
MFAFOE ⟷ HEALTH	0.087	0.140	0.620	0.070
MFAFOE ⟷ HABITAT	0.805	0.474	1.698	0.196
MFAFOE ⟷ RURURB	−0.158	0.177	−0.893	−0.101
HEALTH ⟷ HABITAT	−0.199	0.123	−1.612	−0.183
HEALTH ⟷ RURURB	0.053	0.047	1.132	0.127
HABITAT ⟷ RURURB	−0.412	0.159	−2.586	−0.301

Table 11.8 *Goodness-of-fit values for the height SEM model*

Goodness-of-fit index	Value	
χ^2, 5 df	3.081	$p = 0.688$
GFI	0.990	
AGFI	0.932	
Δ_1	0.964	
ρ_1	0.801	
Δ_2	1.023	
ρ_2	1.183	

purposes. The R^2 for height equals 0.357. As can be seen in Table 11.7, both indicators of height perform well (large standardized estimates). Some of the exogenous variables are strongly correlated such as SES and daily per capita food expenses in the family (DAYFOEX). From the model we can conclude that for these boys height is negatively influenced by daily per capita food expenses. We have positive regressions from SES, monthly family level food expenses and

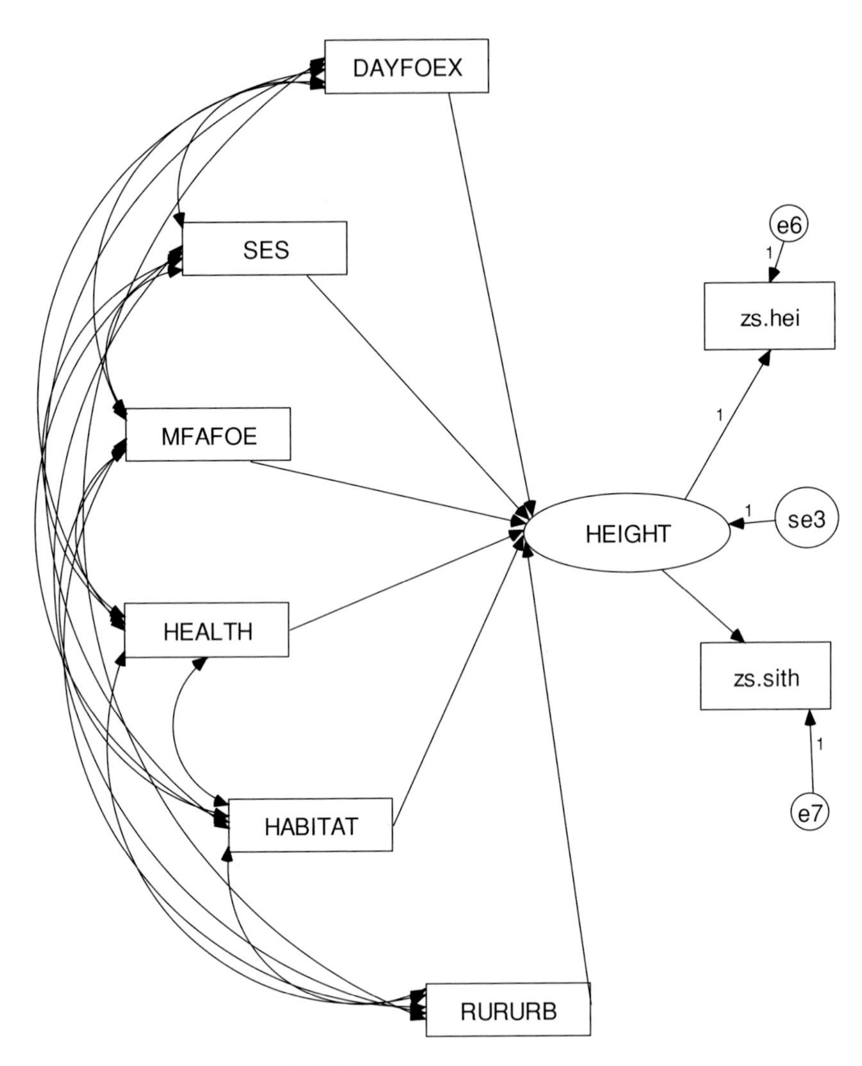

Figure 11.5 SEM for height.

health assurance levels. Housing conditions are marginally positively related and rural or urban conditions cannot explain variance in height.

Data requirements

Since the core of SEM is covariance structure modelling and the most used estimation algorithm, maximum likelihood, requires (in theory) multivariate normal data, the models are not suited for categorical data. However, some authors do include binary predictor variables in the models. Using a more econometric 'dummy coded' perspective, this means for example that gender may be included as an exogenous variable. An alternative for categorical variables may be found in the multiple group models. In this case, a separate model is estimated for each value of the categorical variable. This implies a more complicated approach but is very rich in the sense that for boys and girls in the gender case, two models are estimated simultaneously and can be compared in an explicit way. So, in practice, SEM covariance modelling allows binary, ordinal and metric data.

All commercial software for SEM uses raw data or covariance (or correlation) matrices as input. If a covariance matrix is available, its use is to be preferred to the correlation matrix because it contains the original metrics and allows a more objective means of theory testing by modelling (Hair *et al.*, 1998). For practical purposes, that is, analysing patterns of relationships, the standardized covariance matrix or correlation matrix is suitable. Some packages (e.g. PRELIS: Jöreskog and Sorböm, 1988) allow the estimation of special correlation measures better suited for handling combinations of binary, ordinal and metric data. This means that the correlation matrix to be used as input for SEM contains different types of correlation coefficients: Pearson product–moment correlation (two metric variables), polychoric correlation (both variables are ordinal with at least three categories), tetrachoric (both non-metric measures are binary), polyserial (metric variable with ordinal variable) and biserial (binary with metric). This combination of different correlation measures in one SEM input matrix may yield unstable correlation structures resulting in non-convergence of the estimation algorithm. In order to compute such mixed-type correlation matrices, one needs a data file with no missing values because these methods have been developed for complete data sets. This means that the data set is completely observed for all variables, or that missing values have been replaced in a proper manner (see below).

Any multivariate statistical model requires 'enough data' to ensure reliable estimation of the parameters and the associated standard errors. The usual observed cases/parameters ratio is 10. This means that for each parameter in the model, one needs 10 independent observations. In the literature on sample size requirements for proper modelling, some authors advise to limit the samples to about 200 cases. This 'critical sample size' (Hoelter, 1983) deals with the problem that one encounters when increasing the sample size in order to improve the estimation: as the sample size increases, the probability that we will reject an essentially correct model increases as well, that is, if we use the 'classic' goodness-of-fit χ^2 test. Even the smallest difference between the model-implied covariance structure and the observed one leads to model rejection when there is a lot of data at hand.

As mentioned before, the parameter estimation procedure based on maximum likelihood, requires multivariate normal data. A number of performance studies indicate that, if the sample size is large enough, the maximum likelihood estimation procedure is rather robust against departures from normality. If there are no extreme sharp and/or asymmetric distributed variables to be modelled, but the marginal distributions are not Gaussian, the increase of the cases/parameters ratio to 15, is advised (Boomsma, 1983; Wang *et al.*, 1996).

The collection of growth data by means of surveys and panel studies faces us with problems of missing data. When some cases contain missing data, the researcher needs to pay attention to those. Naive use of SEM software in the presence of missing data may yield biased results (Little and Rubin, 1987; Verleye, 1999). As research shows, it is not so much the amount of but the nature of the missing data that is important in dealing with it. Even if this chapter is not mainly concerned with missing data in multivariate modelling, we will give this topic the attention that is needed in order to achieve proper modelling. We will not consider the case where individuals do not 'deliver' any data at all. Such issues on missing data are treated in works on survey methodology and longitudinal research and require special attention, such as weighting and attrition modelling (Groves *et al.*, 2002). Data analysis procedures that help the researcher to discover the nature of the item missing data process are mainly derived from deeper analyses between (and within) dichotomized variables ($0 =$ observed, $1 =$ missing) and the original observed part (Hair *et al.*, 1998).

In order to make the correct choices for the handling of item non-response in covariance structure modelling, one needs to find answers on the nature of the missing data process. Basically, three processes are possible.

If the 'missingness' (the missing pattern and process) data is 'missing completely at random' (MCAR), as if the raw data matrix was repeatedly hit by a dart, almost every statistical solution yields proper results. List-wise deletion (delete the case if it contains missing data), and pair-wise estimation of the covariance matrix (use all available data for each variable pair) both work. However, we would still advise using more complex techniques such as conditional mean (or regression-based) imputation (replace the missing observations by an estimate given the observed data part), or the EM-based covariance matrix estimation (Schafer, 1997). This latter procedure yields a precise covariance or correlation matrix between the variables in the presence of missing data using the EM algorithm. This algorithm is an important general technique for finding maximum likelihood estimates from incomplete data (Little and Rubin, 1987).

In the case of a 'missing at random' (MAR) process, the missing values of y depend on x, but not on y. This means that the observed part of y constitute a random sample of the y values for each x value, but the observed y part does not necessarily represent a full random sample of all y values. The problem here is that the missing data process is random in the sample but the values (and therefore all inferences) are not generalizable to the population. For example, we fully know the age group of the children (x) but there is missingness in the weight variable (y). We may find that the missing data are random for (within) all groups but occur at a much higher frequency for the oldest age group. In this case, 'passive' missing data handling techniques such as list-wise deletion and even pair-wise deletion cannot be used. Proper imputation techniques (Bayesian data augmentation, hot deck, etc.), EM covariance matrix estimation and full-information maximum likelihood as available in the AMOS SEM software (Arbuckle, 1995) are valid solutions (Verleye, 1999). Bayesian data augmentation is an imputation technique that replaces the missing values through simulation and estimates the missing value using the redundancy between the variables. Hot deck, in general, searches for a case, which is very similar to the case containing missing data, and copies the observed part of that similar case to fill in the missing part.

If the probability of having missing data is related to the values of that same variable, the missing data process becomes non-ignorable. If nutrient data are less available for those children that are poorly fed, a more algebraic formulation of that process needs to be added to the SEM model specification. The uncertainty about the exact nature of the process behind the data makes simulation a necessity (Muthén and Jöreskog, 1983; Rubin, 1987).

Remarks and recommendations

Although SEM allows very complex modelling, we do have to advise that it is better to develop parsimonious models that explain as much as possible with fewer 'arrows'. Practices such as defining correlated errors in the measurement part of a model need a theoretical foundation. Of course, as we add regression paths or covariances in a model, several indices of fit will increase but only due to relaxation of the initial model, merely eating degrees of freedom. So, it is good practice to keep an eye on the parsimonious goodness-of-fit index (PGFI).

Together with the first remark, we argue for a validation of the models obtained from a 'data-driven' analysis. Quite often, the SEM data analyst starts with a theoretical model, which shows an average to bad fit. Using so-called modification indices (estimated drop in χ^2 GFI when adding a path or covariance) one develops a model that is more in line with the data at hand. The only solution to cope with data-driven development instead of a more theory-driven approach is a random split of the original data set into two parts: a development part and a validation part. The latter will serve for a final test of the model derived on the development data set.

An extra note on the possibilities of SEM with multiple groups is also worth mentioning: using such a multiple group approach, we can explicitly test the equivalence of a model for different groups. In the case of the 'Yopal' data analysis with SEM, a model can be derived that shows different strengths of relationships from SES variables with respect to growth and development variables for boys versus girls, or between age groups within gender.

Finally, after using SEM together with more classic analysis procedures such as regression models, we can only advise the reader to learn the massive possibilities of SEM by comparing it to these more traditional procedures. As soon as one is able to produce a regression analysis with SEM, it will be very tempting to explore models beyond standard regression. We do hope that this chapter has helped auxologists to understand the potentials of SEM for fully exploring the interrelationships between biometric variables and genetic and environmental factors. As mentioned in the initial paragraph, social scientists have adopted SEM as a powerful data analysis tool that allows explicit confrontation of theory with data on a conceptual level. We strongly believe that this added value is also interesting for biometric data analysis. Among many interesting books on SEM, we specially recommend Hayduk (1987), Bollen (1989) and Hair *et al.* (1998, especially Chapter 11).

REFERENCES

Arbuckle, J. (1995). *Full Information Estimation in the Presence of Incomplete Data.* Philadelphia, PA: Department of Psychology, Temple University.

Bollen, K. A. (1989). *Structural Equations with Latent Variables.* New York: Wiley Interscience.

Boomsma, A. (1983). On the robustness of LISREL (maximum likelihood estimation) against small sample size and non-normality. Ph.D. thesis, University of Groningen, The Netherlands.

Cole, T. J. (1988). Fitting smoothed centile curves to reference data. *Journal of the Royal Statistical Society A*, **151**, 385–418.

Cronbach, L. J. (1951). Coefficient alpha and the internal structure of tests. *Psychometrica*, **16**, 297–334.

Groves, R. M., Dillman, R. M., Eltinge, J. L. and Little, R. J. (2002). *Survey nonresponse.* New York: John Wiley.

Hair, J. F., Anderson, R. E., Tatham, R. L. and Black, W. C. (1998). *Multivariate Data Analysis.* Englewood Cliffs, NJ: Prentice Hall.

Hauspie, R. C., Chrzastek-Spruch, H., Verleye, G., Kozlowska, M. A. and Susanne, C. (1996). Determinants of growth in body length from birth to 6 years of age: a longitudinal study of Lublin children. *American Journal of Human Biology*, **8**, 21–9.

Hayduk, L. A. (1987). *Structural Equation Modeling with LISREL: Essentials and Advances.* Baltimore, MD: Johns Hopkins University Press.

Hoelter, J. W. (1983). The analysis of covariance structures: goodness-of-fit indices. *Sociological Methods and Research*, **11**, 325–44.

Jöreskog, K. and Sorböm, D. (1988). *PRELIS Software Manual*. Mooresville, IN: Scientific Software.

Little, R. J. and Rubin, D. B. (1987). *Statistical Analysis with Missing Data*. New York: John Wiley.

Muthén, B. and Jöreskog, K. (1983). Selectivity problems in quasi experimental studies. *Evaluation Review*, **7**, 139–74.

Rolland-Cachera, M. F., Brambilla, P., Manzoni, P., *et al.* (1997). Body composition assessed on the basis of arm circumference and triceps skinfold thickness: a new index validated in children by magnetic resonance imaging. *American Journal of Clinical Nutrition*, **65**, 1709–13.

Rubin, D. B. (1987). *Multiple Imputation For Nonresponse in Surveys*. New York: John Wiley.

Schafer, J. L. (1997). *Analysis of Incomplete Multivariate Data*. London: Chapman and Hall.

Sempé, M. and Pavia, C. (1979). *Atlas de maturation squelettique du poignet et de la main*. Paris: Simep/Masson.

Tanner, J. M. and Whitehouse, R. H. (1976). Clinical longitudinal standards for height, weight, height velocity, and stages of puberty. *Archives of Disease in Childhood*, **51**, 170–9.

Verleye, G. (1999). Missing data in linear modeling. *Journal of the Dutch Statistical Society* (Special issue on missing data), **62**, 95–110.

Wang, L. L., Fan, X. and Wilson, V. L. (1996). Effects of non-normal data on parameter estimates for a model with latent and manifest variables: an empirical study. *Structural Equation Modelling*, **3**, 228–47.

12 *Multilevel modelling*

Adam Baxter-Jones
University of Saskatchewan

and

Robert Mirwald
University of Saskatchewan

Introduction

The purpose of this chapter is to assist the practitioner in the use and interpretation of multilevel models in the context of human growth research. The basic concepts of multilevel modelling are discussed and illustrated using a practical example. For detailed technical statistical discussions of multilevel modelling the reader is directed elsewhere (Goldstein, 1995; Kreft and de Leeuw, 1998; Snijders and Bosker 1999).

As we know human physical growth is a highly regulated process. From conception to full maturity the change in size and shape is a continuous process. Many attempts have been made to find mathematical curves that can fit, and thus summarize, the process of human growth. There is considerable literature on the analysis of longitudinal growth data both for linear (Vandenberg and Falkner, 1965; Berkey and Reed, 1987) and non-linear (Jenss and Bayley, 1937; Preece and Baines, 1978) parametric models. Adjusting a mathematical model to a set of growth data is called growth curve fitting or growth modelling. Such growth models have had variable success in describing the pattern of human growth depending on the type of growth variable used, the precision of the measurement, the frequency and age range of the observations and the ability of the model to describe the growth curve (Karlberg, 1998). At the individual level what is required is a curve with relatively few variables, each capable of being interpreted in a biological meaningful way (Tanner, 1989). It is essential that any models used to identify the effects of covariates (e.g. nutrition, physical activity,

Methods in Human Growth Research, eds. R. C. Hauspie, N. Cameron and L. Molinari.
Published by Cambridge University Press.

etc.) on growth incorporate an individual's growth trajectory. Specifically, each individual must have their own growth line; in statistical terms this means that each individual's growth curve has its own intercept and slope. The uniqueness of this approach is that it captures, mathematically, the wide variation shown amongst children's growth parameters at any given age, and in the velocity, or rate of change of these parameters from one age to the next (Tanner, 1962).

Traditionally, the mathematical relationship between a dependent measure and an independent measure of growth has been described using regression equations. For inter-group comparisons analysis of covariance (ANCOVA) is usually used. However, ANCOVA cannot provide or indicate which growth or maturation characteristic(s) within an individual explains the differences between individuals. Repeated-measures analysis of variance (ANOVA), as implemented in SPSS or SAS, is also used to analyse growth data; however, a major disadvantage of this technique is that time-dependent independent covariates cannot be fitted (SAS, 2004; SPSS, 2004). Another disadvantage of using repeated-measures ANOVA is that cells have to be balanced. This is a problem as missing measurement occasions often occur in longitudinal growth studies. The development of multilevel statistical models (Goldstein, 1995), specifically random-effects models and the accompanying computer software (Rasbash *et al.*, 1999), has enabled the incorporation of individual growth characteristics into an analysis of covariates. In these models time-dependent covariates can be fitted. Another advantage of multilevel models is that missing data do not constitute a problem, as long as they are missing at random. In the multilevel model each individual has their own growth trajectory, with intercepts and slope coefficients varying between individuals. Using multilevel models, it is possible to identify independent inter-group effects while simultaneously controlling the effects of growth and maturation within each individual.

Example data

In this chapter we will illustrate the multilevel modeling technique using a data set that investigated the accrual of total body bone mineral content (TBBMC) in normal healthy boys and girls (Baxter-Jones *et al.*, 2002). Subjects were part of the Saskatchewan Bone Mineral Accrual Study (1991–7), which has been described in detail elsewhere (Bailey, 1997). In brief, the study utilized a mixed longitudinal design and incorporated eight age cohorts. At study entry boys and girls were aged between 8 and 15 years (Table 12.1). During seven years of annual data collection the composition of these clusters remained the same.

Table 12.1 *Number of subjects by test year, age group and sex. Initially subjects were recruited using eight age cohorts*

Age group (years)	Test years							Total
	1991	1992	1993	1994	1995	1996	1997	
8	3 (**7**)							3 (**7**)
9	10 (**15**)	3 (**9**)						13 (**24**)
10	19 (**16**)	10 (**18**)	3 (**10**)					32 (**44**)
11	17 (**12**)	18 (**15**)	12 (**18**)	3 (**9**)				50 (**54**)
12	21 (**18**)	17 (**12**)	16 (**15**)	12 (**14**)	3 (**10**)			69 (**69**)
13	17 (**21**)	21 (**16**)	17 (**11**)	14 (**16**)	11 (**14**)	3 (**10**)		83 (**88**)
14	17 (**16**)	15 (**20**)	20 (**15**)	14 (**11**)	14 (**16**)	10 (**12**)	3 (**9**)	93 (**99**)
15	3 (**8**)	17 (**15**)	14 (**20**)	14 (**15**)	12 (**11**)	10 (**16**)	8 (**8**)	78 (**93**)
16		3 (**7**)	17 (**14**)	13 (**12**)	13 (**14**)	11 (**7**)	8 (**11**)	65 (**65**)
17			3 (**7**)	13 (**11**)	13 (**12**)	12 (**12**)	9 (**6**)	50 (**48**)
18				3 (**6**)	13 (**8**)	12 (**11**)	8 (**9**)	36 (**34**)
19					3 (**6**)	8 (**6**)	7 (**8**)	18 (**20**)
20						4 (**3**)	5 (**3**)	9 (**6**)
21							2 (**2**)	2 (**2**)
Total	107 (**113**)	104 (**112**)	102 (**110**)	86 (**94**)	82 (**91**)	70 (**77**)	50 (**56**)	601 (**653**)

n, subjects boys (**girls**).

As there were overlaps in ages between the clusters it was possible to estimate a consecutive 13-year developmental pattern (8 to 21 years of age) over the shorter period of seven years (Table 12.1). To be included in the multilevel model analyses each child required complete measurements of age, height, lean and fat mass, TBBMC, and physical activity at more than one measurement occasion. In addition, all were required to have experienced peak height velocity (PHV). Age at peak height velocity (APHV) is the most commonly used indicator of somatic maturity (biological age) in longitudinal studies of adolescences (Malina, 1978). APHV is the age when the maximum velocity in stature occurs. In girls PHV occurs at approximately 12 years (range 9.5–14.5 years) and in boys 14 years (range 10.5 to 17.5 years) (Tanner, 1989). Eighty-five boys and 67 girls fulfilled the inclusion requirements and comprise our data set for analyses; Figure 12.1 presents a scatter plot of TBBMC against biological age (years from APHV) for all the measurements (862 TBBMC scans) from the 152 children. Within the scatter plot individuals are represented on more than one occasion. Data is plotted against biological rather than chronological age to account for the known effects of pubertal development on bone mineral accrual (Bailey, 1997). From the scatter plot it can be seen that TBBMC increases with increasing biological age. The scatter plot also suggests that the rate of accrual is greater post-PHV compared to pre-PHV.

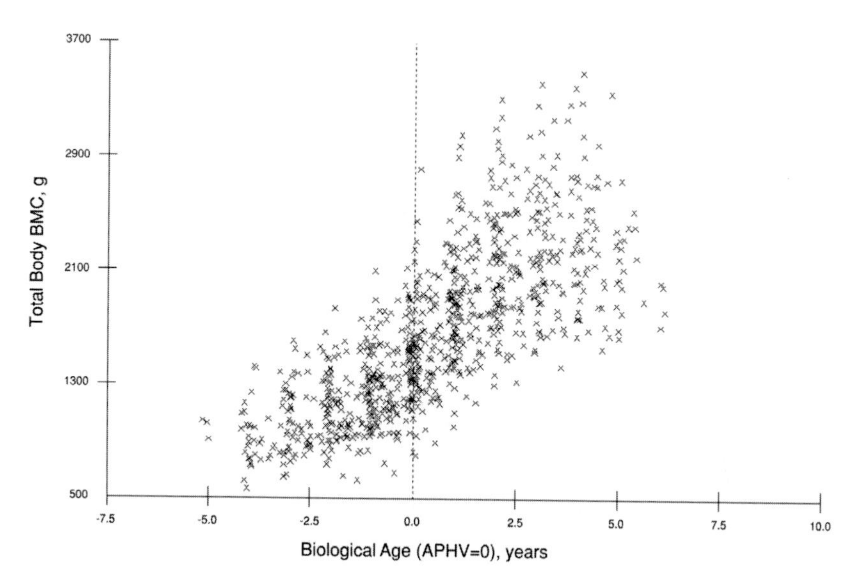

Figure 12.1 Total body bone mineral content (TBBMC) plotted against biological age (years from age at peak height velocity (APHV)) for 862 bone scans from 85 boys and 67 girls.

Cross-sectional versus longitudinal data

In cross-sectional data, only a single dependent variable, y_i ($i = 1, 2, \ldots, n$), is available for each of the n experimental units (e.g. an individual). In contrast longitudinal data are comprised of repeated observations within an individual. The dependent variable, y, therefore has two subscripts, y_{ij} ($i = 1, 2, \ldots, m$; $j = 1, 2, \ldots, k$), subscript ij now refers to measurement occasion i in individual j. Longitudinal data require special statistical methods because the set of observations within an individual (j) tend to be intercorrelated. These correlations must be taken into account to draw valid scientific inferences. A longitudinal study is defined as a study in which variables for each subject are observed on two or more occasions of measurement, thus necessitating the appropriate treatment of correlated data.

The major advantages of a longitudinal study over a cross-sectional study are that they: (1) allow for the assessment of individual variability; (2) provide distance, velocity and acceleration data; (3) allow for the assessment of the timing of increments; and (4) allow for the study of the process of growth. In our example data we want to estimate both the contribution of fixed covariables (e.g. height, lean and fat mass, and physical activity) and random covariable(s) (biological age) to investigate the relationship of TBBMC accrual over time within and between individuals. We want to separate differences among individual children at different ages at a fixed time (cross-sectional) from changes in children over time (longitudinal). The goals of this particular longitudinal growth research are: (1) to characterize patterns of individual's TBBMC accrual over 'time', and (2) to investigate the effects of predictor variables (e.g. sex, biological age, height, lean and fat mass, and physical activity) on these patterns of TBBMC accrual.

Multilevel data structure

To perform a multilevel analysis longitudinally collected data is required; specifically, a repeated-measures study design. This design can be either a true, or a mixed, longitudinal design. A true longitudinal study would study individual's growth from birth to young adulthood. A mixed-longitudinal study is basically a compromise of cross-sectional and longitudinal designs. In the mixed-longitudinal study groups of children are selected into the study at specific ages (age cohorts). Each child is then followed longitudinally at regular intervals and during the course of the study the age cohorts overlap. With special statistical treatment data is fitted between cohorts and thus the whole growth period is covered in a much shorter time period. Our example study used a mixed-longitudinal design, the composition of the age cohorts and the overlapping between age groups over time is illustrated in Table 12.1. The change in *n* within age cohorts illustrates drop-out and retention rates. Since this is a longitudinal study it is possible for subjects to have missed a measurement in one year but to have been measured again in the following year.

Many kinds of data, including repeated-measures data, have a hierarchical or clustered structure. By definition a hierarchy consists of units grouped at different levels. In studies of human growth, the measurement occasions are the level-1 units in a two-level structure, where the level-2 units are the individuals themselves. Figure 12.2 illustrates the clustered structure of TBBMC measurements overtime within individuals. TBBMC measurements (level-1 units) are clustered within individuals (level-2 units); each line in Figure 12.2 represents an individual's growth in TBBMC with increasing biological age. From all these individual lines we can construct an average line to describe TBBMC growth for the group. The mixed-longitudinal study design is illustrated by the fact that individuals' growth lines start and end at different biological ages. The clustering of data points within an individual is an important statistical concept because there is much more variation between individuals, in general, than between occasions within individuals. For example, in the case of children's statural growth once the overall trend with age has been adjusted, the variance

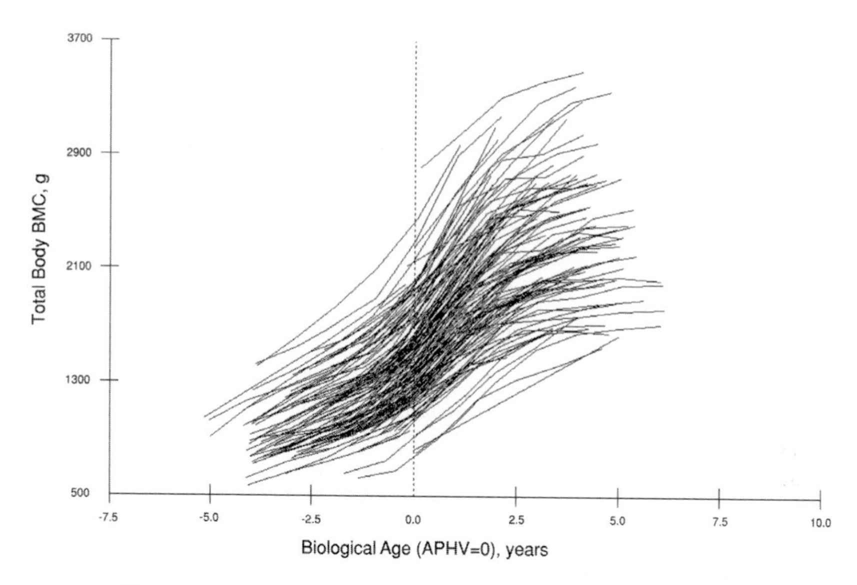

Figure 12.2 Total body bone mineral content (TBBMC) plotted against biological age (years from age at peak height velocity (APHV)). Each line represents an individual's repeated growth trajectory.

between successive measurements on the same individual is generally no more than 5% of the variation in height between children (Goldstein, 1989).

Multilevel data layout

In statistical packages such as SPSS or SAS data are typically stored with each row of data representing an individual, and the repeated measurements as a series of variables within the row. The data structure for a multilevel analysis is generally different. Multilevel models require data to be structured so that each row represents an individual's measurements on a single assessment occasion. Thus each individual will appear in more than one row. For the modelling procedure it is important that each subject has unique identifiers at both level-1 (within subject) and level-2 (between subject). A description of our example data layout is given in Table 12.2. The measurement occasion (sequence number [seq. no.]) is the unique identifier for level-1 units and subject ID is the unique identifier for level-2 units. Each individual (j) has to have complete measurements (y and x_{1-p}) at one or more occasions (i) to be included in the analysis. It is essential that the time period between measurement occasions (i) is the same. In this case measurements were taken approximately 12 months apart.

Table 12.2 *Example of longitudinal data layout for multilevel analysis*

			Independent variables						
Subject ID	Sequence number	Dependent variable y (TBBMC, g)	x_1 (Biological age, years)	x_2 (Sex, 0 = boy, 1 = girl)	x_3 (Height, cm)	x_4 (Lean mass, g)	x_5 (Fat mass, g)	:	x_p (Physical activity, score)
1	1	911.7	−4.95	0	131.9	23688.4	11898.3	:	3.58
1	2	1103.9	−4.07	0	137.3	24909.9	16195.1	:	3.08
:	:	:	:	:	:	:	:	:	:
1	7	2254.2	0.99	0	167.7	54487.1	26775.8	:	2.05
2	1	1525.6	0.37	1	157.1	31955.5	11539.9	:	2.15
2	2	1666.3	1.36	1	162.0	33315.9	12230.5	:	2.35
:	:	:	:	:	:	:	:	:	:
2	7	1974.8	4.30	1	165.2	36173.2	15879.5	:	2.30
:	:	:	:	:	:	:	:	:	:
:	:	:	:	:	:	:	:	:	:
j	i	y_{ij}	x_{1ij}	x_{2ij}	x_{3ij}	x_{4ij}	x_{5ij}	:	x_{pij}

Notes: Subject ID is the unique individual identification number.
Sequence number (Seq. No.) is the unique repeated measurement sequence identification number (1 to 7).
j is the individual identifier (values 1, 2, . . . , 152).
i is the measurement occasion (values 1, 2, . . . , 7).
y_{ij} is the dependent variable for measurement occasion i in individual j; y = total body bone mineral content (TBBMC) in g.
x_{ij} the independent variable ($x_1, x_2, \ldots, x_p$) at measurement occasion i in individual j; x_1 = Biological age (years from age at peak height velocity (APHV)), x_2 = Sex (0 = boy, 1 = girl), x_3 = Height (cm), x_4 = Lean mass (g), x_5 = Fat mass (g) and x_p = Physical activity (1 = low, 5 = high).
Source: Data from Saskatchewan Bone Mineral Accrual Study (Bailey, 1997).

Multilevel Analysis

Analysis models that contain variables measured at different levels of a hierarchy are known as multilevel models. Multilevel regression models differ from conventional regression models in that as well as modelling location parameters they also model the underlying covariance structure of the data.

In the context of linear modelling, the terms fixed and random traditionally apply to three different entities: random or fixed effects, random or fixed variables, and random or fixed coefficients (Kreft and de Leeuw, 1998). Fixed and random effects are concepts commonly used in experimental research where treatments and treatment groups are involved, where data are analysed using analysis of variance. If a treatment(s) or a placebo is applied to a thoroughly controlled laboratory experiment, treatment(s) form a fixed factor. If however, the treatment is applied to a sample, chosen from the population at random, then other factors are likely to influence the treatment. In this case the treatment effect has to be viewed as being random instead of fixed. We would expect more or less different results, due to measurement error, etc., if the experiment were to be repeated in another sample chose at random from the same population.

Fixed and random are also used as prefixes for variables. Again the concept of measurement error applies. A random variable has an expected value (the mean) and a variance. In general, we assume that a random variable is measured with error and differs from measurement to measurement. In our example data, sex would be considered a fixed variable and height a random variable.

In this chapter we use the terms 'fixed' and 'random' to characterize the linear model parameters (the regression coefficients). In traditional regression models the parameters, coefficients for intercept and slope(s), are assumed to be fixed. The variance is therefore dependent on these fixed factors and takes the same value at any point on the regression line. Random coefficients are coefficients where the values are assumed to be distributed according to a probability function (Kreft and de Leeuw, 1998). When variables are allowed to be random within a model, variance will not take the same value at all points on the regression line but will be dependent on the value of the random variable.

In our example data set when TBBMC is repeatedly measured annually for seven consecutive years (Table 12.2; Figures 12.1 and 12.2), if a traditional regression model is fitted all growth effects, changes in biological age, height, and fat and lean mass, are regarded as fixed. The only source of random variation in this model occurs within individuals, no random variation occurs between them. With this type of model it is only possible to show that the growth terms explain how things are. The growth terms only explain group effects, they do not explain why things are, i.e. you can not determine the independent effects of these variables on your outcome variable at each time point. The simplest way to address the why question is to specify random coefficients into the model. This gives rise to random variation in addition to the residual variation and allows for independent variables to have time-dependent effects.

In our example, biological age (our time variable) can be made random at level-2 of the model, thus the variance of TBBMC accrual over increasing biological age will be estimated in two parts. The model now contains two levels of variance (level-1 within individual and level-2 between individual). Level-1 variance is the variance associated with an individual's regression line of TBBMC development on biological age. The second part of the variance in the model (level-2) is variance representing the deviation of each individual's line from the average line for the whole group (Figure 12.2). Multilevel models allow individuals to deviate from the mean solution, either in the intercept or the slope of their TBBMC growth lines. In practice what this means is that each of our individuals has the same predictor variables (sex, biological age, height, lean and fat mass and physical activity) and the same outcome variable (TBBMC), but each individual has different regression coefficients. The individuals' TBBMC growth models are linked together by a level-2 model, in which the regression coefficients of the level-1 models are regressed on the level-2 independent predictor variables.

It is this linking characteristic that determines the nature of the model for the data under study. Within the multilevel model it is possible to specify all the regression coefficients to be fixed (similar to traditional regression), with no random effects specified. In this type of model (a varying effects model (VEM)), the terms just explain how things are; each individual is analysed separately and has their own regression coefficients. However, a model of regression coefficients of level-1 regressed on the level-2 independent predictor variables is not sufficient for specifying a multilevel model. Multilevel modelling involves a statistical integration of different models at different levels. The simplest integration takes place in the random effects model (REM). In REMs level-1 regression coefficients are treated as random coefficients at level-2. Adding level-2 independent predictor variables to the REM makes it more general and more useful to the study of growth (Kreft and de Leeuw, 1998) because individuals have their own regression lines with separate intercepts and slope coefficients. It is essential that any models used to identify the effects of covariates on growth incorporate an individual's growth trajectory (separate intercepts and slopes). Therefore it is imperative that REMs are used to describe a child's growth and not VEMs. The similarities and differences between REM and VEM models are explained in detail elsewhere (Kreft and de Leeuw, 1998) and are discussed further in the section below describing the development of the multilevel model.

Brief history

The idea of attributing random variation to different sources by fitting random effects is not new. In 1925, Fisher outlined the basic method for estimating variance components (ANOVA) by equating the mean squares from an analysis of variance table to their expected values. Balanced data REMs were developed by Henderson (1953) based on the ANOVA table, although subsequent work identified limitations to these methods. In 1977 Harville developed a REM for balanced data. Based on Harville's (1977) work Laird and Ware (1982) introduced a family of two-stage REMs. The key paper introducing multivariate linear models into the growth curve literature was by Pothoff and Roy (1964). In 1965 Rao linked growth curves with REMs. More recently this approach has been pioneered by Goldstein (1989).

The advancement in computer software development has also enabled more people to have access to this type of statistical analysis. There are, currently, many software tools that can be used to analyse hierarchial data. For example, both SAS statistics solutions (e.g. PROC MIXED) and SPSS (e.g. Mixed Models) have options to construct REMs. For an extensive comparison of some of these packages the reader is directed to the work by Kreft and colleagues (Kreft *et al.*, 1994; Kreft and de Leeuw 1998). In this chapter we will illustrate multilevel modeling using the MLwiN package developed and described by Rasbash *et al.* (1999).

Development of the multilevel model

The equation of a straight line is defined in an equation (Eqn 12.1), where y is the dependent variable that we want to predict and x is the independent (predictor) variable.

$$y = \alpha + \beta x + \varepsilon \tag{12.1}$$

In our example data, we may wish to predict a child's TBBMC (y) at a given biological age (x). A straight line can be drawn when the slope of the line and the point at which the line crosses the vertical (y) axis of the graph (the intercept) are known. Regression analysis finds the line that goes through, or as close to, as many of the data points as possible. In Eqn 12.1 (a simple regression equation) β is the slope of the straight line fitted to the data and α is the intercept of that line. There is also a residual term ε, which represents the difference between the predicted TBBMC for an individual and the actual TBBMC for that individual. The line also explains the relationship between x (biological age) and y (TBBMC), i.e. slopes can be positive or negative.

Multiple regression is an extension of simple regression modelling to situations where there are several predictors of the dependent variable. In multiple regression each independent variable (x) has its own coefficient and the dependent variable is predicted from a combination of all the variables multiplied by their respective coefficients plus a residual term. In our example data, the influence of biological age on the growth of TBBMC for an individual can be expressed as follows:

$$y_i = \underbrace{\alpha + \beta x_i}_{\text{fixed}} + \underbrace{\varepsilon_i}_{\text{random}} \tag{12.2}$$

where subscript i takes values from 1 to m (where m is the number of repeated measures of TBBMC within an individual), and y_i and x_i are respectively TBBMC and biological age for the ith assessment. The intercept, α, is where the regression line meets the vertical axis and β is the slope of the line; both parameters are fixed and can be estimated. In Eqn 12.2 ε_i (the residual) is the departure of the ith assessment of actual TBBMC from the predicted TBBMC at this measurement occasion. This is the random part of the model and is the only part of y_i that is not predicted by the fixed part of the regression relationship.

To make Eqn 12.2 a multilevel model we assume that individuals are a random sample taken from the total population and thus we include in the model more than one individual's repeated assessment of TBBMC over time; the model is developed by rewriting Eqn 12.2 as:

$$y_{ij} = \alpha_j + \beta x_{ij} + \varepsilon_{ij} \tag{12.3}$$

where y is TBBMC and subscript i again takes values from 1 to m (where m is the number of repeated measures) and subscript j takes values from 1 to k (where k is the number of subjects in the sample). The coefficient for x (in this case biological age in decimal years) is represented by β and again the subscript i takes values from 1 to m, and subscript j takes values from 1 to k. When an item has two subscripts, ij, the item not only varies between measurement occasions (within-individual variance; level-1 of the model) but it also varies between individuals (between-individual variance; level-2). Where an item has a j subscript only, in this case the intercept (α), it varies across individuals but has the same value for all assessment occasions. Level-1 variance is represented by ε_{ij} and describes the departure of the ith assessment of actual TBBMC, for the jth individual, from their predicted TBBMC at this measurement occasion (within-individual variance). Equation 12.3 differs from Eqn 12.2 because it has a second level of variance, the between-individual variance (level-2 variance). To enable estimation of the α coefficient, Eqn 12.3 is re-expressed as:

$$\alpha_j = \alpha + u_j \tag{12.4}$$

where α is a constant and u_j the departure of the jth individual's intercept from the average value. u_j is the level-2 residual whose value is the same for all measurement of TBBMC in individual j. The model for TBBMC accrual with biological age can now be expressed as:

$$y_{ij} = \underbrace{(\alpha + \beta x_{ij})}_{\text{fixed}} + \underbrace{(u_j + \varepsilon_{ij})}_{\text{random}} \tag{12.5}$$

The two levels of random variation (the random parameters) are represented by u_j and ε_{ij}. The variance of u_j and ε_{ij} can be estimated if we assume that these variables are uncorrelated and that they follow a normal distribution. The quantities α and β (the fixed parameters) are the average intercept and slope, respectively, for all individuals and also need to be estimated. It is the existence of u_j and ε_{ij} in Eqn 12.5, when compared to Eqn 12.2, that identifies this equation as a multilevel model.

Equation 12.5 is the simplest type of multilevel model (a VEM) where the only random parameters are the intercept variances for each individual's model of TBBMC development with biological age (i.e. individuals are allowed to have their own intercepts). This model assumes that the growth of TBBMC with biological age is the same for all individuals; that is, the slopes of individuals' regression lines are the same. However, since we know that there are great variations in tempo of growth between individuals, this assumption is not acceptable. Thus, Eqn 12.5 has to be further developed to allow for the possibility that the slope of TBBMC development with biological age varies from individual to individual. To allow each individual to have their own intercept and slope, Eqn 12.3 is rewritten as:

$$y_{ij} = \alpha_j + \beta_j x_{ij} + \varepsilon_{ij} \tag{12.6}$$

where β_j is the regression slope of biological age on TBBMC for the jth individual. Again we can re-express α_j as $\alpha + u_j$, as in Eqn 12.4, but this time β_j is expressed as $\beta + v_j$. The constants α and β are fixed quantities, the average intercept and slope, which are estimated from the data. u_j and v_j represent the extent to which the jth subject's intercept and slope departs from the average. u_j and v_j are both random variables that vary at level-2 of the model. As they vary at the same level they may be correlated and it is necessary to estimate the covariance between them.

$$y_{ij} = (\alpha + u_j) + (\beta + v_j)x_{ij} + \varepsilon_{ij} \tag{12.7}$$

Equation 12.7 is an example of a REM where level-1 regression coefficients are treated as random variables at level-2. In this example biological age (x) is in both the fixed and random parts of the model. This is seen more clearly when

Eqn 12.7 is rearranged into fixed and random parts:

$$y_{ij} = \underbrace{(\alpha + \beta_j x_{ij})}_{\text{fixed}} + \underbrace{(u_j + \nu_j x_{ij} + \varepsilon_{ij})}_{\text{random}} \quad (12.8)$$

While the regression coefficients in a VEM explains only how things are (only covariate group effects can be examined), the integration of a level-1 independent variable at level-2 of the REM makes the model more general and explains why things are (now covariate individual effects can be examined). By making biological age random at level-2, level-2 variance is now dependent on biological age and thus level-2 variance will change from one biological age to the next. Fitting individual curves to individual's repeated growth measures is the only way of extracting the maximum information about an individual's growth. The ability of the REM to allow individuals not only to have separate intercepts but also their own slopes of growth makes the REM ideal for the development of human growth curves.

The software program MlwiN (Rasbash *et al.*, 1999) uses Eqn 12.8 to estimate both the fixed parameters α and β and the random parameters, u_j, ν_j, ε_{ij}, and the covariance of u_j with ν_j. Equation 12.8 can be further developed by introducing other explanatory terms (x_{2-p}), such as sex, height, etc. as fixed parameters in level-1 of the model; Eqn 12.8 will take the form:

$$y_{ij} = \alpha_j + \beta_j x_{ij1} + \beta x_{j2} + \cdots + \beta x_{ijp} + \varepsilon_{ij} \quad (12.9)$$

where y is TBBMC at assessment occasion i in the jth individual, α_j is the constant for the jth individual, $\beta_j x_{ij1}$ is the slope of TBBMC with biological age for the jth individual, and x_2 to x_{p} are the coefficients of various explanatory variables at assessment occasion i in the jth individual (Table 12.2). ε_{ij} is the level-1 residual (within-individual variance) for the ith assessment of TBBMC in the jth individual. Again, the equations can be rewritten so that we are able to estimate level-2 random variance (u_j, ν_j, ε_{ij}, and the covariance of u_j with ν_j), the fixed parameters α_j (intercept) and β_j (slope) for each individual, and the fixed coefficients of the other explanatory variables ($x_{ij2}, \ldots, x_{ij\mathrm{p}}$).

Curve shaping

The growth pattern of stature, weight and most external dimensions of the body is S-shaped (sigmoid) and has four phases: rapid growth in infancy and early childhood, steady but rather constant growth during middle childhood, rapid growth during the adolescent spurt, and slow increase and eventual cessation of growth after adolescence (Malina and Bouchard, 1991). Figures 12.1 and 12.2 clearly indicate that TBBMC also follows this S-shaped curve of growth.

Since multilevel models are linear models it is important, when applying them to growth data, that the models contain non-linear functions. One way to introduce non-linearity into a linear model is to use polynomial functions. The idea of fitting simple polynomial curves to repeated measurements on individual subjects was systematically proposed by Wishart in 1938 (Wishart, 1938). For a full exposition of a multivariate statistical model base on polynomial curve fitting the reader is directed to the work by Rao (1965). In brief, the S shape can be fitted mathematically by adding 2nd-, 3rd-, 4th- and 5th-order polynomials of the 'time' parameter. In our example, biological age is the 'time' parameter and thus biological age power functions are added to the models to adjust the shape of the curve to its non-linear S shape.

Fit of the model

In a simple regression model the square of the residuals can be used as a gauge of how well a particular regression line fits the raw data. The mathematical technique that estimates the line that best describes the data, is called the least-squares method (LSM). The goodness of fit of the model is assessed by comparing the sum of the squared deviations from the regression line to the sum of their deviations from the mean line. R^2 is the proportion of improvement in deviation due to the regression model, and represents the percentage of the variation in the dependent variable that can be explained by the regression model. In multiple regression goodness of fit is computed by calculating multiple R^2.

The mathematical technique that estimates the line that best describes the multilevel data is iterative generalized least squares (IGLS). Initially, the MLwiN (Rasbash *et al.*, 1999) program derives some starting values for the parameters it is attempting to estimate. It uses these values as input to a calculating procedure, which produces new estimates as output. The new estimates can then be used as inputs for another cycle of calculations. The differences between the input estimates and the output estimates lessen with successive cycles. When these differences become tolerably small the procedure is said to have converged. One cycle of calculations is called an iteration (Woodhouse *et al.*, 1993). In multilevel terms a model where only the dependent variable and an intercept is fitted is called a null model. Table 12.3 displays the parameter estimates for a null model of TBBMC accrual. Table 12.3 indicates that multilevel models have two R^2s (level-1 within-individual variance and level-2 between-individual variance) and therefore the concept of 'explained proportion of variance' for hierarchical linear modelling is somewhat problematic. For a detailed explanation the reader is directed to the work of Snijders and Bosker (1999). Goodness of fit of multilevel models is measured by the

Table 12.3 *The null multilevel model (a) of total body bone mineral content (TBBMC) for 862 measures in 85 boys and 67 girls*

		(a) Null multilevel model
Variables		
Fixed effect	Estimates	Estimates
Constant	α_j	1662.4 ± 33.9
Random effects	*Level-1 (within individuals)*	*Level-1 (within individuals)*
Constant	ε_{ij}	195652.8 ± 10379.7
	Level-2 (between individuals)	*Level-2 (between individuals)*
Constant	u_j	137792.6 ± 20045.4
$-2 \times$ loglikelihood (IGLS)		13190.5

α_j is the intercept for the *j*th individual.
u_j is the departure of the *j*th individual's intercept from the average value.
ε_{ij} describes the departure (variance) of the *i*th assessment of actual TBBMC, for the *j*th individual, from their predicted TBBMC at this measurement occasion.
Fixed effect values are estimated mean coefficients ± SEE (standard error of estimate) (TBBMC, g).
Random effects values are estimated mean variance ± SEE (standard error of estimate) (TBBMC, g^2).
$P < 0.05$ if mean $> 2 \times$ SEE.

deviance between two models, analysed using the likelihood ratio statistic. We compare the difference in likelihood ratios between the model prior to inclusion of a new variable and after the new variable is included. Under the null hypothesis the difference of the likelihoods follows a χ^2 distribution with degrees of freedom equal to the number of new variables. It is therefore possible to look up the probability of obtaining a χ^2 of the difference in likelihood ratio statistics between two models that would be observed by chance alone. If the difference in likelihood ratio statistics between two models is significant, it strongly indicates that the extra parameter should be included in the model, and indicates that this new model is a significant improvement in fit of the data over the preceding model.

Modelling strategy

Initially a null model is fitted; that is, only a dependent variable and no independent variables other than an intercept (Table 12.3), and the likelihood ratio statistic is calculated. Unlike traditional regression models multilevel models do not compute *t*-statistics to investigate significance of β coefficients. The significance of a newly added fixed coefficient is investigated using two approaches. The first is to compare the estimate of the mean β coefficient with the standard error of the estimate (SEE). With a sufficiently large random sample the ratio of a fixed parameter to its standard error should be approximately normally distributed with mean 0 and variance 1. If an estimate is many times its standard error this tells us that the chance of obtaining estimates of this magnitude from sampling fluctuations are vanishingly small, if their true values are zero. We know that 95% of all errors of prediction will be between the estimate $\pm 1.96 \times$ SEE. Thus, for a rule of thumb if the estimate is great than $2 \times$ SEE we can conclude that $P < 0.05$. In Table 12.3 the only fixed coefficient is the intercept; as the estimated value of this intercept (1662.4 g TBBMC) is greater than its estimated standard error (33.9 g TBBMC) we can conclude that if we were to repeat our experiment 100 times, then the true intercept value will be 95 times in the range 1662.4 ± 66.4 g. Table 12.3 also displays the estimated average variance (random effects) around an individual's line (level-1 (within-individual) variance = 195652.8 g^2 TBBMC) and around the average line for all individuals (level-2 (between-individual) variance = 137792.6 g^2 TBBMC). Both these variance values also have the error of the estimates displayed (10379.7 g^2 TBBMC and 20045.4 g^2 TBBMC respectively).

The second approach to investigating the significance of a new variable is to use the likelihood ratio statistic. Within random parameters the distribution of the estimate to its standard error may depart considerably from normality. A better test for random parameters therefore is the use of the likelihood ratio statistic. As previously described, if the difference in the likelihood ratio statistics between two models is significant, it strongly indicates that the extra parameter should be included in the model, because it indicates that this new model is a significant improvement in fit of the data over the preceding model.

The null model is used as a baseline for the estimation of explained versus unexplained variances in comparison to more elaborate models. In this way independent variables are added to the models and are retained if deviance improves and/or if the variances at level-1 and level-2 are reduced. Covariates are accepted as being significant if the estimated mean coefficient is greater than twice the SEE. If retention criteria are not met then the explanatory variable is deleted from the model.

Centring data

In our example data the intercept (biological age = 0) occurs roughly in the centre of our data set. If we were to use chronological age as our time-dependent variable and make this random at level-2 of the model, this would dramatically increase the parameter estimate of the variance at level-2, around the between-individuals intercepts. This is because, as already outlined, all individuals have different growth trajectories in terms of TBBMC accrual. The intercept for each individual's line is the height of that line at $x = 0$, since individuals were never measured at chronological age 0 the model will extrapolate the actual lines back to where it estimates that the line will cross the y axis. Since the growth trajectories are measured between 8 and 16 years of age, and these trajectories are different between individuals, there will be a big vertical spread in these extrapolated lines at chronological age 0 and therefore a large variance. We would therefore be estimating the variance of the intercepts at an age that never occurred in the sample. The easiest way to overcome this problem is to shift the origin of the explanatory variable (in this case chronological age), known as centreing. A convenient place at which to centre a variable is its mean value. Only random variables require centreing; once the variable is centred then the interpretation of the variance of the intercepts indicates the variation in the heights of individual's lines at the mean variable score.

Worked example

All analysis is carried out using MLwiN software (Rasbash *et al.*, 1999). In the models we wish to estimate four quantities (Eqn 12.5), the fixed parameters α and β and the random parameters u_j and ε_{ij}. MLwiN requires every parameter to be associated with an explanatory variable. The fixed parameter β is associated with the predictor variable, but the other three parameters are not associated with any variable. For these we use the same explanatory variable which takes the value 1 for all level-1 units.

Using our example data the first thing we set is the null model (Table 12.3). Our dependent variable is TBBMC (g). The only fixed coefficient in this model (model (a)) is the mean intercept (estimated to be 1662.4 ± 33.9 g TBBMC) for the 152 individual children. The intercept represents the average value of TBBMC when biological age within our individuals is equal to 0, i.e. the age when peak height velocity is attained (APHV).

Table 12.3 also shows the two levels of variance. In the null model (model (a)) only the intercepts are specified as random effects (the constant values); in the models all constants take the value of 1. Level-1, the within-individual variance (estimated to be 195652.8 ± 10379.7 g^2 TBBMC), is identified by the sequence number of the measurement occasion (see Table 12.2) and level-2, the between-individual variance (estimated to be 137792.6 ± 20045.4 g^2 TBBMC), is identified by the individual's ID (see Table 12.2). All coefficient estimates are significant as indicated by the fact that the mean estimates of the coefficients are greater than twice their associated standard errors ($P <$ 0.05). The null model provides a baseline against which to judge the effects of including further terms. The statistic used to do this is the likelihood ratio statistic. In our null model the loglikelihood statistic takes a value of 13190.5.

Model (a) is extended by adding an independent term (biological age); model (b) in Table 12.4. When biological age is added to the model we can see that model (b) is a better fit of the data than model (a). This is indicated by the change in the likelihood ratio statistic. The probability of obtaining a χ^2 of 1607.9 (the difference in likelihood ratio statistic between models (a) and (b)) with the loss of one degree of freedom (one variable added) is $P < 0.001$. When comparisons are made between the two levels of variance, we observe a drop in variance at both levels; again indicating that model (b) is a better model than model (a). Model (b) is an example of a VEM where the only random effects (coefficients) are the intercepts (the constant values). Model (b) is made into a REM by making biological age both a fixed effect (coefficient) and a random effect (coefficient) at level-2 (Table 12.4, model (c)).

By making biological age random at level-2 we create a covariance matrix in the between individual variance (Table 12.4, model (c)). We are now estimating three random (variance) parameters at level-2 (u_j, the variance associated with the intercept (108433.1 × constant × constant); v_j, the variance associated with biological age (2856.2 × biological age × biological age); and the covariance of the u_j with the v_j (2 × 10651.6 × (constant × biological age)). The total variance at level-2 therefore is affected by the biological age of the individual at the time of measurement. When biological age is equal to zero (i.e. at APHV), then total level-2 variance is equal to u_j. For example the total level variance at APHV, at level-2 ($\mathrm{var}(u_j + v_j x_{ij})$), for model (c) is the sum of the following:

$$\begin{aligned}
&\text{Predicted level-2 variance at APHV} \\
&\quad = (108433.1 \times \text{constant}) + (2 \times 10651.6 \times \text{constant} \\
&\qquad \times \text{biological age}) + (2856.2 \times \text{biological age} \times \text{biological age}) \\
&\quad = (108433.1 \times 1) + (2 \times 10651.6 \times 1 \times 0) + (2856.2 \times 0 \times 0) \\
&\quad = 108433.1\ \mathrm{g}^2 \qquad (12.10)
\end{aligned}$$

where constant takes the value of 1 and biological age at APHV the value 0.

By allowing individuals to have separate intercepts and slopes (level-2 random effects) we can see that the goodness of fit of model (c) has significantly improved ($P < 0.05$) from models (b); the likelihood ratio statistic drops by 320 units, with the loss of two degrees of freedom (two new variables added to the random effects in Table 12.4), $P < 0.001$. Level-1 variance decreases by 10962.1 g^2 between models (b) and (c); again indicating that model (c) is superior to model (b). As level-2 variances are now influenced by biological age comparisons between level-2 variance for models (b) and (c) are problematic. For the remainder of this worked example the change in level-2 variance will be

Table 12.4 *Multilevel models of total body bone mineral content (TBBMC): (b) a varying effects model (VEM) and (c) a random-effects model (REM)*

Model	(b) VEM		(c) REM	
Variables				
Fixed effect	Estimates		Estimates	
Constant	1555.3 ± 28.9		1557.3 ± 27.1	
Biological age	216.5 ± 2.8		218.3 ± 4.9	
Random effects	*Level-1 (within individuals)*		*Level-1 (within individuals)*	
Constant	21792.6 ± 1160.6		10830.5 ± 642.0	
	Level-2 (between individuals)		*Level-2 (between individuals)*	
	Constant		Constant	Biological age
Constant	123362.9 ± 14448.6		108433.1 ± 12808.7	10651.6 ± 1853.1
Biological age			10651.6 ± 1853.1	2856.2 ± 419.2
Change in variance from previous model (g^2)	Level-1: 181204.2	Level-2: 14429.7	Level-1: 10962.1	Level-2: at APHV
$-2 \times$ loglikelihood (Change from previous model)	11582.6 (1607.9*)		11262.6 (320*)	
Number of added variables from previous model	1		2	

Fixed effect values are estimated mean coefficients ± SEE (standard error of estimate) (TBBMC, g).
Random effects values are estimated mean variance ± SEE (standard error of estimate) (TBBMC, g^2).
Biological age is years from age at peak height velocity (APHV).
$P < 0.05$ if mean $> 2 \times$ SEE; * $P < 0.05$ for change in loglikelihood between models.

discussed only at APHV, when biological age is equal to zero. The fixed effects in model (c) indicate that the average intercept for an individual's regression line of TBBMC at APHV is equal to 1557.3 ($\pm$ 27) g. The biological age coefficient (218.3 $\pm$ 4.9) indicates that TBBMC increases by 218.3 g per biological age year. For example if biological age = 1 then the model would predict a TBBMC of 1775.6 g (1557.3 + 1 $\times$ 218.3).

Table 12.5 shows the gradual development of the model. Fixed coefficients are added one at a time and their inclusion in the final model is dependent on whether they are biologically or statistically significant. In models (d) and (e) polynomial terms are added (biological age^2 and biological age^3) to allow for the non-linear shape of growth. Although adding biological age^2 as a fixed effect into model (d) does not improve the goodness of fit over model (c) ($-2 \times$ loglikelihood only decreases by 1.5 units), both level-1 and level-2 variances decrease. Including biological age^3 into model (e) improves the goodness of fit of the model (likelihood significantly drops ($P < 0.05$); however the variance at level-2 increases. Throughout statistics practice a common theme is trade-off. Here the trade-off is between increasing the goodness of fit of the model to the data set and decreasing the precision with which the parameters can be estimated, i.e. increasing the variance. Biological age^2 and biological age^3 can also be made random at level-2. In this particular model making biological age^2 and biological age^3 random did not improve the fit of the model, as indicated by no significant change in the likelihood ratio statistic between successive models.

The next stage is to add other fixed effects. When height is introduced as a covariate into the model (Table 12.5, model (f)) the model improves not only in goodness of fit but also by reducing both level-1 and level-2 variance. The inclusion of height not only adjusts the two levels of variance but also the fixed effects of the other coefficients. Model (f) indicates that height is a significant (25.2 $\pm$ 1.0, $P < 0.05$) independent predictor of TBBMC; every cm of stature predicts 25.2 g of TBBMC. The development of the model continues in a similar fashion. Independent variables are added to the model and are retained if deviance of likelihood ratio statistic improves and/or if the variances at level-1 and level-2 are reduced.

Final model of the example data

The goals of the analysis of the example data set were to (1) characterize patterns of individuals' TBBMC accrual, and (2) to investigate the effects of independent variables (e.g. sex, biological age, height, lean and fat mass,

Table 12.5 *Development of the multilevel model of total body bone mineral content (TBBMC) with biological age*

Model	(d)		(e)		(f)	
Variables						
Fixed effect	Estimates		Estimates		Estimates	
Constant	1559.3 ± 27.2		1534.5 ± 27.4		−2475.4 ± 165.9	
Biological age	219.6 ± 5.0		274.0 ± 4.6		104.3 ± 8.0	
Biological age^2	−1.20 ± 0.95		8.11 ± 0.75		15.7 ± 0.7	
Biological age^3			−5.06 ± 0.18		−2.5 ± 0.2	
Height					25.2 ± 1.0	
Random effects	*Level-1 (within individuals)*		*Level-1 (within individuals)*		*Level-1 (within individuals)*	
Constant	10751.6 ± 637.7		4612.6 ± 275.2		3537.0 ± 209.6	
	Level-2 (between individuals)		*Level-2 (between individuals)*		*Level-2 (between individuals)*	
	Constant	Biological age	Constant	Biological age	Constant	Biological age
Constant	108710.9 ± 12834.7	10434.2 ± 1846.4	112751.4 ± 13068.8	10276.4 ± 1629.3	37037.6 ± 4392.5	5623.6 ± 835.0
Biological age	10434.2 ± 1846.4	2859.9 ± 419.0	10276.4 ± 1629.3	2232.7 ± 298.4	5623.6 ± 835.0	1661.1 ± 222.9
Change in variance	Level-1	Level-2 at APHV	Level-1	Level-2 at APHV	Level-1	Level-2 at APHV
from previous model (g^2)	−78.9	+277.8	−6139	+4040.5	−1075.6	−75713.8
2×loglikelihood (Change from previous model)	11261.1 (1.5)		10715.6 (545.5*)		10340.0 (375.6*)	
Number of added variables from previous model	1		1		1	

Fixed effect values are estimated mean coefficients ±SEE (standard error of Estimate) (TBBMC, g).
Random effects values are estimated mean variance ±SEE (standard error of estimate) (TBBMC, g^2).
Biological age is years from age at peak height velocity (APHV); height (cm).
$P < 0.05$ if mean $> 2 \times$ SEE; * $P < 0.05$ for change in loglikelihood between models.

Table 12.6 *Final multilevel model of total body bone mineral content (TBBMC) with biological age*

Variables		
Fixed effect	Estimates	
Constant	−715.3 ± 165.6	
Biological age	69.1 ± 6.7	
Biological age^2	9.08 ± 0.62	
Biological age^3	−1.48 ± 0.17	
Height	6.45 ± 1.20	
Lean mass	0.031 ± 0.001	
Fat mass	0.009 ± 0.001	
Sex	−20.6 ± 18.9	
Physical activity	11.2 ± 5.2	
Random effects	*Level-1 (within individuals)*	
Constant	2823.0 ± 166.0	
	Level-2 (between individuals)	
	Constant	Biological age
Constant	16537.3 ± 1987.6	2166.1 ± 329.1
Biological age	2166.1 ± 329.1	495.4 ± 78.6

Fixed effect values are estimated mean coefficients ± SEE (standard error of estimate) (TBBMC, g).
Random effects values are estimated mean variance ± SEE (standard error of estimate) (TBBMC, g^2).
Biological age is years from age at peak height velocity (APHV).
Height (cm); lean mass (g); fat mass (g); sex (0 = boy, 1 = girl); physical activity score (1 = low, 5 = high).
$P < 0.05$ if mean > 2 × SEE.

and physical activity) on TBBMC accrual. The first goal has been achieved by analysing the data using a random effects model. Table 12.6 summarizes the results of the completed model for TBBMC accrual. The random effects describe the two levels of variance (within individuals (level-1 of the hierarchy) and between individuals (level-2 of the hierarchy)) in g^2 TBBMC. The random effects are significant within individuals, indicating that TBBMC is increasing significantly at each measurement occasion within individuals ($P < 0.05$). The between-individuals variance matrix for the model indicates that individuals had significantly different TBBMC accrual growth curves both in terms of their intercepts (constant/constant, $P < 0.05$) and the slopes of their lines (biological age/biological age, $P < 0.05$). The variance of these intercepts and slopes is positively and significantly correlated (constant/biological age, $P < 0.05$). The variance between individuals is therefore different at different biological ages. Allowing biological age to be random at level-2 of the model has ensured that individuals have separate intercepts and slopes in their individual growth lines of TBBMC accrual.

The coefficients that significantly predicted TBBMC (fixed effects shown in Table 12.6) were biological age, height, lean mass, fat mass and physical activity; their mean estimates were greater than twice their SEE ($P < 0.05$). The power functions of biological age (biological age^2 and biological age^3) were also significant and were included in the linear models to shape the curves. Sex however was not a significant predictor of TBBMC (-20.6 ± 18.9 g, $P > 0.05$). When the confounders of physique and maturation are controlled, no independent sex difference exists. We can also conclude from the final model (Table 12.6) that when the confounders of growth and maturation are controlled, physical activity has a significant independent effect on TBBMC development. For children of the same biological age, height, lean and fat mass TBBMC will be 56 g higher (11.2×5) in the most active child (physical activity = 5) compared to the least active one (physical activity = 1).

Using the regression equation described in Table 12.6 we can predict TBBMC for each individual at each measurement occasion (Eqn 12.11).

$$\begin{aligned} \text{TBBMC} = {} & -715.3 + (69.1 \times \text{ba}) + (9.08 \times \text{ba}^2) - (1.48 \times \text{ba}^3) \\ & + (6.45 \times \text{ht}) + (0.03 \times \text{lm}) + (0.01 \times \text{fm}) + (11.2 \times \text{pa}) \end{aligned} \tag{12.11}$$

where TBBMC is total body bone mineral content (g), ba is biological age (years from APHV), ht is height (cm), lm is lean mass (g), fm is fat mass (g) and pa is physical activity (score 1 to 5).

We can also predict the variance within (random effects level-1) and between (random effects level-2) individuals.

Summary

Many new concepts related to growth curve analysis are being developed. In this chapter we discuss the introduction of multilevel modelling to the growth literature. The latest developments in computer software show that multilevel techniques are becoming more and more accessible and easier to use. In this chapter we compare two multilevel models, the variance effect model (VEM) and the random-effects model (REM). The main conclusion reached is that the REM should be the model of choice for fitting growth curves because it allows each individual to have their own growth trajectories, i.e. separate intercepts and slopes.

REFERENCES

Bailey, D. A. (1997). The Saskatchewan Pediatric Bone Mineral Accrual Study: bone mineral acquisition during the growing years. *International Journal of Sports Medicine*, **18** (Suppl. 3), S191–S194.

Baxter-Jones, A. D. G., Mirwald, R. L., McKay, H. A. and Bailey, D. A. (2003). A longitudinal analysis of sex differences in bone mineral accrual in healthy 8 to 19 year old boys and girls. *Annals of Human Biology*, **30**, 160–175.

Berkey, C. S. and Reed, R. B. (1987). A model for describing normal and abnormal growth in early childhood. *Human Biology*, **59**, 973–87.

Fisher, R. A. (1925). *Statistical Methods for Research Workers*. Edinburgh, UK: Oliver and Boyd.

Goldstein, H. (1989). Flexible models for the analysis of growth data with an application to height prediction. *Reviews of Epidemiology*, **37**, 477–84.

(1995). *Multilevel Statistical Models*, 2nd edn. London: Edward Arnold.

Harville, D. A. (1977). Maximum likelihood approaches to variance component estimation and to related problems. *Journal of the American Statistical Association*, **72**, 320–40.

Henderson, C. R. (1953). Estimation of variance and covariance components. *Biometrics*, **9**, 226–52.

Jenss, R. M. and Bayley, N. (1937). A mathematical method for studying the growth of a child. *Human Biology*, **9**, 556–63.

Karlberg, J. (1998). The human growth curve. In *The Cambridge Encyclopedia of Human Growth and Development*, eds. S. J. Ulijaszek, F. E. Johnston and M. A. Preece, pp. 108–13. Cambridge, UK: Cambridge University Press.

Kreft, I. G. G. and de Leeuw, J. (1998). *Introducing Multilevel Modeling*. London: Sage.

Kreft, I. G. G., de Leeuw, J. and Aiken, L. (1994). Review of five multilevel analysis programs: BMDP-5V, GENMOD, HLM, ML3, VARCL. *American Statistician*, **48**, 324–35.

Laird, N. M. and Ware, J. H. (1982). Random effects models for longitudinal data. *Biometrics*, **38**, 963–74.

Malina, R. M. (1978). Adolescent growth and maturation: selected aspects of current research. *Yearbook of Physical Anthropology*, **21**, 63–94.

Malina, R. M. and Bouchard, C. (1991). *Growth, Maturation and Physical Activity*. Champaign, IL: Human Kinetics.

Pothoff, R. F. and Roy, S. N. (1964). A generalized multivariate analysis of variance model useful especially for growth curve problems. *Biometrika*, **51**, 313–26.

Preece, M. A. and Baines, M. J. (1978). A new family of mathematical models describing the human growth curve. *Annals of Human Biology*, **5**, 1–24.

Rao, C. R. (1965). The theory of least squares when the parameters are stochastic and its application to the analysis of growth curves. *Biometrika*, **52**, 458–64.

Rasbash, J., Browne, W., Goldstein, H., *et al.* (1999). *A Users's Guide to MLwiN*. London: Institute of Education, University of London.

SAS (2004). http://www.sas.com

Snijders, T. A. M. and Bosker, R. J. (1999). *Multilevel Analysis: An Introduction to Basic and Advanced Multilevel Modelling*. London: Sage.

SPSS (2004). http://www.spss.com

Tanner, J. M. (1962). *Growth at Adolescence*, 2nd edn. Oxford, UK: Blackwell Scientific Publications.

(1989). *Foetus into Man: Physical Growth from Conception to Maturity*, 2nd edn. London: Castlemead.

Vandenberg, S. F. and Falkner, F. (1965). Hereditary factors in human growth. *Human Biology*, **37**, 357–65.

Wishart, J. (1938). Growth rate determinations in nutrition studies with the bacon pig, and their analysis. *Biometrika*, **30**, 16–28.

Woodhouse, G., Rasbash, J., Goldstein, H. and Yang, M. (1993). Introduction to multilevel modelling. In *A Guide to ML3 for New User*, ed. G. Woodhouse, pp. 9–54. London: Institute of Education, University of London.

Part IV

Special topics

13 *Methods for the study of the genetics of growth and development*

Stefan A. Czerwinski and Bradford Towne
Wright State University School of Medicine

Introduction

This chapter provides an overview of the methods currently available to study the genetic epidemiology of normal human growth and development. Over the past two decades numerous technological innovations have enabled researchers to investigate issues that were heretofore intractable. These innovations include the advent of relatively low-cost yet powerful computers, the development of sophisticated statistical genetic modelling approaches, and advances in high-throughput genotyping. Progress in these areas has allowed for more thorough genetic investigations of complex traits such as those comprising growth and development. These investigations not only assess the degree of genetic control of a trait, but also identify genes influencing variation in them.

A number of general points can be made regarding past research on the inheritance of growth-related traits. First, almost all studies to date have established that human growth is at least partly influenced by genes. These studies have examined familial resemblance of various growth and development measures including stature, weight and various maturational indicators (e.g. age at menarche, pubertal stage or skeletal age). Estimates of the proportion of variance attributable to the effects of genes (i.e. heritability) vary according to the trait of interest. Stature is among the most highly heritable of growth measures with estimates ranging as high as 0.92, indicating that as much as 92% of the total variation in stature can be attributable to the effects of genes (Wilson, 1976; Kaur and Singh, 1981). Heritability estimates of other growth and development traits generally range between 0.30 and 0.60 (Susanne, 1977; Susanne *et al.*, 1983; Hauspie *et al.*, 1994; Beunen *et al.*, 1998). Second, a common feature of most early genetic studies of growth was the examination

Methods in Human Growth Research, eds. R. C. Hauspie, N. Cameron and L. Molinari.
Published by Cambridge University Press.

of primarily first-degree family members (e.g. twin siblings, parents–offspring or non-twin siblings) for the calculation of familial correlations. In addition, most of these studies were also based solely on cross-sectional data. It is only in recent years that researchers have moved forward from studies investigating familial resemblance in growth traits to studies having the ultimate goal of identifying specific genes related to normal variation in growth and development. And third, to date, extended-pedigree approaches have been utilized very little in examinations of the genetic architecture of growth, despite their relative advantages. Extended-pedigree studies sample not only first-degree relatives, but also family members of varying degrees of relationship to each other and who live in different households. This approach allows researchers to move beyond the now well-established fact that the processes of growth and development are to a large extent controlled by genes, to understanding the changing sources of genetic control during the course of maturation, in constant interaction with environmental conditions, and to ultimately identifying genes and novel pathways controlling physical growth and development.

In this chapter, we present state-of-the-science methods pertaining primarily to extended-pedigree study designs for the investigation of the genetic epidemiology of growth and development. In addition, we focus on one particular analytical approach that is well suited for studies using extended-pedigree data, the variance components approach. Variance component models have been in existence for several decades and have recently regained prominence in genetic epidemiology because of their exceptional flexibility and utility.

Goals of genetic studies of growth

In order to understand the genetic architecture of complex traits such as those comprising physical growth and development, a particular investigatory sequence should be followed. The first step is to quantify and characterize the extent of genetic effects. This is routinely done by performing quantitative genetic analysis on family-based data and estimating the heritability of a trait. Ideally, these family-based data are from members of large extended families or pedigrees. This familial structure offers two key advantages for the study of complex traits. First, it provides increased statistical power by virtue of the thousands of pair-wise combinations of relatives of varying degrees of relationship to each other that are contained in extended pedigrees. And second, it allows for the study of gene × environment interactions because data are gathered from relatives living across different household environments.

Once it has been determined that a trait has a significant genetic basis, it is important to characterize the nature of the genetic effects. (If a trait does not have a significant genetic basis attention should turn to identifying and quantifying the environmental influences on that trait.) Further characterization of genetic effects can be accomplished by quantifying the effects of covariates on the variation observed in the trait, and through multivariate analytical approaches that estimate genetic and environmental correlations as well as interaction effects. Analyses of this nature help to elucidate pleiotropic (i.e. shared genetic) effects between related traits measured at a single time, or the same trait measured at different times. This information provides insight regarding the genetic regulation of complex traits and potential common genetic pathways between biologically related traits. Information gained through simple quantitative genetic analyses can then be used to inform more complex quantitative trait linkage analyses where extensive DNA marker data are incorporated. The ultimate goal of quantitative trait linkage analysis is the identification of chromosomal regions (i.e. quantitative trait loci) harbouring genes that influence variation in a given trait.

Assessing the importance of genes

Much of what we know about the genetic control of measures of physical growth and development has come from family-based studies of correlations between relatives for various anthropometric traits such as stature or weight. The premise underlying these studies is that if a trait has a genetic basis, then more closely related family members will be more similar for that trait. The strength of the correlation between family members of varying degrees of relatedness is therefore indicative of the degree of genetic control exerted on any given trait. For example, for a trait such as adult stature, monozygotic (MZ) twins, who share 100% of their genetic make-up, are more likely to have similar values compared to dizygotic (DZ) twins who share only 50% of their genetic constitution. We therefore expect a higher degree of phenotypic correlation between MZ twin pairs than DZ twin pairs. Through examination of correlations between relatives, heritabilities can be calculated using various equations. Examples of these equations for first-degree relative pairings are presented in Table 13.1. The heritability of a trait is an indication of the degree of genetic control for that trait and ranges from 0 (no genetic effect) to 100% (full genetic effects). It should be noted that although generally robust, heritabilities are population-level estimates and are specific to a particular population in a given environment.

Table 13.1 *Heritability estimation from relative-pair correlations*

Relative pairing	Definitions
Twins	
$h^2 = r_{MZ}$	MZ, monozygotic twins
	DZ, dizygotic twins
$h^2 = (r_{MZ} - r_{DZ})/(1 - r_{DZ})$	
Parent–child	
$h^2 = 2r_{pc}/(1 + r_{sp})$	pc, parent–child
	sp, spouse–spouse
Sibling	
$h^2 = 2r_{sib}$	sib, siblings

Concept of heritability: definition and utility

The concept of heritability is central to understanding the nature of genetic influences on any trait. According to classical quantitative genetics theory (e.g. Falconer and Mackay, 1996) the observed phenotypic variance (σ_P^2) in a trait can be expressed as the sum of both genetic (σ_G^2) and random environmental effects (σ_E^2). Mathematically, this can be expressed as:

$$\sigma_P^2 = \sigma_G^2 + \sigma_E^2 \qquad (13.1)$$

Although simplistic, this model provides a starting point for understanding the quantitative genetics of growth and development. It is acknowledged that variation in a trait is also influenced by environmental factors, and by interactions between different genes and between genes and the environment. These interactions can be modelled using this variance components approach, but for now the discussion will focus on the simplest model comprised of just genetic and environmental components. From this simple model the heritability (h^2) of any trait can be easily calculated as the proportion of the phenotypic variance explained by additive genetic effects. This is mathematically expressed as:

$$\sigma_G^2/\sigma_P^2 \qquad (13.2)$$

The interpretation of heritability estimates from this model depends upon a number of assumptions that may or may not be valid for a given trait in a particular population. For example, the expression above refers to what is often considered a narrow-sense heritability. In this case, all of the genetic effect is considered to be additive in nature, that is, no gene × environment or gene × gene interactions or dominance effects are accounted for in the model. Interaction and dominance effects are considered as noise and not accounted for in the genetic effect component. Ignoring interaction and dominance effects in a model often results in an underestimation of the overall genetic effect. Generally, models that incorporate interaction and dominance effects in the genetic component allow for the calculation of a broad-sense heritability. In practice, the narrow-sense heritability has been extremely useful in characterizing the genetic effects of continuously distributed traits such as stature, body mass index or blood pressure because inheritance of these traits is most likely to be influenced by a number of genes with small to moderate effects.

Familial study designs

Prior to embarking on a genetic epidemiologic study of growth and development, a number of practical considerations should be addressed. Apart from the usual study design issues of determining whether a study will be cross-sectional (i.e. observations at one point in time) or longitudinal (i.e. multiple observations over time) and what traits should be measured, genetic epidemiologic studies require several additional considerations. The most important of these considerations is the sampling scheme. Most of the traits of interest in a study of growth and development (e.g. stature, weight, age at menarche, etc.) are complex traits. By this is meant that these traits are influenced by numerous genes, environmental factors, and their interactions. In this regard, these traits are also often referred to as polygenic or, more properly, oligogenic, since variation in them is due to several genes, some with small effects and some with moderate effects. Variation in common complex traits is usually measured on a continuous scale. This distinction is in contrast to discrete traits that are measured on a 'presence' or 'absence' scale, and which often have a single-gene inheritance pattern. Traits inherited in this manner are usually considered to be monogenic (i.e. one gene having very large measurable effects on the trait). Examples of monogenic traits include cystic fibrosis and phenylketonuria. There are a number of monogenic disorders that influence growth. For example, Turner syndrome and achondroplasia are both single-gene disorders that have pronounced effects on stature and general development. For the purpose of this chapter, however, we will consider indicators of normal growth to be common complex sets of traits and will limit our discussion to issues concerning the genetics of oligogenic inheritance. Although there are a number of possible study designs available for genetic studies, the study designs discussed in the following section are most appropriate for the study of complex traits.

Twin studies

Studies of twins have been useful in establishing the existence of familial resemblance for many complex traits. The classic twin model compares phenotypic correlations between two classes of twins, monozygotic and dizygotic. As stated previously, MZ twins share 100% of their genetic make-up, while DZ twins, like full siblings from singleton births, share on average half of their genetic make-up. Because of this, any observed differences between MZ twins are assumed to be the result of environmental factors only, while phenotypic differences between DZ twins are considered to be due to differences in both genes and environmental exposure. Thus, by observing phenotypic differences between twin groups, assertions can be made about the nature of genetic control for different traits.

The twin study design has a number of strengths, especially when viewed in the context of studies of growth and development. First, data collection can proceed fairly rapidly because data can be collected simultaneously from twin pairs. This is especially beneficial if the study has a longitudinal component where a number of measures are taken over time. In addition to the relative ease of data collection, another strength of twin studies is that data are most often collected from each twin at the same age, thus there is no need to incorporate age-adjustments in data analysis. And finally, because data are collected for individuals at the same age and at the same time there are no generational or secular effects that can sometimes pose difficulties in samples containing multiple generations.

Despite some of the apparent advantages of twin studies, there are a number of weaknesses as well. For example, one important assumption in the classical twin study design is that both MZ and DZ twin pairs are equally likely to share a common environment. This assumption may not necessarily be valid, however, because MZ twins are often more likely to share common activities, foods and other aspects of the environment than DZ twins. It is for this reason that studies of heritability in twins often yield inflated estimates with some studies finding heritability estimates greater than 1, an estimate that is biologically impossible. In general, heritability estimates obtained from studies of twins are best regarded as the upper limit of the true biological value.

The twin study design can be especially problematic if the focus of a study is a growth-related outcome. Twin births are physiologically different from singleton births due to competition for maternal resources during pregnancy. Hence, even at birth, growth rates among twins may be considerably divergent. In fact, compared with siblings of singleton births, twins may actually be more different. These early differences may influence subsequent growth patterns (e.g. 'catch-up' growth), and ultimately more distant health-related outcomes as has been suggested by studies investigating the so-called 'foetal origins' hypothesis.

Nuclear families

Another important study design that is commonly employed in genetic epidemiology is the nuclear family study design. With this study design correlations between the various classes of first-degree relatives in a family are estimated. These correlations include parent–offspring, sibling–sibling and spouse–spouse. From these correlations heritabilities can be estimated. Heritability estimates calculated from studies of nuclear families are often subject to inflation due to the effects of shared environmental factors such as diet and lifestyle among family members living in a single household. As a consequence of this, heritability estimates are often adjusted by accounting for the degree of spousal correlation in the family. It is assumed, in the absence of inbreeding, that any correlation found between spouses is the result of shared environmental factors, although this assumption ignores possible positive assortative mating.

There are a number of practical considerations to be taken into account in nuclear family studies apart from the problems encountered from shared environmental effects. For example, it is sometimes difficult to obtain information about certain life events because they are often separated generationally in time. Consequently, it may require 20 to 30 years of waiting in order to collect growth measures of the children of parents who, as children themselves, were originally enrolled in a growth study. Also, generational differences in growth parameters may exist due to secular trends over time. This may effectively reduce the heritability of certain traits by diminishing the degree of phenotypic correlation observed between parents and offspring.

Extended pedigrees

The study design that offers considerable promise for estimating the heritabilities of complex traits is the extended family approach. This approach involves collecting information from all available family members and estimating phenotypic correlations between all relatives of varying degrees of relationship to one another. By sampling members outside of the immediate nuclear family, many of the problems encountered with shared environmental effects in other study designs are minimized because family members come from a number of different households. This results in a more reliable estimation of heritability that is less likely to be overestimated. By sampling members in different households who live under potentially different environmental conditions, one is able to rigorously investigate gene × environment interactions. Although the methods involved in calculating heritabilities are especially computationally intensive, advances in computer technology make this less of an obstacle today than in the past.

There are a few drawbacks to this approach, however. As with nuclear family studies, there are logistical considerations when collecting data from individuals of varying ages. Many of these issues are compounded today by the unpredictable nature of health research funding. Above all, such studies need to be well conceived prior to data collection with particular attention paid to the outcomes of interest and the methods of analysis required to accomplish the specific goals of the study.

Characterizing the nature of genetic effects on growth-related traits

Quantitative genetic analysis

The first step in characterizing the effects of genes on a trait is to perform quantitative genetic analysis. Generally speaking, quantitative genetic analyses should proceed from simple univariate quantitative genetic models that estimate heritabilities and covariate effects to more complex multivariate quantitative genetic models that assess pleiotropy and various kinds of interaction. Once again, quantitative genetic analysis is based on the ability to relate quantitative phenotypic variation (e.g. stature at a given age) among individuals in a family to their genetic relatedness to one another. The purpose of these analyses is to decompose the phenotypic variance of a trait into variance components due to genetic and environmental effects. For the simplest case where we assume additive polygenic inheritance, the covariance matrix, Ω, for a quantitative trait in a pedigree of n members has the form:

$$\Omega = 2\Phi\sigma_{\mathrm{G}}^2 + I_n\sigma_{\mathrm{E}}^2 \tag{13.3}$$

where Φ is the kinship matrix for the pedigree, σ_{G}^2 is the variance due to additive genetic effects, I_n is the $n \times n$ identity matrix, and σ_{E}^2 is the variance due to random, individual-specific environmental effects. The parameters estimated in these models include, minimally, the mean of the trait, phenotypic variances and the heritability of the trait (Hopper and Mathews, 1982).

Initial analyses are intended to quantify the genetic and non-genetic sources of variation in the study traits. Additionally, from these models the relative contribution of important covariates can be estimated. For example, in an analysis of the heritability of birthweight, mother's age, height, weight and parity can all be entered as covariates. By constructing a series of reduced models, the significance and relative effects of all covariates on the trait variance can be estimated.

Principles of maximum-likelihood estimation theory guide the evaluation of specific hypotheses (Hopper and Mathews, 1982; Lange and Boehnke, 1983). Several hypotheses or models may be compared to each other. Each model specifies different assumptions regarding genetic and non-genetic influences on the trait. For example, one may hypothesize that age at menarche is significantly heritable. In this case, a model where the heritability of the trait is estimated from the data is compared to another model in which the heritability of the trait is constrained to zero (i.e. a null model with no genetic effect). Likelihoods can then be calculated for these various models and evaluated for significance using the likelihood ratio test statistic. The significance of individual covariates can also be evaluated in the same manner.

Using the statistical genetic tools described above, genetic and non-genetic variance components can be estimated for any growth- and development-related traits of interest, and the relative importance of pertinent covariates can also be established. All of this information informs the subsequent construction of increasingly more complex genetic models.

Multivariate quantitative genetic analysis

Multivariate extensions of the simple variance components method discussed above have also been developed (Lange and Boehnke, 1983; Lange *et al.*, 1988). Using the genetic information contained in the kinship coefficients between members of extended pedigrees, and the maximum-likelihood variance decomposition techniques outlined above, the phenotypic correlations between different traits can be partitioned into additive genetic and random environmental components. The multivariate phenotype of an individual can be modelled as a linear function of the measurements of the individual's traits, the means of these traits in the population, the covariates and their regression coefficients, plus the additive genetic and random environmental standard deviations. From such a model, a phenotypic variance–covariance matrix can be obtained. From this, the additive genetic and random environmental components can be estimated, given the kinship coefficients obtained from the pedigree structure and standard quantitative genetic theory. Mathematically this can be expressed as:

$$\Omega = G \otimes 2\Phi + E \otimes I \tag{13.4}$$

where ⊗ is the Kronecker product operator, Φ is the kinship coefficient, G is the additive genetic covariance matrix, E is the environmental covariance matrix and I is an identity matrix. From the genetic and environmental variance–covariance matrices, the additive genetic correlation, ρ_G, and the random environmental correlation, ρ_E between trait pairs can be estimated.

Just as the genetic and environmental components of the phenotypic variance and covariance matrices for the pedigree are additive, so too are the components of the phenotypic correlation matrix. Therefore, by using maximum-likelihood estimates of additive genetic and random environmental correlations, an estimate of the total phenotypic correlation between two traits, ρ_P, can then be obtained by:

$$\rho_P = \sqrt{h_1^2}\sqrt{h_2^2}\rho_G + \sqrt{(1 + h_1^2)}\sqrt{(1 - h_2^2)}\rho_E \tag{13.5}$$

The additive genetic correlation may, theoretically, range from –1.0 to 1.0 and is a measure of the shared genetic basis of the two traits. An additive genetic correlation of 1.0 indicates complete positive pleiotropy, (i.e. the same genes are affecting variation in both traits). An additive genetic correlation of zero between the traits indicates that different genes control each of them. And a genetic correlation of −1.0 indicates that genes that act to increase the value of one trait act to decrease the trait value of the other. Finally, a genetic correlation of a value intermediate between zero and 1.0 or −1.0 indicates incomplete pleiotropy. That is, the two traits are influenced to some extent by the same genes, but each trait also has, to some extent, a genetic basis unique from the other. Similarly, the random environmental correlation is estimated and serves as a measure of the strength of the correlated response of the traits to non-genetic factors.

Genetic correlations of a single trait at different times

Of particular interest in the study of growth and development is elucidating the nature of genetic control of a trait over time. In order to conduct these types of analyses, it is necessary to have serial measurements (or repeated measures) of the trait(s) of interest. In classic auxological theory, canalization refers to the tendency of a growth-related trait to follow a certain course or trajectory. From a genetic perspective, traits that are highly canalized show a relative insensitivity to changes in environmental conditions and are likely to be under strong genetic control with relatively high heritabilities.

Serial measurements of traits separated by time are normally correlated to some degree, with the traits measured more closely in time being most highly correlated on the phenotypic level. The more highly canalized a trait is, the higher the degree of phenotypic correlation between measurements. To test hypotheses concerning the genetic control of growth at different ages, and to investigate issues of growth canalization, more elaborate extensions of the simple variance components model illustrated above can be constructed. In this case, the variance terms of each measurement can be decomposed and genetic and environmental correlations between repeated measures of the trait at different age intervals can be calculated. The utility of this approach is that it allows for disentangling shared genetic effects from shared environmental effects when examining correlations between the traits measured over time. This approach provides an added level of information that is obscured when examining simple Pearson product–moment correlations.

The strength of a genetic correlation for a single trait with repeated measures is indicative of the degree of uniformity in genetic control over time. For example, if we observe a genetic correlation equal to 1.0 in a comparison of stature at 8 and 18 years of age, then we can infer that the genes controlling stature at each of these ages are likely to be the same. Squaring the genetic correlation gives a measure of the proportion of the genetic variance between the two measures that is shared in common (in the same manner that squaring a standard phenotypic correlation would yield the proportion of variance explained). In this vein, a genetic correlation of zero indicates that no genes are shared in common at the different ages, and a correlation between zero and 1.0 indicates an intermediate proportion of shared genetic variance. If a genetic correlation is significantly different from 1.0, then there is evidence that a different suite of genes influences the trait at different ages. A negative genetic correlation in this context is generally not considered because it would be a very unusual circumstance in where a set of genes would influence a trait positively at one age and negatively at another.

Genetic correlations between different traits

It is also important to understand the nature of the shared genetic underpinnings of traits. To examine the genetic relationship between different traits measured at the same time, a similar extension of the variance components model can be employed. The method is the same as for the repeated-measures analysis, only the interpretation of the results is slightly different. If significant genetic correlations exist between two traits, then there is evidence for pleiotropy (i.e. shared genetic effects). Some proportion of the genes that control one trait are likely to also influence the other trait. The strength of pleiotropy between traits can be evaluated with the same logic employed for the repeated-measures analyses. However, in between-trait analysis, the sign of the genetic correlation is an important indicator of the direction of the pleiotropic effect. For example, a negative genetic correlation between two traits indicates that the genes involved in increases of one trait are involved in decreases of the other trait. This type of analysis is useful for investigating the relationships between different, but related, aspects of growth and development, for example, understanding the common genetic basis between growth in stature and maturational outcomes such as skeletal development or age at menarche.

Gene × environment interaction

Similarly, more elaborate models can be developed that incorporate interaction effects such as gene × environment interaction or gene × sex interaction (Blangero *et al.*, 1990; Blangero, 1993). Understanding how genes interact with aspects of one's physical or physiological environment is essential for a complete understanding of the genetic architecture of complex traits. In studies where relatives live in different environments, gene × environment interaction can be modelled by examining the genetic variance of a trait as a function of differing environmental conditions. There are two underlying hypotheses that need to be evaluated in order to test for these types of interactions. First, does the partitioned genetic variance of a trait differ significantly in different environments? And second, does the same gene or suite of genes influence a trait in different environments? Gene × environment interaction exists when the genetic variance of a trait is different in different environments, and/or there are different genes influencing variation in a trait across environments.

The multivariate quantitative genetic models discussed above are able to incorporate and test for interaction effects. Blangero *et al.* (1990) developed a maximum-likelihood approach for the study of gene × environment interaction using related individuals in different environments. In this framework, additional parameters are modelled including environment-specific genetic variances and the genetic correlation between individuals living in different environments. Gene × environment interaction is indicated by significantly different additive genetic variances for individuals in different exposure groups, and/or by having a genetic correlation that is significantly less than 1.0 between groups as defined by exposure status. These two conditions indicate that there are differences either in gene expression with differing degrees of environmental exposure, or that there are different genes influencing variation in trait levels with differing environmental exposure.

Gene × environment interactions can be investigated using the general approach described above where the environment is now viewed as a traditional environmental factor (e.g. smoking or alcohol use), or a characteristic of the individual (e.g. age, sex or birthweight) that is known or hypothesized to influence the trait of interest. For example, given the current research interest involving hypothesized foetal programming, one could use a series of models to investigate the possibility of gene × birthweight interaction on subsequent growth during childhood. In these models, one could evaluate whether gene action is different between groups of related individuals who differ by birthweight status. In the simplest case, environmental factors can be modelled as categorical traits (e.g. high vs. low). For this particular example, the expected genetic covariance between a pair of relatives consisting of an individual with an arbitrarily defined birthweight as either high or low ($\mathrm{COV}(G_{\mathrm{high}}, G_{\mathrm{low}})$) is therefore defined as:

$$\mathrm{COV}(G_{\mathrm{high}}, G_{\mathrm{low}}) = 2\Phi\rho_{G_{(\mathrm{high,low})}}\sigma_{G_{\mathrm{high}}}\sigma_{G_{\mathrm{low}}}, \tag{13.6}$$

where subscripts high and low refer to high or low birthweight status, respectively; Φ is as defined previously the coefficient of kinship between two individuals; and $\rho_{G_{(\mathrm{high,low})}}$ is the genetic correlation between trait values in high and low birthweight individuals. In the absence of gene × birthweight interaction (i.e. the null hypothesis in this instance), each trait measured in high and low birth-weight individuals will have a $\rho_{G_{(\mathrm{high,low})}}$ of 1.0, and high and low birthweight individuals will have the same genetic variance. If gene × birthweight interaction exists, then each trait measured in high and low birthweight individuals will have a $\rho_{G_{(\mathrm{high,low})}} < 1.0$ and/or there will be different genetic variances between the groups. To conclude that gene × birth weight interaction exists, either one or both of the null hypotheses need(s) to be rejected.

Gene × sex interaction

Another example of gene × environment interaction that can be provisioned for in the same manner is gene × sex interaction (Robertson, 1959; Eisen and Legates, 1966). There are significant sex differences in measures of growth and development (e.g. stature, sexual maturity indicators), but the genetic basis of these sexual dimorphisms is poorly understood. The internal, hormonal, physiological environments of males and females differ considerably, and the expression of autosomal genes controlling a quantitative trait is likely to be influenced by sex steroid hormones. In this case, separate genetic variances can be estimated for male and female relatives and interaction effects can be evaluated.

Gene × age interaction

Extensions of these methods allow for the modelling of continuous environmental factors as well. Another special case of gene × environment interaction is gene × age interaction. For example, in one study we found the genetic variance of fasting insulin levels to increase during adulthood in a sample of individuals drawn from the Fels Longitudinal Study. These results indicated significant gene × age interaction with gene expression changing as a function of age (Towne *et al.*, 2000). The modelling of continuous environmental factors is slightly more complicated, but is easily incorporated into the variance component model (Almasy *et al.*, 2001). One way to model gene × age interaction, in the absence of serial data, is to use a matrix of age differences between related individuals to structure the additive genetic variance component (σ_g^2):

$$\Omega = 2\Phi \odot \Psi \sigma_g^2 + I\sigma_e^2 \tag{13.7}$$

where Ψ is a matrix of scaled similarities among individuals with regard to age, Φ is a kinship matrix, I is an identity matrix, σ_e^2 is a residual environmental component, and $\odot$ is the Hadamard product operator. Ψ can be modelled in a number of ways depending on the nature of the hypothesis to be examined. For example, one can use a simple exponential decay function where Ψ: $\nu_{ij} = \exp(-\lambda|(\text{age}_i - \text{age}_j)|)$. Gene × age interaction is evaluated by testing whether λ is significantly different from zero using likelihood ratio tests.

Localizing genes influencing growth-related traits

Once it has been determined through quantitative genetic analyses that there is a significant genetic influence on a trait, attention turns toward identifying the actual genes influencing variation in the trait. Over the past 50 years our understanding of a number of monogenic growth disorders has increased dramatically, but until recently our understanding of the oligogenic nature of normal variation has proceeded at a slow pace. Advances in molecular and statistical genetic methodology over the last two decades, however, have enabled us to begin to search for genes influencing common complex traits related to growth and development. Methodologies have evolved that improve the probability of finding influential genes of small to moderate effect. These innovations hinge on relatively new laboratory methods that are able to generate extensive genetic marker data. Using polymerase chain reaction (PCR) methods and gel electrophoresis, tens of thousands of genetic polymorphisms have been identified throughout the human genome. Using genetic marker data, the search to identify specific genes involved in the regulation of growth and development can take two forms: population-based association studies or family-based quantitative trait linkage studies.

Population association studies of candidate genes

The first approach to gene discovery is the candidate gene association approach, where genes suspected to be physiologically involved in growth are screened for possible relationships with various growth-related outcome measures. Association studies generally sample a population of unrelated individuals, and compare the trait values of individuals with different genotypes. If trait values differ significantly among individuals with different genotypes, then there is evidence for a population association. Association studies depend upon the existence of linkage disequilibrium between the genetic marker and the quantitative trait locus. Two loci are in equilibrium when alleles at the two loci are randomly associated with each other. If the relationship between the loci is not random then linkage disequilibrium is present. Linkage disequilibrium can occur for a number of reasons including new mutations, genetic drift and the presence of selection. For these reasons interpretations of association studies must be made with caution. Additionally, significant associations can be observed due to heterogeneity in the population sampled. This occurs when population subgroups differ systematically in both allele frequencies and levels of the quantitative trait of interest.

Quantitative trait linkage analysis

In recent years there have been many advances in linkage analysis as applied to complex traits. Most recently, allele-sharing methods have gained prominence for the detection of genetic loci controlling continuously distributed traits (Lander and Schork, 1994). The key premise behind allele-sharing methods is the concept of identity by descent (IBD). In comparisons between relatives, two alleles that are structurally identical are said to be identical by state (IBS); alleles that are structurally identical and inherited from a common ancestor (e.g. two siblings getting the same allele from their mother) are further classified as IBD. A pair of relatives can share either 0, 1 or 2 alleles IBD at any given marker locus. The likelihood of their sharing 0, 1 or 2 alleles IBD is contingent upon their coefficient of kinship. Linkage between a quantitative trait locus (QTL) and a marker exists in chromosomal regions when pairs of relatives who are more phenotypically similar share more alleles at a marker locus than pairs of relatives who are more phenotypically dissimilar (Terwilliger and Goring, 2000).

One strategy that has become increasingly popular over the past decade is using quantitative trait linkage analysis to scan the entire genome for QTL influencing various traits. This is commonly referred to as a whole genome scan. This strategy has been relatively successful in genetic epidemiologic studies of obesity and cardiovascular disease, and other common complex diseases. This approach entails genotyping many genetic markers for each individual. Traditionally, markers are selected so that they are spaced at fairly uniform intervals across every chromosome. For humans, it requires approximately 400 genetic markers to construct a genetic marker map with 10 cM resolution, the marker map density most commonly used in studies of complex disorders. Once all family members are genotyped quantitative trait linkage analyses can be performed that look for linkage between a trait and marker loci spread across the entire human genome.

Variance components linkage analysis

Variance components approaches to quantitative trait linkage have been extensively developed in recent years (Schork, 1993; Amos, 1994). Extensions of these approaches can accommodate large complex pedigree structures (Almasy *et al.*, 1996; Almasy and Blangero, 1998). In similar fashion to the quantitative genetic variance components models presented above, the pedigree-based variance component linkage method extends these strategies to estimate the genetic variance attributable to the region around a specific genetic marker (Schork, 1993; Amos, 1994). This approach is based on specifying the expected genetic covariances between arbitrary relatives as a function of the IBD relationships at a QTL. The variance component method is more powerful for the analysis of quantitative traits than the widely used sib-pair methods, and can provide at least a provisional estimate of the variance of a trait attributable to an underlying major locus (Amos *et al.*, 1996; Williams and Blangero, 1999; Goring *et al.*, 2001). For a simple model in which one QTL and residual polygenes influence a quantitative trait, one can assume that each individual's phenotype is a function of three general components: the QTL effect, an additive polygenic effect (which is assumed to be due to a large number of loci acting additively), and a random environmental effect. The covariance matrix (Ω) for a pedigree is then given by:

$$\Omega = \Pi\sigma_{\mathrm{q}}^2 + 2\Phi\sigma_{\mathrm{a}}^2 + I_n\sigma_{\mathrm{e}}^2 \qquad (13.8)$$

where σ_{q}^2 is the genetic variance in the trait (e.g. stature, age at menarche) attributed to the QTL and Π is a matrix whose elements specify the proportion of genes that individuals *j* and *k* share IBD at that QTL that is linked to a genetic marker locus. As in simple variance component models described previously, σ_a^2 is the genetic variance due to residual additive genetic factors, Φ is the kinship matrix, σ_e^2 is the variance due to individual-specific environmental effects and I_n is an $n \times n$ identity matrix.

Using the variance component model, one can test the null hypothesis that σ_q^2, (the genetic variance due to the QTL) equals zero (no linkage) by comparing the likelihood of this restricted model with that of a model in which σ_q^2 is estimated. The difference between these two $\log_{10}$ likelihoods produces a logarithm of difference (LOD) score that is the equivalent of the classical LOD score of linkage analysis. Twice the difference in the log-likelihoods yields a test statistic that is asymptotically distributed as a mixture of χ^2 distributions. Furthermore, in this framework one can follow a sequential strategy in which the genome is scanned for linkage and the chromosomal location that yields the largest marginal LOD score is retained for subsequent conditional analyses. The genome can then be scanned once more examining the conditional LOD scores. The benefits of conditional analyses are that they yield estimates of QTL effect size, diminish the risk of false positives and increase power to detect linkage. That is because it maximizes the relative signal-to-noise ratio of true linkages.

Multipoint extension of linkage analysis

In order to better localize linkage signals, the variance components approach also permits multipoint analysis (Almasy and Blangero, 1998). Using this method, the IBD allele-sharing at any location along a chromosome is predicted using a constrained linear function of observed IBD probabilities for genetic markers at known locations. Thus, use of multipoint analysis improves the resolution of the chromosomal location of the QTL. Initially, marker-specific IBD probability matrices are estimated and IBD probabilities between markers imputed. These matrices are stored for future use in the maximum-likelihood variance components estimation routines. Simulations have shown that this procedure produces unbiased estimates of QTL location and an estimate of effect size (Amos *et al.*, 1996; Williams and Blangero, 1999; Goring *et al.*, 2001).

Multivariate linkage analysis

After the initial detection and localization of QTL that influence individual traits of interest, one can use multivariate quantitative trait linkage analysis to test whether, and to what extent, these loci also have effects on other potentially related traits. Results from multivariate quantitative genetic analyses performed prior to linkage analysis can then be used to inform which sets of traits should be examined in multivariate linkage analyses. The multivariate extension of the procedures described above takes into consideration multiple phenotypes (Almasy *et al.*, 1997, 1999), thereby allowing one to test these hypotheses. This model, once again, is an extension of the multivariate quantitative genetic models presented above with the inclusion of additional parameters. In the case of t traits, the phenotypic covariance matrix (Ω) for a pedigree can be expressed as:

$$\Omega = M \otimes \Pi + G \otimes 2\Phi + E \otimes I \tag{13.9}$$

where $\otimes$ is the Kronecker product operator, M is the $t \times t$ additive covariance matrix due to the QTL, G is the residual additive genetic covariance matrix, E is the environmental covariance matrix and I is an identity matrix. Testing for linkage involves the examination of the diagonal elements of M for deviation from zero. Trait-specific estimates of σ_q^2, σ_a^2 and σ_e^2 can be obtained as well as the following correlations between traits: $\rho_q, \rho_a \ and \ \rho_e$. The parameters ρ_a and ρ_e represent the shared residual additive genetic and environmental influences on the traits, respectively. Parameter ρ_q is a measure of the shared QTL effects near the region for which linkage is being assessed. Using this model, one can test the null hypothesis that the σ_q^2 term equals zero by comparing the likelihood of a restricted model to that of a model in which σ_q^2 is estimated for the traits. When evidence is obtained that QTL for related traits are linked to the same chromosomal region, one can consider the possibility of a single gene with pleiotropic effects on both traits. Because colocalization of traits to the same region of the genome could also be the result of coincident linkage to separate, closely linked QTL (Almasy *et al.*, 1997, 1999), however, one must furthermore compare the likelihood of a model in which ρ_q is constrained to zero (indicating no shared major locus effects for the two QTL in that region) to a model in which ρ_q is constrained to 1.0 (i.e. complete pleiotropy). Rejection of the model in which ρ_q is constrained to zero, but not in which ρ_q is constrained to 1.0 suggests the existence of a single gene with pleiotropic effects.

Through use of the tools described above, researchers can identify chromosomal regions harbouring genes influencing various growth- and development-related traits. Although initially these regions may be relatively large, further laboratory-intensive methods such as fine-mapping and positional cloning can be performed in order to identify the functional mutations influencing phenotypic variation. As in any discipline that is advancing rapidly, genetic methodologies continue to evolve and many of the new challenges faced today involve the statistical treatment of the abundant genetic sequence data that are rapidly being generated. As these challenges are addressed, our knowledge of the

Table 13.2 *Internet resources relevant to research in the genetics of growth and development*

Site	Site address
General resources	
Center for Medical Genetics (Marshfield, WI, USA)	http://research.marshfieldclinic.org/genetics/
GENATLAS QUERY	http://www.dsi.univ-paris5.fr/genatlas/
OMIM Home Page – Online Mendelian Inheritance in Man	http://www3.ncbi.nlm.nih.gov/Omim/
The Genome Database	http://gdbwww.gdb.org/
The Human Genetic Analysis Resource	http://darwin.cwru.edu/
National Human Genome Research Institute (NHGRI)	http://www.genome.gov/
National Center for Biotechnology Information	http://www.ncbi.nlm.nih.gov/
Analytical resources	
Genetic Linkage Analysis (a comprehensive analytical resource)	http://linkage.rockefeller.edu/
ACT (variance components software)	http://www.epigenetic.org/Linkage/act.html
Cyrillic (pedigree drawing software)	http://www.cyrillicsoftware.com/
Fisher Genetic Analysis Software	http://www.biomath.medsch.ucla.edu/faculty/klange/software.html
Merlin (linkage analysis software)	http://www.sph.umich.edu/csg/abecasis/Merlin
Pedsys (pedigree management software)	http://www.sfbr.org/sfbr/public/software/pedsys/pedsys.html
Pedigree/Draw (pedigree drawing software)	http://www.sfbr.org/sfbr/public/software/pedraw/peddrw.html
SOLAR (variance components software)	http://www.sfbr.org/sfbr/public/software/solar/index.html

genetic architecture of normal growth and development will continue to be advanced.

Summary and conclusion

The methods presented in this chapter are intended to be a starting point for students interested in pursuing genetic epidemiological studies of growth and development. Like many research endeavours today, a genetic epidemiological study is by its nature multidisciplinary. Therefore, it is rare that any one individual will have the skills and resources necessary to pursue such an involved undertaking alone. It is therefore vital to seek collaborative interactions with individuals who have the varied skills and training necessary to conduct such research. For example, to conduct the type of large-scale whole genome scan linkage studies described above, it is necessary to organize researchers from a wide variety of fields including: data collection specialists, molecular biologists, population biologists, as well as statisticians. Finally, in an effort to help students find the necessary resources available to conduct genetic epidemiological studies of growth and development, we have provided the addresses of a number of research-based websites and a brief description of their content (Table 13.2). These websites provide numerous resources including free analytical software as well as further reference material.

REFERENCES

Almasy, L. and Blangero, J. (1998). Multipoint quantitative trait linkage analysis in general pedigrees. *American Journal of Human Genetics*, **62**, 1198–211.

Almasy, L., Hixson, J. E., VandeBerg, J. L., MacCluer, J. W. and Blangero, J. (1996). Multipoint quantitative-trait linkage analysis in general pedigrees. *American Journal of Human Genetics*, **59**, A173.

Almasy, L., Dyer, T. D. and Blangero, J. (1997). Bivariate quantitative trait linkage analysis: pleiotropy versus co-incident linkages. *Genetic Epidemiology*, **14**, 953–8.

Almasy, L., Hixson, J. E., Rainwater, D. L., *et al.* (1999). Human pedigree-based quantitative-trait-locus mapping: localization of two genes influencing HDL-cholesterol metabolism. *American Journal of Human Genetics*, **64**, 1686–93.

Almasy, L., Towne, B., Peterson, C. and Blangero, J. (2001). Detecting genotype × age interaction. *Genetic Epidemiology*, **21** (Suppl. 1), S819–S824.

Amos, C. I. (1994). Robust variance-components approach for assessing genetic linkage in pedigrees. *American Journal of Human Genetics*, **54**, 535–43.

Amos, C. I., Zhu, D. and Boerwinkle, E. (1996). Assessing genetic linkage and association with robust components of variance approaches. *Annals of Human Genetics*, **60**, 143–60.

Beunen, G., Maes, H., Vlietinck, R., *et al.* (1998). Univariate and multivariate genetic analysis of subcutaneous fatness and fat distribution in early adolescence. *Behavior Genetics*, **28**(4), 279–88.

Blangero, J. (1993). Statistical genetic approaches to human adaptability. *Human Biology*, **65**, 941–66.

Blangero, J., Williams-Blangero, S. and Konigsberg, L. W. (1990). Analysis of genotype–environment interaction using related individuals in different environments. *American Journal of Physical Anthropology*, **81**, 195–208.

Eisen, E. J. and Legates, J. E. (1966). Genotype–sex interaction and the genetic correlation between the sexes for body weight in *Mus musculus*. *Genetics*, **54**, 611–23.

Falconer, D. S. and Mackay, T. F. C. (1996). *Introduction to Quantitative Genetics*. Harlow, UK: Longman.

Goring, H. H., Terwilliger, J. D. and Blangero, J. (2001). Large upward bias in estimation of locus-specific effects from genomewide scans. *American Journal of Human Genetics*, **69**(6), 1357–69.

Hauspie, R. C., Bergman, P., Bielicki, T. and Susanne, C. (1994). Genetic variance in the pattern of the growth curve for height: a longitudinal analysis of male twins. *Annals of Human Biology*, **21**, 347–62.

Hopper, J. L. and Mathews, J. D. (1982). Extensions to multivariate normal models for pedigree analysis. *Annals of Human Genetics*, **46**, 373–83.

Kaur, D. and Singh, R. (1981). Parent–adult offspring correlations and heritability of body measurements in a rural Indian population. *Annals of Human Biology*, **8**, 333–9.

Lander, E. and Schork, N. (1994). Genetic dissection of complex traits. *Science*, **265**, 2037–48.

Lange, K. and Boehnke, M. (1983). Extensions to pedigree analysis. VI. Covariance components models for multivariate traits. *American Journal of Medical Genetics*, **14**, 513–24.

Lange, K., Weeks, D. and Boehnke, M. (1988). Programs for pedigree analysis: Mendel, Fisher, and dGene. *Genetic Epidemiology*, **5**, 471–2.

Robertson, A. (1959). The sampling variance of the genetic correlation coefficient. *Biometrics*, **15**, 469–85.

Schork, N. J. (1993). Extended multipoint identity-by-descent analysis of human quantitative traits: efficiency, power, and modelling considerations. *American Journal of Human Genetics*, **53**, 1306–19.

Susanne, C. (1977). Heritability of anthropological characters. *Human Biology*, **49**, 573–80.

Susanne, C., Defirse-Gussenhoven, E., Van Wanseele, P. and Tassin, A. (1983). Genetic and environmental factors in head and face measurements of Belgian twins. *Acta Genetica Medicae et Gemellologiae*, **32**, 229–38.

Terwilliger, J. and Goring, H. (2000). Gene mapping in the 20th and 21st centuries: statistical methods, data analysis, and experimental design. *Human Biology*, **72**, 63–132.

Towne, B., Guo, S. S., Chumlea, W. C., Roche, A. F. and Siervogel, R. M. (2000). Genetics of age-related changes in cardiovascular disease risk factors. *American Journal of Physical Anthropology* (Suppl.) **30**, 304.

Williams, J. T. and Blangero, J. (1999). Comparison of variance components and sibpair-based approaches to quantitative trait linkage analysis in unselected samples. *Genetic Epidemiology*, **16**, 113–34.

Wilson, R. (1976). Concordance in physical growth for monozygotic and dizygotic twins, *Annals of Human Biology*, **3**, 1–10.

14 *Prediction*

Noël Cameron
Loughborough University

Introduction

Prediction within the context of research in human growth and development may be defined as the estimation of the magnitude of some unmeasured dimension of interest. Prediction may take place in both cross-sectional and longitudinal scenarios. In cross-sectional scenarios one may wish to predict a dimension that cannot, for whatever reason, be measured at the time of assessment of other dimensions. For instance, one may wish to estimate total body fat when it is not possible to take a direct measure but when anthropometric estimations of subcutaneous fat (skinfolds) are available. In longitudinal situations one is interested in predicting the magnitude and/or timing of some future event, such as birthweight, or the timing of menarche, or adult stature.

The process of human growth and development lends itself particularly well to prediction methods for two reasons; first, almost all physical dimensions are associated with each other and do not vary independently, and second, human somatic growth is a process of change over time and has, for the most part, distinct and measurable end points. The end points also do not usually vary independently of earlier size and shape. However, the magnitude and timing of human growth and development are also variable and thus whilst it may be known with certainty that a particular event will occur, e.g. the achievement of adult stature, it is not known precisely at what time the event will occur or what magnitude it might be. It is not particularly surprising, therefore, that within research on human growth interest in prediction has been a fascination of research scientists since the earliest longitudinal studies of the twentieth century.

A cursory review of the literature on prediction and growth will reveal numerous approaches that include the use of morphological dimensions to predict: (1) other morphological dimensions, e.g. body fat from body mass index

Methods in Human Growth Research, eds. R. C. Hauspie, N. Cameron and L. Molinari.
Published by Cambridge University Press.

(Deurenberg *et al.*, 2001); (2) body areas and volumes, e.g. head circumference to predict brain volume (Bartholomeusz *et al.*, 2002); (3) functional dimensions, e.g. age, height and weight to predict respiratory function (Perez-Padilla *et al.*, 2003); (4) biochemical composition, e.g. total body potassium from age, height and weight (Larsson *et al.*, 2003) and (5) the timing and magnitude of future events, e.g. birthweight from foetal characteristics during pregnancy (Schild *et al.*, 2000; Stetzer *et al.*, 2002). The majority of methods have in common the fact that they use anthropometrically determined somatic dimensions to predict less easily determined variables. Indeed, the comparative ease with which size and shape may be assessed makes somatic dimensions primary candidates for non-invasive procedures to predict other less easily assessed dimensions. Invariably regression procedures are used to derive the prediction formulae.

The accepted technique to derive prediction formulae is quite simple. A sample of normal individuals is identified, all of whom are assessed for the variables of interest. The individuals are then randomly assigned to one of two subsamples, the prediction sample and the validation sample. Power analysis will provide an estimation of the required sample size for each group. The statistical association between the variables of interest is investigated in the prediction group using multiple regression. The exact form of the regression (linear, non-linear, etc.) will be dictated by minimization of the sums of squares of deviations, i.e. the goodness of fit. Those variables that combine to produce the smallest residual standard deviation become the predictor variables. The regression equation so produced is validated on the validation sample to obtain an estimate of the lowest prediction error. The resulting error will be lowest because the validation sample represents a random sample from precisely the same population as the prediction sample. The prediction equations are presented in the form of the constant and parameters with their standard errors.

Clearly any physical or physiological dimension or quality is capable of being predicted. The choice of what to predict and when to predict it will depend on the reasons for prediction. In most cases the purpose is to identify individuals who will fall above or below some cut-off point for normality/abnormality. Thus birthweight prediction is to identify possible low birthweight infants who characteristically present with risk factors and will probably need special neonatal care. Similarly, body mass index prediction is to identify individuals who will be at risk of overweight and obesity. Adult height prediction is to identify individuals who will fall within acceptable limits for normal stature, and so on. Often the status of the predicted variable will mean that intervention is considered. For example, the prediction of an abnormally low adult stature may instigate the use of hormonal treatment to improve the eventual outcome, or the prediction of a low birthweight may herald an intervention to improve the growth status of the foetus.

Given the variety of possible uses for prediction the question of what constitutes an acceptable accuracy of prediction becomes important. In other words, at what point does the researcher decide to accept a predicted outcome as accurate enough for intervention? There are no accepted guidelines. Researchers strive to arrive at an outcome that minimizes the residual standard deviation in regression, and usually simply present the prediction equations and report their validation error. It seems reasonable, however, to attempt to provide guidelines for the researcher in human growth. Our acceptability of the accuracy of any measurement is based on our knowledge of the reliability of measurement. It is usually accepted within research on human growth that the technical or standard error of measurement (TEM; SE_{meas}) provides acceptable statistics by which to judge reliability (Cameron, 1984; Lohman *et al.*, 1988). There seems to be no theoretical reason why prediction accuracy should not be comparable to measurement accuracy and that a desirable accuracy for the prediction of an anthropometric dimension should be to within 95% confidence limits for the reliability of anthropometric measurement or $\pm 1.96 \times$ TEM or SE_{meas}. In practice it is highly unlikely that prediction can ever be as good as that. The best, i.e. smallest, residual standard deviations that could be achieved when validating the latest Tanner–Whitehouse (TW3) prediction equations for adult height (Tanner *et al.*, 2001) was 1.4 cm for postmenarcheal girls at 14.0 to 14.9 years of age, i.e. 95% confidence limits of ±2.75 cm. Height can be measured to a TEM or SE_{meas} of about 0.3 cm (Ulijaszek and Lourie, 1994) and 95% confidence limits of ±0.59 cm, i.e. four times more accurately than the best height prediction. Thus other tests of acceptability need to be considered. The sensitivity, specificity, and positive and negative predictive values for predicting whether a validation individual will be outside or within the cut-off points of interest, e.g. birthweight greater or less than 2500 g, are relevant indicators of the success of prediction but each carries a different implication for the researcher. An acceptable proportion of individuals misclassified can be decided on depending on the outcome of the misclassification. Naturally if misclassification results in the absence of some life-sustaining intervention then sensitivity and a high positive predictive value would be the priority, i.e. it would be more serious not to treat an affected child than to offer treatment to an unaffected child. On the other hand the financial cost of intervention may persuade the researchers that it is necessary to treat only those who are definitely affected by the condition of interest. In this scenario specificity and negative predictive value will be the priority statistics.

This chapter will use adult height prediction as the basis to discuss prediction within research in human growth because the principles developed and used within this quest are universally applicable to all prediction situations.

Predicting adult stature

Early fascination with prediction focused on the prediction of adult stature because of the clinical importance of being able to estimate, with low (or at least clinically insignificant) errors, the adult height of an individual from data obtained at pubertal or prepubertal examinations and the information that the errors of prediction provide with regard to the relative importance of the variables thought to influence adult stature. Clearly adult height prediction is a special case of prediction in that it seeks to predict the magnitude of a physical dimension at some time in the future rather than to predict the size of another dimension that is not being measured concurrently. However, concurrent prediction is the simplest case of prediction and is adequately covered in the previous discussion. Adult height prediction on the other hand is more complex involving longitudinal data and the necessity to predict with an acceptable error from the earliest possible age using a minimum of easily and reliably collected variables. The lessons learnt by reviewing the historical and current approaches to adult height prediction can be applied to all other prediction situations.

Orthopaedic surgeons Gill and Abbott (1942) were interested in the effect of leg-lengthening operations on adult height. Such operations demanded accurate methods of estimating the future rate of growth, the ultimate length of the extremities, and final height. The technique developed by Gill and Abbot (1942) was based on the beliefs that canalization was a strong enough phenomenon to allow accurate prediction of adult height and that the accuracy would increase when an estimate of bone maturation was also considered. In addition, they determined that the relative proportions of the length of the femur and tibia to stature are maintained with only small age variations throughout the adolescent period. Gill and Abbot (1942) took the age-adjusted proportions of the femur and tibia to stature and multiplied the final proportions by predicted final height. The method was validated by investigating the serial predictions of 33 girls and 12 boys using the percentile (canalization) method. A child's present height was entered on a growth chart, drawn up by Gill and Abbott from the Whitehouse Conference data of 1932, at either the appropriate chronological age or, if the skeletal age and chronological age differed by more than 6 months, at the skeletal age, and the centile followed to a final height estimation at 18.5 years for boys and 16.5 years for girls. This technique produced an average error of ±2.5 cm and 90% confidence limits of ±5.08 cm.

Frank Shuttleworth of Yale University's Institute of Human Relations was the first to predict 'mature height' for biological interest rather than clinical necessity (Shuttleworth, 1939). Data from the Harvard Longitudinal Growth Study were used to develop two types of tables presented separately for boys and girls and for northern Europeans and Italians. The first used the variables of initial height and chronological age and the second included, in addition to the previous variables, the annual height increment for ages 10.5 to 14.5 years in girls and 12.5 and 16.5 years in boys. Shuttleworth (1939) emphasized that these latter tables should be used with 'some discrimination if their value is to be realized. The interval on which the gain (annual increment) is based should be within two or three weeks of the exact one-year period. The original measurements must be taken with care, by the same method, at the same time of day, and to the nearest quarter of an inch. Finally, these tables must not be used outside the specified age ranges.' Shuttleworth also included the probable error of estimate for each age specific column of his tables, ±4 cm at worst.

These two early methods highlight the original clinical and biological interest in prediction techniques and both relied on the phenomenon of canalization and the use of a maturity variable to infer the possibility of deviations from the genetically determined growth canal; Gill and Abbott (1942) used skeletal maturity, and Shuttleworth (1939) used height velocity as the refining variable.

Almost without exception the variables of present height and chronological age have been used on all prediction methods since these two pioneering studies. The major, and very important, difference between most techniques is the way in which other variables have been used to cater for individual differences in growth rate. It is these predictor variables that have provided interesting and important information on the process of growth and the major influences that affect final height.

Approaches to prediction

Prediction techniques have been developed using three approaches. In the first and simplest case single cross-sectional variables have been used to predict adult height. The use of parental height is a prime example of this method. However, in all other methods for predicting a future event in the growth and development of a child the current age of the child (chronological age), is a necessary variable. Any measured variable associated with a child during growth implicitly carries the information of chronological age. Thus height is actually height-for-age, and its use becomes an example of the second case in which combinations of cross-sectionally obtained variables have been used. Shuttleworth's (1939) technique is an example in which present height and age are the initial variables. The third case involves longitudinally obtained data. These data relate to the changing rate of somatic growth and/or maturation over a set period of time. Shuttleworth's (1939) second technique is such a situation in which height velocity from one age to the next is used in conjunction with cross-sectionally derived variables.

The prediction of adult height will be examined from the perspective of these three different approaches in order to highlight how individual variables, such as present height and parental height, can be assessed for usefulness, and to demonstrate the effect of combining variables. Almost without exception the more recent prediction techniques are multivariate techniques employing regression equations. This fact alone indicates that multivariate techniques reduce the errors of prediction. The problem involved in multivariable techniques is the variety of data required to accomplish a prediction. Some authors (e.g. Roche *et al.*, 1975a) have tried to overcome this problem by allowing the substitution of population mean values for the missing data. Others (Tanner *et al.*, 1975, 1983a, 2001) have provided a number of alternative equations depending on the available data.

Parental size

In many respects it is a misnomer to describe parental height as a predictor variable. The main reason for this is that prediction is usually thought of as determining a single, discrete, adult value rather than a range of values. It is, of course, true that a range of error is always associated with a predicted height but when using parental height alone only a range of possible heights, the target range, can be determined. The method is constant and involves finding the average of the heights of the parents after having corrected the mother's or father's height for the gender of the subject. Obviously a range exists around this value. The size of the range is determined by within-family variance and, for British families, is 17.0 cm (Tanner, 1989). It is important not to confuse the midparent height with a predicted height. The difference is that when predicting a child's height one is making an estimate of the likely adult height of that specific child. Midparent height is the average height expected from the offspring of a specific set of parents. Welch (1970) demonstrated that prediction improves with a combined estimate of parental stature. Standard deviations of errors of prediction reduced, in males, from 6.06 cm using father's height, to 5.29 cm using mother's height and 5.08 cm using midparent stature. In girls the reduction was from 5.83 cm to 5.21 cm and 5.17 cm respectively. An earlier method of correcting parental stature for the gender of the subject was to multiply father's stature by 0.923 or mother's stature by 1.08 (Gray, 1948). This slightly enlarges the difference between British adult male and female mean heights but only by the order of 0.5 cm when increasing mother's height and 1.2 cm when decreasing father's height. Except as a rule-of-thumb method to estimate a likely range for children's heights the use of parental height is not very satisfactory. It has been suggested, however, that this technique may have particular usefulness when used in conjunction with growth charts to determine whether a child is achieving his or her genetic potential for growth (Goldstein and Tanner, 1980). Or indeed, in developing countries where there are no growth standards or charts, a target range determined from parental heights may have advantages over using international references.

Single predictors

Roche *et al.* (1975a) investigated the usefulness of nine single predictors of adult stature; current stature, stature at menarche or puberty, tibial length, pattern of growth in stature, body shape, secondary sex characters, age at menarche, age at peak height velocity and skeletal age. Apart from current stature variables that are partitions of stature, e.g. long-bone lengths, provide little information.

The pattern of growth in stature requires some explanation; Roche *et al.* (1975a) took this description to mean a child's centile position at one moment in time or the rate of change in stature from one time to a subsequent time, or a series of incremental values covering a longer period of time. The former interpretation as a cross-sectionally derived estimate of height in relation to the rest of the population takes us back to the percentile method used by Gill and Abbot (1942) and Shuttleworth (1939). Certainly in the first few years of growth centile position is not a good indicator of eventual adult centile position. Tanner *et al.* (1956), using data from the Aberdeen study of Alexander Low, demonstrated the relatively low correlations between height in the first 3 to 5 years of life and eventual adult height in the same individuals. Simmons (1944) demonstrated that most children stay within a particular quartile range of height for most of the growth but, as Roche *et al.* (1975a) point out, this does not imply that the assumption of an invariate growth canal is suitable to provide reliable prediction information. Centile position, however, in combination with other predictor variables is a useful part of a multiple-variable predictor situation.

It is difficult, if not impossible, to crystallize 'body shape' into a meaningful variable for predictive purposes. Body shape is generally taken to mean a particular somatotype derived from either the Sheldon (1940) or Heath–Carter (Heath and Carter, 1967) methods. Whilst methods do exist to predict adult somatotype from childhood ages (Walker and Tanner, 1980) height per se is not an inherent variable of body shape analysis except within a weight/height index such as body mass index.

The age at which certain pubertal events such as menarche or peak height velocity occur has been used within multiple regression techniques to predict adult stature (Onat, 1975). Age at menarche alone, however, accounts for only 65% of total variance in adult stature when calculated from a multiple correlation coefficient (Simmons, 1944). Age at menarche has none the less been used in a number of prediction techniques as has the presence or absence of a regular menstrual cycle (Onat, 1975; Tanner *et al.*, 1975). The problem of obtaining an exact age of a particular event is that of the subject's inability to recall accurately their exact age at the time of the event. Historical clues can be used to prompt the subject (e.g. the season of the year, whether she was attending school or on vacation, etc.) but the best that can usually be achieved is a 'guesstimate' to within 3 months.

Empirical evidence obviously requires serial data and thus as a cross-sectional single-predictor variable age at peak height velocity (PHV) is not useful. The presence or absence of PHV is possibly useful but one needs to use other characteristics, e.g. the ossification of the ulnar sesamoid, whether menarche has occurred or the development of secondary sexual characteristics, to determine whether the subject is pre- or post-PHV.

The suitability and importance of skeletal maturity as a predictor variable has been recognized since the early twentieth century (Long and Caldwell, 1911). Methods of accurately determining skeletal maturity were, however, not widely available until Wingate Todd, William Walter Greulich and Idell Pyle pioneered the atlas techniques of the 1940s and 1950s (Todd, 1937; Greulich and Pyle, 1950, 1959), which are still probably, the most commonly used skeletal maturity methods. Once this technique was available all future prediction methods made use of skeletal maturity as a predictor variable. In contrast to chronological age, skeletal age provides a developmental time base to growth without the necessity of serial data to investigate relative position within the growth curve. Used alone, however, skeletal age has been shown by Bayley (1943) and Simmons (1944) to account for only 40–80% of the percentage of adult stature achieved by males and females between the ages of 10 and 17 years. One of the problems may be that the hand and wrist, commonly used as the preferred site for skeletal age determination, may not provide the required information that would more closely relate skeletal maturity to adult stature. It could be that the maturity of the knee, obtained from long-bones (Roche *et al.*, 1975b), would provide more sensitive information. None of these single predictor variables will, therefore, predict adult height with reasonable errors when used in isolation.

Multiple predictors

The reason for this lack of accuracy is that the prediction of adult height is not simply a process of predicting a one-dimensional variable – height; it is, in fact, a process of predicting two dimensions – height and maturity, i.e., the height of the individual at full maturity. Thus it is necessary to use indicators of both height and maturity in a prediction technique. When this is done errors of prediction are dramatically reduced and the errors remaining are due to inaccuracy in our maturity assessments plus an inability to estimate correctly the exact genetic contribution of the parents of the subject. It follows that the most accurate variables for assessing maturity will be of most use in a prediction situation. It is unlikely that presence or absence variables, e.g. the presence or absence of menarche, PHV, the ulna sesamoid, etc., will be as useful or more useful than time-based variables such as skeletal age. It also follows that continuous variables will be of more use than discrete variables. It is, therefore, not surprising that prediction methods that have survived the tests of clinical and biological usefulness are the ones that incorporate a multitude of variables that relate to adult height and maturity and that are sensitive in their assessment of childhood maturity.

Predicting adult size by determining the percentage of completed growth

The Bayley–Pinneau tables, which allow the prediction of adult height, were first produced in 1946 and later modified and republished in 1952. The theoretical basis for the method is the determination of the percentage adult height at a particular skeletal age. This theoretical basis was developed in Bayley's 1943 publication in which she investigated skeletal maturing in adolescence as a basis for determining the percentage of completed growth. Bayley studied the radiographs of the left hands and knees of a group of 177 children (87 girls) who were examined as part of the adolescent growth study of the University of California at Berkeley's Institute of Child Welfare. Radiographs were taken every 6 months beginning when the subjects averaged 14 years of age and continuing for 4.5 years. The importance of skeletal age as the major determinant of growth status was seen from the following results: (1) standard deviations of percentage of mature size attained at successive ages were smaller when the subjects were grouped according to skeletal age than when grouped according to chronological age, (2) correlations between different measures at each age group gave higher correlations with skeletal age, (3) correlations for consistency in height over a period of years were greater between skeletal age groups.

Bayley (1943) concluded that 'It appears that growth in size is closely related to the maturing of the skeleton. At a given skeletal age we may say that a child has achieved a given proportion of his eventual adult body dimensions. Consequently, mature size can be predicted with fair accuracy if a child's present size and skeletal age are known.' Three years later these findings had been distilled into the Bayley prediction technique using the original Todd atlas (Todd, 1937). Tables were presented that facilitated estimates of adult stature for boys and girls between the ages of 7 years and maturity. The tables were constructed from two sources: Massachusetts children from the Harvard growth study of Shuttleworth (1939) and Californian children from the Berkeley study mentioned above. Bayley (1946) describes the error as a mean and standard deviation from the actual mature height of a validating group of Berkeley girls and Harvard boys. Sixty per cent of both males and females had errors within 2.5 cm of actual height after the age of 11 years in boys and 9 years in girls when skeletal age is used. Chronological age delayed these ages to 15 and 14 years respectively before more than 60% of the subjects are within 2.5 cm of actual height predicted. The usefulness of Bayley's (1946) method was thus during adolescence when variations in the tempo of human growth deviate a child from the genetically determined growth canal.

The revision of Todd's *Atlas of Skeletal Maturation* by Greulich and Pyle (1950) necessitated a revision of Bayley's (1946) tables. Rather than simply update the existing tables Bayley, with Samuel Pinneau, completely reworked her data using 192 subjects from the Guidance Study of the Institute of Child Welfare at Berkeley and validated them on 46 children from the Berkeley Growth Study. The new tables were supplemented by tables allowing for accelerated and retarded subjects during adolescent years.

These Bayley–Pinneau (1952) tables were the only prediction tables available to the clinical and biological practitioner until Alex Roche and James Tanner, working independently, published their techniques almost 25 years later (Roche *et al.*, 1975a; Tanner *et al.*, 1975).

Multivariate prediction

Roche's 1975 monograph on predicting adult stature for individuals was a landmark in prediction literature (Roche *et al.*, 1975a). Here, for the first time, the subject of prediction as a scientific theory and practical exercise was examined in depth as an introduction to a new method of predicting adult stature and is fundamentally important reading for the researcher in human growth and development. The Roche–Wainer–Thissen method grew out of a detailed consideration of previous prediction methods and current knowledge concerning the factors influencing present growth status and future mature height. Roche *et al.* (1975a) examined possible single and multiple predictors before embarking on the development of their own method. The need for a new method had grown out of Roche's dissatisfaction with Bayley–Pinneau's technique when applied to 'deviant' children. In particular Roche and Wettenhall (1969) had investigated the prediction of adult stature in tall girls and, in the discussion to the paper, outlined a series of criticisms:

- The Bayley–Pinneau method makes no allowance for possible future changes in the rates of statural growth and skeletal maturation.
- The measurement of height and the assessment of skeletal maturity (as used by Bayley and Pinneau (1952)) were 'inexact.'
- The carpal bones need not be assessed when determining skeletal age and an arithmetic mean of the other bones of the hand–wrist should be accepted instead.
- Successive predictions on the same subject may differ markedly.
- There was a 'slight tendency' for errors to be smaller for predictions made at more advanced skeletal ages, or at longer intervals after menarche, or when there were smaller actual potentials for growth in height.
- At skeletal ages of 14.0 years or more there was a tendency to overpredict by a mean of 7 mm.

Roche and his colleagues did not expect to solve all these problems. Indeed, Roche describes their technique as a 'partial solution'. Clearly the achievement of a perfect prediction method was, and still is, outside the scope of human biology.

Roche *et al.* (1975a) chose their sample from the long-running Fels Longitudinal Growth Study. They chose the age of 18 years as their prediction target age and excluded any subjects whose skeletal maturation was complete or 'nearly complete'. In practice this meant that children were excluded at those ages when 50% or more of the bones of either the hand–wrist, foot–ankle or knee were adult. Roche maintains that this exclusion was necessary because a skeletal age cannot be assigned to a mature individual. This is true of all skeletal maturity systems; at full maturity, regardless of chronological age, one is 'mature' even though in an assessment made 24 hours earlier one may have had a skeletal age of 17.99 bone age 'years'. Naturally the exclusion of such children means that late maturers bias the older age groups. Roche maintains that the inclusion of older, mature children would increase the accuracy of prediction because they would have ceased growing. Correlations between present and adult heights would be increased and the weighting given to bone age would be 0.0 to exclude the effect of maturity. The children were born between 1929 and 1954 but the Fels participants showed little or no positive secular trends in height over that time period. They were reported as a fairly representative sample of American children of the time coming from social backgrounds equally spread amongst labourers, farmers, small businessmen, white-collar workers and professionals. The subjects were seen at 3-monthly intervals up to 12 months and then at 6-monthly intervals to 18 years, thereafter every 2 years. Radiographs were taken of the hand and wrist, foot and ankle, and knee. Skeletal maturities of the regions, specific bones within the regions and parts of specific bones, e.g. lateral femoral condyle in addition to femoral skeletal age, were obtained from standard radiographic atlases using the bone specific technique. (Greulich and Pyle, 1959; Hoerr *et al.*, 1962; Pyle and Hoerr, 1969). Bone-specific skeletal ages from the hand and wrist were combined to produce 11 further estimates of maturity. Thus 75 different estimates of skeletal maturity were available for analysis – 64 from the individual bones or parts of bones and 11 area skeletal ages. In addition the measurements of recumbent length, stature, weight and parental heights were added to the serial data set for each child. This collection of 78 possible predictor variables for each child was obviously far too large to contemplate as a usable predictor set. Roche thus set about reducing the number of variables.

Two strategies were applied; the first involved examining the interrelationships among the possible predictors and the second involved close examination of a reduced set of predictors within sex and age. Six conclusions were reached by Roche and his colleagues following these procedures:

1. Composite or area skeletal ages were more effective than bone-specific skeletal ages.
2. Although means and medians were approximately equal in utility medians were more stable in the presence of outlying values.
3. Whilst the knee and hand–wrist skeletal ages were of about equal value the hand–wrist conformed with criteria of reduced irradiation to the subject, ease of positioning and ease of measurement.
4. Almost all information about skeletal maturity needed for prediction is present in the hand–wrist complex.
5. Maternal and paternal stature aided equally in prediction thus midparent stature was chosen.
6. Present recumbent length and present weight were the only other desirable variables.

Thus the total of 78 variables was reduced to four; present recumbent length (or adjusted stature), present weight, midparent adult stature and median skeletal age of the hand–wrist.

These four predictor variables having been selected, their regression weightings were determined for each age. Not all of the predictors were useful at all ages. In boys, for example, all variables were only useful at 13 to 14 years of age, i.e. during the period of rapid pubertal growth and development. Up to the age of 3 years midparent stature and recumbent length were most useful and weight contributed significantly from 3 to 12.5 years. The squared multiple correlation coefficients for boys and girls demonstrated that on average over 70% of the variance in adult stature was being accounted for by this combination of predictor variables.

Experienced researchers in human growth may well take the view that a logical subjective assessment of the predictor variables would have resulted in length, midparent stature and skeletal maturity as being the most appropriate variables without going through the laborious process of variable selection that Roche and his colleagues felt necessary. However, it is of signal importance that Roche and his colleagues chose to investigate 75 possible predictor variables from the maturing skeleton because they provided empirical evidence that the hand–wrist was the most suitable area and that combinations of bone maturities were better than single bone indicators.

The prediction methods developed by Tanner and his colleagues in the UK used a less rigorous but equally successful approach to prediction. They applied standard regression techniques to predict adult stature from the variables of current height, chronological age and skeletal age determined by the Tanner–Whitehouse radius, ulna and short bones (RUS) method (see Chapter 5). No other cross-sectional variables were included but greater prediction accuracy was obtained by using information on whether or not girls had attained menarche. The source data came from two longitudinal growth studies: the Harpenden (HGS) and the International Children's Centre, London (ICC) studies. In both studies a relatively large number of children had been followed until growth in height had diminished to less than 1.0 cm/year. This was true for 79 boys and 56 girls from the Harpenden study and 37 boys and 39 girls from the ICC study. While the ICC children were measured annually the Harpenden children were measured following a similar regime to the Fels study; 6-monthly until puberty, 3-monthly during puberty, yearly until age 20 years and then 5-yearly. In addition to standard anthropometric measurements radiographs of the hand–wrist were obtained on each measurement occasion and were rated following the TW2 system (Tanner *et al.*, 1983b). Tanner applied classical regression analyses to these data within yearly age groups, e.g. 8.000 to 8.999 years, with one source of sampling error; because subjects were measured more than once a year they were included in the regressions more than once within any particular age group – indeed Tanner states that 'most appear two or more times in each group'. This inclusion had the effect of reducing the accuracy of the residual standard deviations of regression but according to Tanner *et al.* (1975), the residuals were not biased.

A variety of combinations of the predictor variables were regressed on adult height to find the best combination. The quality of the prediction was determined from the residual standard deviation and the correlation coefficient. The successive inclusion of chronological age and RUS bone age with present height reduced the size of the residuals. In addition interaction variables, e.g. chronological age × height were tried but the best combination was height, chronological age and RUS bone age. Of particular interest is that Tanner *et al.* (1975) found the carpal bone age to be detrimental to the accuracy of prediction. The exclusion of the carpal bones reduced the residual standard deviation of the full TW2 (20) system. Roche *et al.* (1975a) also found the carpal bones to be unhelpful in this prediction situation. It will also be noted that interaction terms of bone age with height could not improve the prediction beyond that already achieved by the RUS system in isolation. Tanner *et al.* (1975) found that bone age was important over a greater age span than Roche *et al.* (1975a) and described its inclusion as 'essential' from 11 to 16 years in boys and from 9 to 14 years in girls. The knowledge of whether a female was pre- or post-menarcheal significantly improved predictions but the actual age of menarche made no significant improvement.

Like Roche *et al.* (1975a) Tanner and his colleagues smoothed the partial regression coefficients, not by using a curve-fitting approach but by using 'graphical' smoothing. The final equations were presented in half-year age bands to reduce the total number of equations. Tanner *et al.* (1975) computed the regressions at each year of age between the errors of estimate and midparent height from the ICC study. The relationship was significant over the ages 9 to 14 years in boys and 7 to 13 years in girls, suggesting that an allowance should be made. This problem was solved by recommending that one-third of the difference between midparent height and mean midparent height should be added to the final predicted height of the subject. Clinical experience of the UK team showed there to be a variety of children who were beyond the upper age limit of the chronological-age-based equations, i.e. 17.99 years, and yet were still growing. This is particularly true of boys classified as 'small-delayed'. Because of this situation Tanner *et al.* (1975) provided an additional set of equations based on RUS bone age rather than chronological age. In this way an individual of 18+ years with a delay of 2 or 3 years in bone age could still be included in a prediction situation.

During the next few years Tanner's colleagues, who were using prediction techniques on children attending the Growth Disorder Clinic at the Hospital for Sick Children in London, became increasingly concerned that they were applying a prediction method based on normal children to children suspected of abnormality. Aware of the dangers of extrapolating regression equations beyond the characteristics of the original source sample the team decided to revise the original equations including in the new source sample numbers of very tall and very short children in the standardizing group (Tanner *et al.*, 1983a). In addition a complete revamp of the original predictor variables was carried out because of the realization that clinicians often had data available of a longitudinal nature that might be of use in prediction. The new source sample was composed of 69 boys from the HGS, 41 from the ICC study and 34 boys from the Growth Disorder Clinic who had growth delay, genetic short stature or both. Fifty-two females from the HGS, 41 from the ICC study and 57 girls from the Royal Ballet School were included. (Pupils at the Royal Ballet School were required to have their adult heights predicted because they were expected to be within certain prescribed height ranges as adults within the corps de ballet.) The girls at the Royal Ballet School had been the subjects of height predictions throughout the 1970s and were now adult according to the criteria of a velocity less than 1.0 cm/year. Nineteen tall girls and 10 short girls from the Growth Disorder Clinic were also included. Thus the new sample was not only greater in numbers but also included more extreme subjects. In addition to the previously tested predictor variables of chronological age, present height and RUS bone age, two longitudinal variables were added to the analysis; the increment of height (ΔH) and the increment of RUS bone age (ΔRUS) in the year preceding measurement. Thus five variables were included in the standard multiple regressions in the form of four equations with each incorporating the relevant interaction terms:

1. Height, chronological age and RUS.
2. Height, chronological age, RUS and ΔH.
3. Height, chronological age, RUS and ΔRUS.
4. Height, chronological age, RUS, ΔH and ΔRUS.

In girls these equations were calculated for pre- and postmenarcheal subjects separately.

The first equation revealed the standard relationship of RUS bone age being important from 8 years in boys and 6 years in girls. ΔH improved the prediction significantly at ages 12 and 13 years in boys and at 8 to 11 years in girls. In postmenarcheal girls ΔH improved the prediction from 12 to 15 years but this was not the case for premenarcheal girls. ΔRUS had no positive effect on male predictions but reduced the residual standard deviation in premenarcheal girls at 13 and 14 years and postmenarcheal girls at 12 and 13 years. The inclusion of both ΔH and ΔRUS was no advantage to male predictions but in girls this combination was the best, having important effects on all girls at 13 years.

The third version of the Tanner–Whitehouse height prediction equations (TW3) (Tanner *et al.*, 2001) incorporated bone maturity score rather than bone age as a predictor variable. Tanner and his colleagues posited that skeletal age was conditioned by environmental factors such as nutrition, health and socio-economic status that caused secular trends. Actual maturity was represented more accurately by the actual bone maturity scores rather than bone age and thus bone score was used in the regressions rather than bone age. The equations are of the form:

$$\text{predicted adult height} = \text{present height} + a \text{ RUS score} + b$$

where a and b are constants which vary according to the chronological age of the child. Once again height increment is used at certain ages (12.0 to 14.9 in boys and 11.0 to 13.9 in girls) and knowledge as to whether the girls are pre- or postmenarcheal improves prediction. The source for this new series of equations was the Zürich Longitudinal Study of Prader *et al.* (1989) and the equations were validated using the ICC, HGS and University of Virginia growth studies.

Validation

Both Roche *et al.* (1975a) and Tanner *et al.* (1975) validated their prediction equations by using them on other suitable study samples; Roche *et al.* (1975a) against the Fels, the Child Research Center, Denver and the School of Public Health, Harvard samples and Tanner *et al.* (1975) against girls from the Royal Ballet School. Roche's testing was, perhaps, more rigorous in that he tested R^2 and relative and absolute errors compared to Tanner's testing of relative error only. But both investigations are of interest. The Roche–Wainer–Thissen method was compared to the Bayley–Pinneau technique and generally produced a smooth error profile across all age bands compared to the jagged appearance of the Bayley–Pinneau error profile. The Bayley–Pinneau method tended to underpredict compared to that of Roche–Wainer–Thissen but, apart from the older age groups, produced greater prediction errors than the latter method. The original Tanner–Whitehouse method (TW1), tested on 62 girls from the Royal Ballet School, resulted in all children falling within ±8 cm of actual final height. Sixty (97%) were within ±6 cm, 54 (87%) were within ±4 cm and 32 (52%) were within ±2 cm of actual final height. Cameron *et al.* (1985) investigated the application of the updated TW prediction equations to a group of Canadian boys aged 11 years. These investigators wished to test whether the increased heterogeneity of the source sample and the generally larger residual standard deviations of the TW2 equations would adversely affect the accuracy of prediction in normal children. In addition they tested, on normal short and tall boys, whether the TW2 equations predicted better than the older TW1 equations. The TW2 equations significantly improved the accuracy of prediction and particularly so in short boys but had no real increased accuracy in tall boys. Harris *et al.* (1980) compared the 'relative accuracies' of these prediction techniques, excluding TW2, on a sample of 22 males and 24 females from the Child Research Center, Denver, longitudinal growth study. Six ages were tested within each sex: 5.0, 8.0, 12.0, 13.0, 14.0 and 15.0 years. The Bayley–Pinneau technique tended to overpredict compared to consistent underpredictions for Roche–Wainer–Thissen and TW1. There were significant differences in the methods' accuracies, but inspection of the data by age revealed that the major discrimination occurred during adolescence. The major finding was that the method of skeletal age assessment, i.e. Greulich–Pyle or TW2, was more critical to accurate height prediction than the choice of method per se. Harris *et al.* (1980) correctly interpreted this as a reflection of the inter-population differences in the 'rate and pattern of progress towards maturity'. Tanner *et al.*'s latest (TW3) technique produced unbiased estimates of adult stature in the ICC and Virginia boys with mean differences of less than 2 cm and none that exceeded twice the standard error.

Implications for general prediction in human growth

There is no doubt that the approaches of Bayley, Roche and Tanner advanced, in no small degree, an understanding of the relative importance and age dependence of certain predictor variables. Whilst the need for 'clinically insignificant' error is evident the achievement of that goal is difficult to assess. Mainly because the significance of the error depends upon the importance of the need for clinical assessment of final height. If the predicted final height signals invasive, and potentially dangerous, treatment regimes then the clinician and patient must be sure that the gain is worth the risk. The 'significance' of the error is not, of course, a statistical significance but a subjective significance determined by the clinician and patient.

The principles of prediction

The principles of prediction that emerge from this discussion and review are:

1. Future morphological status can be predicted using any single associated variable or combination of associated variables obtained at any earlier time. These variables are known as 'predictor variables'.
2. The inclusion of indicators of growth increment, maturity, maturity increment and heritability will increase prediction accuracy.
3. The usefulness of predictor variables will change over time such that those useful at one time may be less useful at another.
4. The closer the prediction target is to the time of prediction the more accurate will be the outcome.

REFERENCES

Bartholomeusz, H. H., Courchesne, E. and Karns, C. M. (2002). Relationship between head circumference and brain volume in healthy normal toddlers, children, and adults. *Neuropediatrics*, **33**(5), 239–41.

Bayley, N. (1943). Skeletal maturing in adolescence as a basis for determining percentage of completed growth. *Child Development*, **14**, 1–46.

(1946). Tables for predicting adult height from skeletal age and present height. *Journal of Pediatrics*, **28**, 49–64.

Bayley, N. and Pinneau, S. R. (1952). Tables for predicting adult height from skeletal age: revised for use with Greulich–Pyle hand standards. *Journal of Pediatrics*, **40**, 423–41.

Cameron, N. (1984). *The Measurement of Human Growth*. London: Croom-Helm.

Cameron, N., Mirwald, R. L., Bailey, D. A. and Davies, P. S. W. (1985). The application of new height prediction equations (Tanner–Whitehouse Mark 2) to a sample of Canadian boys. *Annals of Human Biology*, **12**, 233–9.

Deurenberg, P., Andreoli, A., Borg, P., *et al.* (2001). The validity of predicted body fat percentage from body mass index and from impedance in samples of five European populations. *European Journal of Clinical Nutrition*, **55**(11), 973–9.

Gill, G. G. and Abbott, L. C. (1942). Practical method of predicting the growth of the femur and tibia in the child. *Archives of Surgery, Chicago*, **45**, 286–315.

Goldstein, H. and Tanner, J. M. (1980). Ecological considerations in the creation and use of child growth standards. *Lancet*, **i**, 582–7.

Gray, H. (1948). Predictions of adult stature. *Child Development*, **19**, 167–75.

Greulich, W. W. and Pyle, S. I. (1950). *Radiographic Atlas of the Skeletal Development of the Hand and Wrist*. Stanford, CA: Stanford University Press.

(1959). *Radiographic Atlas of the Skeletal Development of the Hand and Wrist*, 2nd edn. Stanford, CA: Stanford University Press.

Harris, E. E., Weinstein, S., Weinstein, L. and Poole, A. E. (1980). Predicting adult stature: a comparison of methodologies. *Annals of Human Biology*, **7**, 225–34.

Heath, B. H. and Carter, J. E. L. (1967). A modified somatotype method. *American Journal of Physical Anthropology*, **27**, 57–74.

Hoerr, N. L., Pyle, S. I. and Francis, C. C. (1962). *Radiographic Atlas of Skeletal Development of the Foot and Ankle: A Standard of Reference*. Springfield, IL: C. C. Thomas.

Larsson, I., Lindroos, A. K., Peltonen, M. and Sjostrom, L. (2003). Potassium per kilogram fat-free mass and total body potassium: predictions from sex, age, and anthropometry. *American Journal of Physiology, Endocrinology and Metabolism*, **284**(2), E416-E423.

Lohman, T. G., Roche, A. F. and Martorell, R. (1988). *Anthropometric Standardization Reference Manual*. Champaign, IL: Human Kinetics.

Long, E. and Caldwell, E. W. (1911). Some investigations concerning the relation between carpal ossification and physical and mental development. *American Journal of Diseases of Childhood*, **1**, 113–38.

Onat, T. (1975). Prediction of adult height of girls based on the percentage of adult height at onset of secondary sexual characteristics, at chronological age, and skeletal age. *Human Biology*, **47**, 117–30.

Perez-Padilla, R., Regalado-Pineda, J., Rojas, M., *et al.* (2003). Spirometric function in children of Mexico City compared to Mexican-American children. *Pediatric Pulmonology*, **35**(3), 177–83.

Prader, A., Largo, R. H., Molinari, L. and Issler, C. (1989). Physical growth of Swiss children from birth to 20 years of age: First Zürich Longitudinal Study of Growth and Development. *Helvetica Paediatrica Acta, Supplementum*, **52**, 1–125.

Pyle, S. I. and Hoerr, N. L. (1969). *A Radiographic Standard of Reference for the Growing Knee*. Springfield, IL: C. C. Thomas.

Roche, A. F. and Wettenhall, H. N. B. (1969). The prediction of adult stature in tall girls. *Australian Paediatric Journal*, **5**, 13–22.

Roche, A. F., Wainer, H. and Thissen, D. (1975a). Predicting adult stature for individuals. *Monographs in Pediatrics*, **3**, 1–114.

(1975b). *Skeletal Maturity: The Knee Joint as a Biological Indicator*. New York: Plenum Press.

Schild, R. L., Fimmers, R. and Hansmann, M. (2000). Foetal weight estimation by three-dimensional ultrasound. *Ultrasound Obstetrics and Gynecology*, **16**(5), 445–52.

Sheldon, W. H. (1940). *The Varieties of Human Physique*. New York: Harper and Row.

Shuttleworth, E. K. (1939). The physical and mental growth of girls and boys age six to nineteen in relation to age at maximum growth. *Monographs of the Society for Research in Child Development*, **4**, 1–291.

Simmons, K. (1944). The Brush Foundation Study of child growth and development. II. Physical growth and development. *Monographs of the Society for Research in Child Development*, **9**, 1–87.

Stetzer, B. P., Thomas, A., Amini, S. B. and Catalano, P. M. (2002). Neonatal anthropometric measurements to predict birth weight by ultrasound. *Journal of Perinatology*, **22**(5), 397–402.

Tanner, J. M. (1989). *Foetus into Man*, 2nd edn. London: Castlemead.

Tanner, J. M., Healy, M. J. R., Lockhart, J. D., MacKenzie, J. D. and Whitehouse, R. H. (1956). Aberdeen Growth Study. I. The prediction of adult body measurements from measurements taken each year from birth to 5 years. *Archives of Disease in Childhood*, **31**, 372–81.

Tanner, J. M., Whitehouse, R. H., Marshall, W. A. and Carter, B. S. (1975). Prediction of adult height from height, bone age, and occurrence of menarche, at ages 4 to 16 with allowance for midparent height. *Archives of Disease in Childhood*, **50**, 14–26.

Tanner, J. M., Landt, K., Cameron, N., Carter, B. S. and Patel, J. (1983a). Prediction of adult height from height and bone age in childhood: a new system of equations (TW Mark 11) based on a sample including very tall and very short subjects. *Archives of Disease in Childhood*, **58**, 767–76.

Tanner, J. M., Whitehouse, R. H., Cameron, N., *et al.* (1983b). *Assessment of Skeletal Maturity and Prediction of Adult Height (TW2 Method)*, 2nd edn. London: Academic Press.

Tanner, J. M., Healy, M. J. R., Goldstein, H. and Cameron, N. (2001). *Assessment of Skeletal Maturity and Prediction of Adult Height (TW3 Method)*, 3rd edn. London: Academic Press.

Todd, T. W. (1937). *Atlas of Skeletal Maturation*. St Louis, MO: C. V. Mosby.

Ulijaszek, S. J. and Lourie, J. A. (1994). Intra- and inter-observer error in anthropometric measurement. In *Anthropometry: The Individual and the Population*, eds. S. J. Ulijaszek and C. G. N. Mascie-Taylor, pp. 30–55. Cambridge, UK: Cambridge University Press.

Walker, R. N. and Tanner, J. M. (1980). Prediction of adult Sheldon somatotypes I and II from ratings and measurements at childhood ages. *Annals of Human Biology*, **7**, 213–24.

Welch, Q. B. (1970). Fitting growth and research data. *Growth*, **34**, 293–312.

15 *Ordinal longitudinal data analysis*

Jeroen K. Vermunt and Jacques A. Hagenaars
Tilburg University

Introduction

Growth data and longitudinal data in general are often of an ordinal nature. For example, developmental stages may be classified into ordinal categories and behavioural variables repeatedly measured by discrete ordinal scales. Consider the data set presented in Table 15.1. This table contains information on marijuana use taken from five annual waves (1976–80) of the National Youth Survey (Elliot *et al.*, 1989; Lang *et al.*, 1999). The 237 respondents were 13 years old in 1976. The variable of interest is a trichotomous ordinal variable 'Marijuana use in the past year' measured during five consecutive years. There is also information on the gender of the respondents.

Ordinal data like this is often analysed as if it were continuous interval level data, that is, by means of methods that imply linear relationships and normally distributed errors. However, the data in Table 15.1 is essentially categorical and measured at ordinal, and not at interval level. Consequently, a much better way to deal with such an ordinal response variable is to treat it as a categorical variable coming from a multinomial distribution; the ordinal nature of the categories is then taken into account by imposing particular constraints on the odds of responding, i.e. of choosing one category rather than another. As will be further explained below, an ordinal analysis can be based on cumulative, adjacent-categories, or continuation-ratio odds (Agresti, 2002). The constraints are in the form of equality or inequality constraints on one of these types of odds.

In this chapter, we will discuss the three main approaches to the analysis of longitudinal data: transition models, random-effects or growth models and marginal models (Diggle *et al.*, 1994; Fahrmeir and Tutz, 1994). Roughly speaking, transition models like Markov chain models concentrate on overall gross changes or transitions between consecutive time points, marginal models investigate net changes at the aggregated level and random-effects or growth models

Methods in Human Growth Research, eds. R. C. Hauspie, N. Cameron and L. Molinari.
Published by Cambridge University Press.

Table 15.1 *Data on marijuana use in the past year and gender, taken from five yearly waves of the National Youth Survey*

Gender[a]	1976[b]	1977	1978	1979	1980	Frequency	Gender	1976	1977	1978	1979	1980	Frequency
0	1	1	1	1	1	63	1	1	1	1	3	1	1
0	1	1	1	1	2	10	1	1	1	1	3	3	1
0	1	1	1	1	3	3	1	1	1	2	1	3	1
0	1	1	1	2	1	4	1	1	1	2	2	1	2
0	1	1	1	2	2	2	1	1	1	2	2	2	2
0	1	1	1	3	1	1	1	1	1	2	2	3	1
0	1	1	1	3	2	1	1	1	1	2	3	3	5
0	1	1	1	3	3	3	1	1	1	3	1	2	1
0	1	1	2	1	1	2	1	1	1	3	2	2	1
0	1	1	2	1	2	2	1	1	1	3	3	3	3
0	1	1	2	2	1	3	1	1	2	1	1	2	1
0	1	1	2	2	2	7	1	1	2	1	2	1	1
0	1	1	2	2	3	1	1	1	2	2	1	1	2
0	1	1	2	3	3	1	1	1	2	2	2	1	1
0	1	2	1	1	1	1	1	1	2	2	1	3	1
0	1	2	1	1	2	2	1	1	2	2	3	3	1
0	1	2	2	1	2	1	1	1	2	3	2	2	1
0	1	2	2	2	1	1	1	1	2	3	2	3	1
0	1	2	2	3	3	2	1	1	2	3	3	2	1
0	1	2	3	1	2	1	1	1	2	3	3	3	4
0	1	2	3	3	2	1	1	1	3	1	3	3	1
0	1	2	3	3	3	1	1	1	3	2	2	2	1
0	1	3	3	2	2	1	1	1	3	3	3	3	2
0	2	1	1	3	3	1	1	1	3	3	2	2	1
0	2	1	2	2	2	1	1	2	1	1	1	1	3
0	2	1	3	3	3	1	1	2	2	2	2	2	1
0	2	3	3	3	3	1	1	2	2	3	3	3	1
0	2	3	3	3	2	1	1	2	3	2	1	1	1
0	3	3	3	2	3	1	1	2	3	2	3	3	1
1	1	1	1	1	1	48	1	2	3	3	3	3	2
1	1	1	1	1	2	8	1	3	1	1	1	1	1
1	1	1	1	1	3	4	1	3	2	3	3	3	1
1	1	1	1	2	1	2	1	3	3	3	3	1	1
1	1	1	1	2	2	4	1	3	3	3	3	3	1
1	1	1	1	2	3	1							

[a] 0, female; 1, male.
[b] 1, never; 2, not more than once a month; 3, more than once a month.

study developments at the individual level. Variants of each of these have been developed for ordinal categorical variables (Agresti, 2002).

There are various complicating issues that have to be dealt with when analysing longitudinal data in general and ordinal longitudinal data in particular. The first is the issue of misclassification or measurement error (Hagenaars, 1990, 2002; Bassi *et al.*, 2000). Measurement of developmental stages is almost never perfect. For example, even in situations in which this is theoretically impossible, backward transitions will be observed and usually different indicators of the same phenomenon will provide partly inconsistent information. An important way to cope with measurement error is to introduce latent variables into the analysis. When dealing with categorical ordinal data that are assumed to measure ordinal true states, it is most natural to use categorical latent variables and latent class methods to this purpose.

Another important complicating issue, especially potentially harmful in transitional and growth models is the problem of unobserved heterogeneity (Vermunt, 1997, 2002). Random-effects approaches may be used to solve this problem. With ordinal variables, because of their non-linear nature, these models are somewhat more complicated and their estimation is more time consuming than with continuous outcome variables. A possible way out is to use a non-parametric random-effects approach based on latent class or finite mixture modelling.

The third issue is the presence of partially observed data. Methods for longitudinal data analysis are less useful if they can only deal with complete records. Fortunately, transition and growth models for ordinal variables can easily be adapted to deal with missing data (Hagenaars, 1990; Vermunt, 1997).

These complicating issues will be further discussed at the end of this chapter. First, logit models for ordinal response variables will be introduced. They form the basic building blocks for models for categorical longitudinal data. The main ways of analysing longitudinal data, viz. marginal, transitional and random-effects models for ordinal categorical variables, are also discussed.

Ordinal logit models

Given a dichotomous or polytomous ordinal outcome variable, the most popular model is the logit model. The logit model is a regression model in which the odds of choosing a particular category (or categories) of the response variable rather than another category (other categories) are assumed to depend on the values of certain independent variables (Agresti, 2002). Note that an odds is simply a ratio of two probabilities. The term logit comes from log odds, which refers to the fact that a logit model is a linear model for log odds.

The measurement level of the independent variables is also often ordinal. This ordinal character of the variables involved frequently leads to the assumption that the log odds are a linear function of predictors, which implies certain *equality* constraints on the odds ratios. As shown below, another important way of dealing with the ordinal nature of the variables of interest is to specify *inequality* restrictions on the odds ratios instead of *equality* restrictions. Various

Table 15.2 *Cross-tabulation of year and marijuana use in the past year based on data in Table 15.1*

	Time (X)				
Marijuana use (Y)	1. 1976	2. 1977	3. 1978	4. 1979	5. 1980
1. Never	218	195	167	156	138
2. No more than once a month	14	27	41	41	52
3. More than once a month	5	15	29	40	47

types of ordinal logit models can be defined depending on the type of odds that are being used (as mentioned above). It is also possible to use another link function than the logit link (using the generalized linear modelling jargon), and formulate probit, log-log, or complementary log-log models; Agresti provides an excellent overview of these possibilities (Agresti, 2002).[1] Here, we will only deal with logit models.

In order to make the discussion more concrete, assume that we have a two-way cross tabulation of X – time (or age) and Y – marijuana use, as shown in Table 15.2 (derived from Table 15.1). Variable X with category index i has $I = 5$ levels or categories and variable Y with category index j has $J = 3$ levels. It is assumed that X (time/age) serves as independent or predictor variable, that Y (marijuana use) is the dependent or response variable, and that we are interested in the conditional distribution of Y given X. The probability of giving response j at time i is denoted by $P(Y = j|X = i)$.

The substantive research question of interest is whether there is an increase of marijuana use with age; that is, whether the proportion of respondents in the highest categories increases over time. Note that this hypothesis does not imply any specific parametric form for the relationship between X and Y: we only assume that if X increases, Y will increase as well.

The most common way of modelling relationships between ordinal categorical variables is by means of a (linear) logit model that imposes equality constraints on certain odds ratios. Four types of odds can be used for this purpose (Agresti, 2002): cumulative odds denoted here as $O(\text{cum})_{i,j}$, adjacent-category (or local) odds $O(\text{adj})_{i,j}$, or one of two types of continuation-ratio odds

[1] The term link refers to the transformation of the dependent variable yielding the linear model. In a logit model, the response probabilities are transformed to log odds. It is, however, possible to work with other types of transformations. A probit link, for example, involves transforming the response probabilities to z values using the cumulative normal distribution.

$O(\text{conI})_{i,j}$, or $O(\text{conII})_{i,j}$. These are defined as

$$\begin{aligned}
O(\text{cum})_{i,j} &= P(Y \leq j|X = i)/P(Y \geq j+1|X = i) \\
O(\text{adj})_{i,j} &= P(Y = j|X = i)/P(Y = j+1|X = i) \\
O(\text{conI})_{i,j} &= P(Y = j|X = i)/P(Y \geq j+1|X = i) \\
O(\text{conII})_{i,j} &= P(Y \leq j|X = i)/P(Y = j+1|X = i)
\end{aligned}$$

for $1 \leq j \leq J-1$ and $1 \leq i \leq I$. Below, the symbol $O_{i,j}$ will be used as a generic symbol referring to any of these odds. As follows from these formal definitions, for an ordinal variable with four categories, the 'first' cumulative odds will be the odds of choosing category 1 rather than one of the other categories 2 or 3 or 4: $(1)/(2+3+4)$ and the other cumulative odds denoted in a similar way are $(1+2)/(3+4)$ and $(1+2+3)/(4)$. Using the same shorthand notation, the adjacent odds are $(1)/(2)$, $(2)/(3)$, and $(3)/(4)$ and the first type of continuation odds are $(1)/(2+3+4)$, $(2)/(3+4)$, and $(3)/(4)$.

In practical research situations, one has to make a choice between these four types of odds, that is, one has to specify a model for the type of odds that fits best to the process assumed to underlie the individual responses. The cumulative odds are the most natural choice if the discrete ordinal outcome variable is considered as resulting from a discretization of an underlying continuous variable. The adjacent category odds specification fits best if one is interested in each of the individual categories; that is, if one perceives the response variable as truly categorical. The continuation-ratio odds correspond to a sequential decision-making process in which alternatives are evaluated from low to high (type I) or from high to low (type II).

If X and Y are both treated as nominal level variables, no restrictions will be imposed on the odds. The logit model for this *nominal–nominal* case can be expressed as

$$\log O_{i,j} = \alpha_j - \beta_{ij}$$

which is just a decomposition of the log odds. Here, α_j is an intercept parameter and β_{ij} is a slope parameter.[2] As is the case in most models for categorical dependent variables, the intercept is category specific (for the categories of the dependent variable). Typical for the nominal–nominal case is that the slope depends on both the category of X and the category of Y.

[2] Note that the index j goes from 1 to $J-1$ and the index i from 1 to I. This means that no further restrictions need to be imposed on the $J-1$ α_j parameters. On the other hand, there are $(J-1) \times (I)$ free β_{ij} parameters, but we can identify only $(J-1) \times (I-1)$ of them. For identification, one can, for instance, assume that $\beta_{1j} = 0$ for each j, which amounts to treating the first category of X as reference category.

The most restricted specification is obtained if both the dependent and the independent variable are treated as ordered. This yields the *ordinal–ordinal* model

$$\log O_{i,j} = \alpha_j - \beta x_i$$

where x_i denotes the fixed score assigned to category i of X, and α_j and β are the intercept and the slope of the logit model. In most cases, x_i will be equal-interval scores (for example, 1, 2, 3, etc., or 1976, 1977, 1978, etc.), but it is also possible to use other scoring schemes for the X variable. As can be seen, the slope is assumed to be independent of the categories of X and Y.

Because of the restrictions involved, a better, although not common, name for the ordinal–ordinal would be the *interval–interval* model. For the adjacent-categories odds, this model is also known as the linear-by-linear association model (Goodman, 1979; Clogg and Shihadeh, 1994; Hagenaars, 2002). In fact, we do not only specify scores for the categories of X, but implicitly also assume that the categories of Y are equally spaced with a mutual distance of 1; for example, 1, 2 and 3, or 0, 1 and 2.

The other two, intermediate, cases are the *nominal–ordinal* specification (with Y nominal and X ordinal) and the *ordinal–nominal* case (with Y ordinal and X nominal). These are defined as follows:

$$\log O_{i,j} = \alpha_j - \beta_j x_i$$
$$\log O_{i,j} = \alpha_j - \beta_i$$

As can be seen, in the nominal–ordinal case, the slope is category specific for Y, but does not depend on X. In the ordinal–nominal case, the slope does not depend on Y.

These models are also known, for the adjacent categories odds, as row- or column-association models (Goodman, 1979; Clogg and Shihadeh, 1994; Hagenaars, 2002). In our case, the nominal–ordinal model is a row-association model because the nominal variable (marijuana use) serves as the row variable in Table 15.1. In a row-association model, the scores of the categories of the column variable are fixed and the scores of the categories of the row variable are unknown parameters to be estimated. In our parameterization, β_j can be interpreted as the distance between the unknown scores of categories $j+1$ and j. For equivalent reasons, the ordinal–nominal model is a column-association model, where β_i represents an unknown column score. The row scores are treated as fixed and assumed to have a mutual distance of 1.

Except for the nominal–nominal specification, each of these specifications implies certain equality constraints on the odd ratios $O_{i,j}/O_{i+1,j}$. When we use equal-interval x_i, the ordinal–ordinal model implies that the log-odds increase or decrease linearly with X and that, in other words, the log odds ratios between adjacent levels of X are assumed to be constant, to be equal to β:

$$\log(O_{i,j}/O_{i+1,j}) = \log O_{i,j} - \log O_{i+1,j} = \beta$$

for $1 \leq j \leq J-1$ and $1 \leq i \leq I-1$. The fact that these differences between log odds do not depend on the values of X and Y can also be expressed by the following two sets of equality constraints:

$$(\log O_{i,j} - \log O_{i+1,j}) - (\log O_{i,j'} - \log O_{i+1,j'}) = 0, \qquad (15.1)$$

$$(\log O_{i,j} - \log O_{i+1,j}) - (\log O_{i',j} - \log O_{i'+1,j}) = 0, \qquad (15.2)$$

where $i' \neq i$ and $j' \neq j$.

In the nominal–nominal specification of the logit model, the log odds ratio between adjacent categories of X equals $\beta_{i+1,j} - \beta_{i,j}$ and none of the two equality constraints (15.1) and (15.2) are valid. In the nominal–ordinal case, the log odds ratio equals β_j and equality constraints (15.2) apply: the odds ratios vary with Y but do not depend on X. In the ordinal–nominal case, the log odds ratio equals $\beta_{i+1} - \beta_i$ and equality constraints (15.1) are applicable: the odds ratios vary with X and do not depend on Y.

What can be observed is that the ordinal nature of the predictor variable is dealt with by assuming that the odds ratio is the same for each pair of adjacent categories. This amounts to assuming that the distances between all adjacent categories are equal, which is a much stronger assumption than ordinal, and essentially the interval-level assumption. The constraints related to the outcome variable imply that the effect of the predictor variable is assumed to be the same (constant) for each of the category-specific odds, which is also a more restrictive assumption than ordinal. An advantage of imposing the ordinal logit constraints is, however, that they force the solution to be in agreement with an ordinal relationship. Moreover, they provide a parsimonious and easy-to-interpret representation of the data: the effect of an ordinal predictor variable on an ordinal outcome variable is described by a (very) small number of parameters.

However, in many situations, the ordinal–ordinal specification is too restrictive even if the relationship between the variables of interest is truly ordinal. In such cases, one may consider staying closer to the definition of ordinal measurement. In terms of odds, the purest definition of a positive, (weakly) monotonically increasing relationship between two ordinal variables involves the following set of inequality constraints:

$$\log O_{i,j} - \log O_{i+1,j} \geq 0, \qquad (15.3)$$

for $1 \leq j \leq J-1$ and $1 \leq i \leq I-1$ (Vermunt, 1999). As can be seen, we are assuming that all log odds ratios are at least zero, or, equivalently that all odds ratios are larger than or equal to 1. Such a set of constraints is often referred to as simple stochastic ordering, likelihood ratio ordering, or uniform stochastic ordering for cumulative, adjacent category, and continuation odds, respectively (Dardanoni and Forcina, 1998).

The inequality constraints (15.3) are equivalent to the following constraints on the slope parameters of the nominal–nominal model

$$\beta_{i+1,j} - \beta_{i,j} \geq 0$$

This shows that this 'non-parametric' ordinal approach can be seen as a nominal–nominal approach with an additional set of constraints. As long as these constraints are not violated, the order-restricted solution will be the same as the nominal–nominal solution. Similar types order constraints can be defined for the nominal–ordinal and ordinal–nominal models to guarantee that the solution is ordered in the sense of a monotonic relationship. The order constraints corresponding with a positive association are $\beta_j \geq 0$ in the nominal–ordinal case and $\beta_{i+1} - \beta_i \geq 0$ in the ordinal–nominal case.

Another type of model for odds is a class of logit models with bi-linear terms. When working with adjacent category odds, these are called row–column association models (Goodman, 1979; Clogg and Shihadeh, 1994). The model of interest has the form

$$\log O_{i,j} = \alpha_j - \beta_j \gamma_i$$

where the γ_i are unknown parameters to be estimated. These can be seen as free scores for the categories of *X*. This model is, in fact, a restricted version of the nominal–nominal model. It can also be seen as a less restricted variant of the nominal–ordinal model (free instead of fixed scores for *X*) or of the ordinal–nominal model (non-constant slope). If $\beta_j \geq 0$ and $\gamma_{i+1} - \gamma_i \geq 0$ for all i and j, the solution is in agreement with an ordinal relationship.

In order to illustrate the equality and inequality constraints implied under the various specifications, we applied the models to the data in Table 15.2 with an adjacent-category odds formulation. Table 15.3 reports the estimated odds ratios obtained with the estimated models. Note that an odds ratio larger than one is in agreement with the postulated positive relationship between time and marijuana use. As can be seen from the outcomes of the nominal–nominal model, the data contains only one violation of an ordinal relationship. The test results indicate that this can be attributed to sampling fluctuation. In the estimation and testing, we treated the observations at different time points as independent samples, which is not correct. Results should therefore be treated with some caution. The next section discusses methods that take the dependence between observations into account. Moreover, the testing of models with inequality constraints is not straightforward. Because the number

Table 15.3 *Adjacent-category odds ratios under various models for the data in Table 15.3*

Model[a]	L-sq (df)[b]	Marijuana use (Y)	Time/Age (X) 1/2	2/3	3/4	4/5
Nominal–nominal	0.00 (0)	1/2	2.16	1.77	1.07	1.43
		2/3	1.56	1.27	1.38	0.93
Ordinal–ordinal	11.37 (7)	1/2	1.36	1.36	1.36	1.36
		2/3	1.36	1.36	1.36	1.36
Nominal–ordinal	8.74 (6)	1/2	1.48	1.48	1.48	1.48
		2/3	1.19	1.19	1.19	1.19
Ordinal–nominal	1.88 (4)	1/2	2.00	1.57	1.19	1.19
		2/3	2.00	1.57	1.19	1.19
Row–column	0.72 (3)	1/2	2.05	1.65	1.27	1.27
		2/3	1.75	1.41	1.08	1.09
Order-restricted	0.06 (1)	1/2	2.16	1.77	1.09	1.38
		2/3	1.56	1.27	1.32	1.00

[a] The models were estimated with the LEM program (Vermunt, 1997).
[b] L-sq is the likelihood-ratio statistic, which is defined as twice the difference between the log-likelihood of the data and the log-likelihood of the model concerned. The number of degrees of freedom is denoted by df. In the order-restricted model, df refers to the number of odds ratios that are equated to 1.

of degrees of freedom is a random variable, the asymptotic distribution of the test statistics is a mixture of χ^2 distributions, which is usually denoted as $\bar{\chi}^2$ distribution (Vermunt, 1999; Galindo-Garre *et al.*, 2002). Especially the order-restricted model using only inequality restrictions fits almost perfectly, but also the ordinal–ordinal model, using equality restrictions, fits well and provides an excellent very parsimonious description of the data. However, comparison of the ordinal–ordinal with the ordinal–nominal model (L-sq = 9.5 with df = 3) indicates that the ordinal constraint is somewhat too restrictive for the column variable time. (Nested models can be compared by a likelihood-ratio test. For this purpose we subtract their L-sq and df values, which yields a new asymptotic χ^2 test.)

Three approaches to longitudinal data

There are three main approaches to the analysis of longitudinal data: transitional models, random-effects models and marginal models (Diggle *et al.*, 1994; Fahrmeir and Tutz, 1994). Transitional models such as Markov-type models concentrate on changes between consecutive time points. Marginal models can be used to investigate changes in univariate distributions, and random-effects or growth models study development of individuals over time. (Here, we concentrate on situations in which there is a single response variable. With multiple response variables, one may wish to study change in multivariate distributions (e.g. Croon *et al.*, 2000)). These three approaches do not only differ in the questions they address, but also in the way they deal with the dependencies between the observations. Because of their structure, transitional models take the bivariate dependencies between observations at consecutive occasions into account. Growth models capture the dependence by introducing one or more latent variables. In marginal models, the dependency is often not modelled, but dealt with as found in the data and in general is taken into account in a more ad hoc way in the estimation procedure. In this section, we present marginal, transition, and random-effect models for ordinal outcome variables.

Before describing the ordinal data variants of the three approaches, we first extend our notation to deal with the longitudinal character of the data. The total number of time points is denoted by T, and a particular time point by t, where $1 \leq t \leq T$. Moreover, we denote the response variable at time point t by Y_t, a particular value of Y_t by j_t, and the number of levels of Y_t by J. Notice that number of levels of the response variable is assumed to be the same for each time point. Predictor k is denoted by X_{kt}, where the index t refers to the fact that a predictor may change its value over time. When referring to a vector of random variables, we use boldface characters. For example, the conditional distribution of the time-specific responses given a particular covariate pattern is denoted by $P(\mathbf{Y} = \mathbf{j}|\mathbf{x})$.

The models of interest will be illustrated with the data set reported in Table 15.1. This means that we have a trichotomous response variable measured at five occasions; that is, $J = 3$ and $T = 5$. There is a single time-constant predictor gender, whose value is denoted by x, where $x = 1$ for males and $x = 0$ for females.

Marginal models

The analysis presented in the previous section is an example of a marginal analysis. We studied the trend in the age- or time-specific marginal distributions of the response variable of interest. However, when estimating the model parameters, we assumed that observations at different time points are independent, which is, of course, unrealistic with longitudinal data. The purpose of the marginal modelling framework is to test hypotheses like the one discussed in the previous section, while taking the dependencies between the observations into account. Parameters can be estimated by maximum likelihood (ML), but also by other methods, such as generalized estimating equations (GEE) or weighted least squares (WLS).

The ML approach takes the full multidimensional distribution of the response variables, $P(\mathbf{Y} = \mathbf{j}|\mathbf{x})$, as the starting point. In addition, to the marginal model of interest, a model has to be specified for the joint distribution. This is not necessarily a restricted model and often the saturated model is simply used. The ML estimates of the probabilities in the joint distribution should be in agreement with both the model for the joint distribution and the model for the marginal distribution. A disadvantage of the ML approach is that it is not practical with more than a few time points. Although theoretically inferior to ML, the GEE approach has the advantage that it can also be applied with larger numbers of time points.

Using the ordinal logit formulation introduced in the previous section, a marginal model for the data displayed in Table 15.1 could, for instance, be of the form

$$\log O_{jt} = \alpha_j - \beta_1 t - \beta_2 x$$

Here, α_j is the intercept for the log odds corresponding to category j, β_1 is the time effect and β_2 is the gender effect. (In the example $J = 3$, which means that there are two sets of odds since $1 \le j \le J - 1$). The fact that the time and gender effects do not depend on the category of the outcome variable shows that we are using an ordinal logit model specification for the outcome variable. Moreover, the log odds are assumed to change linearly with time or age, which amounts to using an 'ordinal' specification for the time effects. Since gender is a dichotomous variable, for this variable there is no difference between a nominal or ordinal specification.

In the previous section, we showed how to relax the 'ordinality' assumptions. For example, the assumption that the time trend is linear can be relaxed by replacing the term $\beta_1 t$ by β_{1t}; that is, by introducing a separate parameter for each time point. Moreover, inequality constraints can be used to transform such a nominal specification for the time effect into ordinal. The ordinal-logit assumption for the dependent variable can be relaxed by having a separate set of effect parameters for $j = 1$ and $j = 2$.

Another possible modification of the above model is the inclusion of an interaction effect between time and gender; that is,

$$\log O_{jt} = \alpha_j - \beta_1 t - \beta_2 x - \beta_3 tx$$

This model relaxes the assumption that the (linear) time trend is the same for males and females.

What should be clear from this example is that marginal models are very much similar to the logit models for ordinal response variables presented in the previous section. The only fundamental difference appears in the estimation procedure in which the dependencies between the time-specific observations have to be taken into account. As was already mentioned, ML estimation involves estimating the cell probabilities in the joint distribution of the time-specific response variables given gender. These cell probabilities should be in agreement with the marginal model of interest and the model that is specific for the joint distribution. As shown by Lang and Agresti (1994) and Bergsma (1997), this estimation problem can be defined as a restricted ML estimation problem.

Transitional models

Typical for transitional models is that a regression model is specified for the conditional distribution of the response variable Y_t given the responses at previous time points (Y_{t-1}, Y_{t-2}, Y_{t-3}, etc.) and predictor values. The fact that Y_t is regressed on a person's state at previous occasions distinguishes transitional from marginal and growth models. Further, a transitional model implies a model for the joint distribution of the time-specific responses. The most popular transitional model is the first-order Markov model in which Y_t is assumed to depend on the state at $t - 1$, but not on responses at earlier occasions. For our example, a first-order Markov model implies the following structure for the joint distribution $P(\mathbf{Y} = \mathbf{j}|x)$:

$$P(\mathbf{Y} = \mathbf{j}|x) = P(Y_1 = j_1|x)P(Y_2 = j_2|Y_1 = j_1, x)P(Y_3 = j_3|Y_2 = j_2, x) \\ \times P(Y_4 = j_4|Y_3 = j_3, x)P(Y_5 = j_5|Y_4 = j_4, x)$$

Further restrictions may be imposed on the initial and transition probabilities. For example, one might assume that the transition probabilities are time homogeneous, yielding what is called a stationary first-order Markov model. The ordinal nature of the response variable can be exploited by restricting the model probabilities by means of an ordinal logit model. An example of a restricted ordinal logit model for the transition probabilities $P(Y_t = j_t|Y_{t-1} = j_{t-1}, x)$ is

$$\log O_{jt} = \alpha_j - \beta_1 y_{t-1} - \beta_2 t - \beta_3 x$$

Except for the presence of Y_{t-1} as a predictor, this transitional logit model is similar to the marginal logit model presented above. Note that y_{t-1} denotes the fixed score corresponding to category of the response at time point $t - 1$. As in a marginal model, ordinal specifications can be changed into a nominal specification, inequality constraints can be imposed, and interaction terms can be included.

ML estimation of transitional models is straightforward. One makes use of a log-likelihood based on a multinomial density with probabilities $P(\mathbf{Y} = \mathbf{j}|x)$. By including states at previous occasions as predictors, the dependence between the time-specific observations is automatically taken into account. More precisely, observations are assumed to be independent given these previous states.

Random-effects growth models

The model structure of a growth model is similar to that of a marginal model. The probability of being in a certain state at occasion t is assumed to be a function of time and predictors. A difference is, however, that the dependence between observations is dealt with in another way. More specifically, the dependence between observations is attributed to systematic differences between individuals. This unobserved heterogeneity is captured by the introduction of random effects in the regression model. A random effect is a parameter that takes on a different value for each individual and that is assumed to come from a particular distribution. Random-effect terms are, in fact, latent variables, which means that a random-effects model is a latent variable model.

An ordinal logit model with a linear growth structure, a gender effect, and a random intercept has the form

$$\log O_{ijt} = \alpha_j + u_i - \beta_1 t - \beta_2 x$$

Here, u_i is the random intercept for individual i. The most common specification is to assume that u_i comes from a normal distribution with a mean equal to zero and variance equal to σ^2; that is, $u_i \sim N(0,\sigma^2)$. Introducing such a random effect amounts to specifying that the intercept is person specific, where person i's intercept equals $\alpha_j + u_i$. It should be noted that apart from the random effect, this logit model is the same as the marginal logit model presented above.

Not only the intercept can be specified to be person specific, but also the time or predictor effects can be assumed to vary across individuals. For example, a model with a random time effect is obtained by

$$\log O_{ijt} = \alpha_j + u_{1i} - \beta_1 t - u_{2i} t - \beta_2 x$$

In this case, the joint distribution of the two random effects u_{1i} and u_{2i} has to be specified. A common choice is bivariate normal, which means that besides the variances also the covariance between the two random effects has to be estimated.

ML estimation is based on the log-likelihood function derived from the multinomial density with probabilities $P(\mathbf{Y} = \mathbf{j}|x)$. Similarly to a transitional model, a random-effects model implies a particular model for the joint distribution

$P(\mathbf{Y} = \mathbf{j}|x)$; that is,

$$P(\mathbf{Y} = \mathbf{j}|x) = \int_{\mathrm{u}} P(Y_1 = j_1|x, \mathbf{u})P(Y_2 = j_2|x, \mathbf{u})P(Y_3 = j_3|x, \mathbf{u}) \times P(Y_4 = j_4|x, \mathbf{u})P(Y_5 = j_5|x, \mathbf{u})f(\mathbf{u})\,d\mathbf{u}$$

As can be seen, occasion-specific responses are assumed to be independent given the random effects **u**. In order to obtain $P(\mathbf{Y} = \mathbf{j}|x)$, we have to integrate out the unobserved random effects. However, contrary to the case of a linear model with normal errors, in our case this integral cannot be solved analytically. Two possible ways to solve the integral are numerical integration by Gauss–Hermite quadrature or integration by simulation methods. Both methods can become quite time consuming with more than a few random-effects terms.

An alternative to the above random-effects approach in which a parametric form is specified for the mixing distribution $f(\mathbf{u})$ is to use a non-parametric specification for the mixing distribution (Laird, 1978; Vermunt and Van Dijk, 2001; Agresti; 2002). The distribution of the random effects is then approximated by a small number of mass points (or latent classes), whose locations and weights are unknown parameters to be estimated. This approach, which is usually referred to as latent class regression or finite mixture regression, has several advantages over using a multivariate normal mixing distribution. One advantage is that it not necessary to make non-testable assumptions about the form of the distribution of the random effects. Another advantage is the much smaller computation burden resulting from the fact that $P(\mathbf{Y} = \mathbf{j}|x)$ can be obtained by summing over a small number of latent classes instead of a large number of quadrature points.

Let the index c refer to a latent class or mixture component and let C be the number of latent classes. A non-parametric specification of the ordinal logit model with a random intercept and a random time effect is

$$\log O_{jt} = \alpha_j + u_{1jc} - \beta_1 t - u_{2c} t - \beta_2 x$$

Rather than assuming that each individual has its own intercept and time effect, we now say that each individual belongs to one of C latent classes, each of which has its own set of logit parameters. A more common way to express this is to index the regression coefficients by c; that is,

$$\log O_{jt} = \alpha_{jc} - \beta_{1c} t - \beta_2 x$$

where $\alpha_{jc} = \alpha_j + u_{1jc}$ and $\beta_{1c} = \beta_1 + u_{2c}$. The implied model for the joint distribution is now

$$P(\mathbf{Y} = \mathbf{j}|x) = \sum_{\mathrm{c}} P(Y_1 = j_1|x, c)P(Y_2 = j_2|x, c)P(Y_3 = j_3|x, c) \times P(Y_4 = j_4|x, c)P(Y_5 = j_5|x, c)P(c)$$

which is much simpler than in the case of the parametric random effects. Note that $P(c)$ is the probability that an individual belongs to latent class c.

Both in the parametric and the non-parametric model it is possible to compute individual-level effects. The most popular are expected a posteriori (EAP) estimates.

Combining the three approaches

Above, we described the three approaches for dealing with longitudinal data as if those provided three mutually exclusive options. However, in some situations, to answer certain questions, one may wish to combine approaches. As explained above, an ML approach to marginal modelling involves the specification of a model for the joint distribution. This could be a restricted model, for example, a first-order Markov model, a random-effects model, or a combination of the two. Vermunt *et al.* (2001), for instance, proposed a combination of the three approaches, in which the random-effects part of the model had the form of a log-linear Rasch model.

Another interesting (and popular) combination is that between a transitional and a random-effects model. Each of these models makes a very specific assumption about the dependence structure of the repeated measures. In a first-order Markov model, for example, it is assumed that dependencies between observations can be fully described by means of first-order autocorrelation terms. In a random effects model, on the other hand, it is assumed that after controlling for the random effects (or latent class memberships), there is no autocorrelation. It is very important when using transitional models to take unobserved heterogeneity into account since failure to do so may result in a strong negatively biased time dependence (Vermunt, 1997). A possible model that combines the two approaches is a first-order Markov model with a non-parametric specification of the random effect:

$$\begin{aligned} P(\mathbf{Y}=\mathbf{j}|x) = \sum_{c} & P(c)P(Y_1=j_1|x,c)P(Y_2=j_2|Y_1=j_1,x,c) \\ & \times P(Y_3=j_3|Y_2=j_2,x,c)P(Y_4=j_4|Y_3=j_3,x,c) \\ & \times P(Y_5=j_5|Y_4=j_4,x,c) \end{aligned}$$

This transitional model with unobserved heterogeneity is usually referred to as mixed Markov model (Langeheine and Van de Pol, 1994; Vermunt, 1997).

And of course, also in these combined models, ordinal logit models can be specified to further restrict the model probabilities. A simple example is

$$\log O_{jt} = \alpha_{jc} - \beta_1 y_{t-1} - \beta_2 t - \beta_3 x$$

in which the intercept is assumed to vary across latent classes.

Table 15.4 *Test results for adjacent-category ordinal logit models estimated with the data in Table 15.1*

Longitudinal type	Specification for Y and X	L-sq[a]	Degrees of freedom
Marginal	M1. nominal–nominal	5.33	8
	M2. nominal–ordinal	19.46	14
	M3. nominal Y and no effect of X	96.09	16
	M4. ordinal–nominal	9.27	13
	M5. ordinal–ordinal	25.06	16
	M6. ordered-restricted	5.40	9[c]
Transitional	T1. nominal–nominal	208.47	466
	T2. nominal–ordinal	258.50	470
	T3. nominal Y and no effect of X	307.45	472
	T4. ordinal–nominal	237.39	473
	T5. ordinal–ordinal	274.84	475
	T6. order-restricted	209.37	468[c]
Random effects (parametric)	R1. nominal–nominal	216.65	470
	R2. nominal–ordinal	230.07	476
	R3. nominal Y and no effect of X	409.36	478
	R4. ordinal–nominal	221.67	476
	R5. ordinal–ordinal	235.36	479
	R6. order-restricted model	216.65[b]	470[c]
Random effects (three-class mixture)	L1. nominal–nominal	204.79	466
	L2. nominal–ordinal	211.57	472
	L3. nominal Y and no effect of X	388.01	474
	L4. ordinal–nominal	208.61	471
	L5. ordinal–ordinal	214.85	474
	L6. order-restricted model	204.79[b]	466[c]

[a] L-sq is the likelihood-ratio statistic, which is defined as twice the difference between the log-likelihood of the data and the log-likelihood of the model concerned. It is sometimes referred to as the deviance statistic.

[b] The fact that the L-sq of the order-restricted model has the same value as the one of the nominal–nominal model indicates that the latter was already in agreement with the order restrictions.

[c] The number of order-restricted parameters that is equated to zero is added to the degrees of freedom.

Application to data set on marijuana use

Table 15.4 reports the test results for a number of models that were estimated for the data in Table 15.1. As can be seen in Table 15.4, marginal, transitional and two types of random-effects models were estimated using different types of specifications for the response and time variables. As in the previous section, only adjacent-category logits have been applied. Many models contained a time–gender interaction term. But as in none of these models the interaction effects were significant, they are not reported here. Obviously, the development of boys and girls over time with respect to marijuana use is the same according to this data set. In the transitional models, we conditioned on the state at the first occasion, which means that no logit restrictions were specified for the response at the first time point. The random-effects models are parametric and three-class mixture models with a random intercept: no more than three latent classes were needed in the non-parametric models and the random time effect was not significant.[3]

The test results show that in the marginal model and the two types of random-effects models, there is no problem if the response variable is treated as ordinal: the difference in L-sq between the ordinal–nominal and nominal–nominal specification is small given the difference in degrees of freedom. In the transitional model, this is somewhat more problematic. In each of the models, there is clear evidence for a time effect since the models without a time effect have much higher L-sq values than the models with a time effect, given the differences in degrees of freedom.[4] Although the ordinal (linear trend) specification captures the most important part of the time dependence, this specification is only satisfactory in the three-class mixture model.

In order to give an impression of the differences in parameter estimates between the four longitudinal data approaches, we present the parameters obtained for the ordinal–ordinal specification in Table 15.5. As can be seen, the signs of the time and gender effects are the same in each of the approaches: marijuana use increases over time and males are more likely to use marijuana than females. The time effects are significant in each of the models. The estimates for the gender effects are on the borderline of significance in all models, except for the transitional model, in which it is clearly significant. A well-known phenomenon that can also be observed in this application is that effect sizes are generally larger in random-effects than in marginal models. The autocorrelation term in the transitional model shows that there is a strong dependence between responses at consecutive time points. The variance of random intercept shows that there are large differences in marijuana use among children.

[3] The non-parametric random-effects models were estimated with the Latent GOLD (Vermunt and Magidson, 2000; www.latentclass.com), which is a very user-friendly Windows program for latent class analysis. The other models were estimated with LEM (Vermunt, 1997; www.uvt.nl/mto), a general program for categorical data analysis. There are several software packages available for the estimation of parametric random-effects model for ordinal variables.

[4] The overall goodness of fit of most models is very good. The number of degrees of freedom is usually larger than the value of the likelihood-ratio statistic L-sq. The number of degrees of freedom equals the number of independent cells in the table that is analysed, minus the number of parameters to be estimates plus the number of constraints that are imposed.

Table 15.5 *Parameter estimates for the ordinal–ordinal models*

Model[a]	Autocorrelation	Variance[b]	Time	Gender
Marginal			0.28	0.28
Transitional	1.27		0.40	0.33
Random effects (parametric)		4.17	0.69	0.60
Random effects (three-class mixture)		3.37/1.44	0.68	0.39

[a] Standard errors are obtained by computing the observed information matrix, the matrix of second-order derivates of the log-likelihood function towards all parameters. The square root of diagonal elements of the inverse of this matrix contains the estimated standard errors.

[b] The three-class mixture model contains two variance terms, one for the log odds between categories 1 and 2 and one for the log odds between categories 2 and 3.

Special issues

It has been mentioned that longitudinal data analysis methods should be able to deal with three main problems: unobserved heterogeneity, measurement error and incomplete responses. The issue of unobserved heterogeneity was already addressed above within the context of random-effects modelling.

Longitudinal data is often incomplete. When using ML estimation it is always possible to use cases with missing data on some of the occasions in the analysis. The assumption generally made is that the missing data are missing at random (MAR). ML under missing at random is straightforward in transitional and random-effects models since only the observed time points contribute to the likelihood function. Although in marginal models things are a bit more complicated, the missing data problem can be dealt with, for example, by ML estimation using an expectation maximization (EM) algorithm.

Another serious problem in longitudinal data analysis, especially in the transitional modelling approach, is measurement error in the response variable. As a result of measurement error in the response variable, the number of observed transitions will be much larger than the true number of transitions, a phenomenon that has the highest impact for the smallest response category (Hagenaars, 1990; Bassi *et al.*, 2000). Another effect of measurement error in the dependent or the independent variable is that covariate effects may be biased. Measurement error can easily be taken into account when using a transitional approach. The transition model is then specified for the true unobserved states Φ_t, which are connected to observed stated Y_t by means of probabilities $P(Y_t | \Phi_t)$. This yields a model that is usually referred to as hidden Markov, latent class Markov, or latent transition model (Langeheine and Van de Pol, 1994; Vermunt, 1997).

The latent Markov model can also be used to deal with multiple response variables. Each of the response variables will then be linked to the latent states by a set of conditional probabilities $P(Y_{tp} | \Phi_t)$, where Y_{tp} denotes response variable p at time point t. With multiple response variables, it is also possible to specify random-effects models in which measurement error is taken into account (Vermunt, 2003). Also the marginal modelling approach could be extended to deal with measurement error in the response variable.

Essentially, the introduction of (partially) unobserved or latent variables is a powerful means to overcome many of the most important problems and complexities in longitudinal analysis. Incorporated into one of the basic approaches towards longitudinal analysis and in combination with flexible kinds of ordinal restrictions to take the ordered nature of the categories into account, it provides the developmental researcher with excellent tools for answering the relevant research questions. And maybe the best news for the researcher is: many easy-to-use computer programs are available to carry out the job.

REFERENCES

Agresti, A. (2002). *Categorical Data Analysis.* New York: John Wiley.

Bassi, F., Hagenaars, J. A., Croon, M. A. and Vermunt, J. K. (2000). Estimating true changes when categorical panel data are affected by uncorrelated and correlated errors. *Sociological Methods and Research*, **29**, 230–68.

Bergsma, W. (1997). *Marginal Models for Categorical Data.* Tilburg, The Netherlands: Tilburg University Press.

Clogg, C. C. and Shihadeh, E. S. (1994). *Statistical Models for Ordinal Data.* Thousand Oaks, CA: Sage.

Croon, M., Bergsma, W. and Hagenaars, J. A. (2000). Analyzing change in discrete variables by generalized log-linear models. *Sociological Methods and Research*, **29**, 195–229.

Dardanoni, V. and Forcina, A. (1998). A unified approach to likelihood inference or stochastic orderings in a nonparametric context. *Journal of the American Statistical Association*, **93**, 1112–23.

Diggle, P. J., Liang, K. Y. and Zeger, S. L. (1994). *Analysis of Longitudinal Data.* Oxford, UK: Clarendon Press.

Elliot, D. S., Huizinga, D. and Menard, S. (1989). *Multiple Problem Youth: Delinquence, Substance Use and Mental Health Problems.* New York: Springer-Verlag.

Fahrmeir, L. and Tutz, G. (1994). *Multivariate Statistical Modelling Based on Generalized Linear Models.* New York: Springer-Verlag.

Galindo-Garre, F., Vermunt, J. K. and Croon, M. A. (2002). Likelihood-ratio tests for order-restricted log-linear models: a comparison of asymptotic and bootstrap methods. *Metodología de las Ciencias del Comportamiento*, **4**, 325–37.

Goodman, L. A. (1979). Simple models for the analysis of association in cross-classifications saving ordered categories. *Journal of the American Statistical Association*, **74**, 537–52.

Hagenaars, J. A. (1990). *Categorical Longitudinal Data: Log-Linear Analysis of Panel, Trend and Cohort Data*. London: Sage.

(2002). Directed loglinear modeling with latent variables: causal models for categorical data with nonsystematic and systematic measurement errors. In *Applied Latent Class Analysis*, eds. J. A. Hagenaars and A. L. McCutcheon, pp. 234–86. New York: Cambridge University Press.

Laird, N. (1978). Nonparametric maximum likelihood estimation of a mixture distribution. *Journal of the American Statistical Association*, **73**, 805–11.

Lang, J. B. and Agresti, A. (1994). Simultaneously modeling joint and marginal distributions of multivariate categorical responses. *Journal of the American Statistical Association*, **89**, 625–32.

Lang, J. B., McDonald, J. W. and Smith, P. W. F. (1999). Association modeling of multivariate categorical responses: a maximum likelihood approach. *Journal of the American Statistical Association*, **94**, 1161–71.

Langeheine, R. and Van de Pol, F. (1994). Discrete time mixed Markov latent class models. In *Analyzing Social and Political Change: A Casebook of Methods*, eds. A. Dale and R. B. Davies, pp. 171–97. London: Sage.

Vermunt, J. K. (1997). *Log-Linear Models for Event Histories*. Thousand Oaks, CA: Sage.

(1999). A general non-parametric approach to the analysis of ordinal categorical data. *Sociological Methodology*, **29**, 197–221.

(2002). A general latent class approach to unobserved heterogeneity in the analysis of event history data. In *Applied Latent Class Analysis*, eds. J. A. Hagenaars and A. L. McCutcheon, pp. 383–407. New York: Cambridge University Press.

(2003). Multilevel latent class models. *Sociological Methodology*, **33**.

Vermunt, J. K. and Magidson, J. (2000). *Latent GOLD User's guide*. Belmont, MA: Statistical Innovations Inc.

Vermunt, J. K. and Van Dijk, L. (2001). A nonparametric random-coefficients approach: the latent class regression model. *Multilevel Modelling Newsletter*, **13**, 6–13.

Vermunt, J. K., Rodrigo, M. F. and Ato-Garcia, M. (2001). Modeling joint and marginal distributions in the analysis of categorical panel data. *Sociological Methods and Research*, **30**, 170–96.

Index